Studies in Fuzziness and Soft Computing

Volume 417

The series "Studies in Fuzziness and Soft Computing" contains publications on various topics in the area of soft computing, which include fuzzy sets, rough sets, neural networks, evolutionary computation, probabilistic and evidential reasoning, multi-valued logic, and related fields. The publications within "Studies in Fuzziness and Soft Computing" are primarily monographs and edited volumes. They cover significant recent developments in the field, both of a foundational and applicable character. An important feature of the series is its short publication time and world-wide distribution. This permits a rapid and broad dissemination of research results.

Indexed by SCOPUS, DBLP, WTI Frankfurt eG, zbMATH, SCImago.

All books published in the series are submitted for consideration in Web of Science.

Rasoul Sabetahd · Ardashir Mohammadzadeh · Ommegolsoum Jafarzadeh · Hamid Taghavifar · Rathinasamy Sakthivel · Chunwei Zhang

Applied Type-3 Fuzzy Logic Systems and Controllers

Springer

Rasoul Sabetahd
Department of Civil Engineering, Sofian Branch
Islamic Azad University
Sofian, Iran

Ardashir Mohammadzadeh
Department of Electrical and Electronics Engineering
Sakarya University
Sakarya, Türkiye

Ommegolsoum Jafarzadeh
Department of Civil Engineering
National University of Skills (NUS)
Tehran, Iran

Hamid Taghavifar
Department of Mechanical, Aerospace and Automotive Engineering
Concordia University
Montreal, QC, Canada

Rathinasamy Sakthivel
Department of Applied Mathematics
Bharathiar University
Coimbatore, Tamil Nadu, India

Chunwei Zhang
Multidisciplinary Center for Infrastructure Engineering
Shenyang University of Technology
Shenyang, China

ISSN 1434-9922 ISSN 1860-0808 (electronic)
Studies in Fuzziness and Soft Computing
ISBN 978-3-031-75987-1 ISBN 978-3-031-75988-8 (eBook)
https://doi.org/10.1007/978-3-031-75988-8

This Springer imprint is published by the registered company Springer Nature Switzerland AG
The registered company address is: Gewerbestrasse 11, 6330 Cham, Switzerland

Preface

Type-2 fuzzy systems are extensively utilized in key sectors of the economy, particularly in industrial and application domains, including services, health, military, and others. Recent studies and research have concentrated on type-2 fuzzy systems, which allow for managing uncertainty in several applications, including intelligent control, robotics, pattern recognition, time series prediction, and medical diagnosis. Recently, there has been a growing interest in the usage of interval type-3 fuzzy systems, and several successful applications have been developed in the past year. These concerns were considered when determining the scope of this book, as we recognized the necessity to provide the fundamental theoretical principles of type-3 fuzzy logic theory, together with techniques for designing, constructing, and executing type-3 fuzzy systems. This book examines the fundamental principles and their application in the development and execution of interval type-3 fuzzy systems, which are novel models for representing uncertainty and imprecision. This work primarily focuses on the fundamental justifications for the utilization of interval type-3 fuzzy systems in various application domains. It then presents the suggested definitions for interval type-3 fuzzy sets and relations, as well as interval type-3 inference and systems. Furthermore, we elucidate techniques for formulating interval type-3 fuzzy systems and exemplify this through various instances and simulations. Additionally, we offer a comparison between the type-3 fuzzy approach and type-2 fuzzy systems. This book serves as a valuable resource for scientists and engineers who wish to utilize interval type-3 fuzzy logic techniques to address difficulties in various fields of application. This book can serve as a reference for graduate courses such as fuzzy logic, soft computing, hybrid intelligent systems, and other related subjects. The present context can serve as a valuable resource for generating innovative ideas for new study directions or for furthering the research indicated by the writers. Furthermore, we elucidate techniques for formulating interval type-3 fuzzy systems and exemplify this through various instances and simulations in modern aspects of technology such as renewable energy, microgrids, etc. In addition, we offer a comparison between the type-3 fuzzy method and type-2 fuzzy systems.

Sofian, Iran — Rasoul Sabetahd
Sakarya, Türkiye — Ardashir Mohammadzadeh
Tehran, Iran — Ommegolsoum Jafarzadeh
Montreal, Canada — Hamid Taghavifar
Coimbatore, India — Rathinasamy Sakthivel
Shenyang, China — Chunwei Zhang

Introduction

The application of artificial intelligence (AI) theory and practice has been extensively utilized across a wide range of human activities. During a period of strong competition among different AI approaches, Prof. Lotfi Aliasker Zadeh recognized that cooperation would yield greater benefits than engaging in claims and counterclaims of supremacy. There is a pressing demand for a novel approach, theory, and technology to create intelligent systems with enhanced machine intelligence quotient (MIQ).

L. Zadeh proposed the use of soft computing (SC) technology to address this question. SC is a collection of intelligent paradigms, including neurocomputing (NC), fuzzy logic (FL), probabilistic computing (PC), evolutionary computing (EC), and chaotic computing (CC). These paradigms have successfully solved numerous significant real-world problems that traditional artificial intelligence methods were unable to tackle. These approaches have the characteristic of being more accommodating towards imprecision, partial truth, and uncertainty compared to their equivalents based on hard computing.

A fundamental attribute of soft computing is that its component techniques are primarily complimentary rather than competitive. The fundamental principle of soft computing is that, in most cases, superior outcomes can be attained by utilizing the constituent techniques in combination rather than individually. Currently, the most prominent combination is that of neuro-fuzzy systems, which have the highest level of visibility. Another such combo that is gaining prominence is neuro-fuzzy-genetic systems.

Neuro-fuzzy hybridization creates a hybrid intelligent system combining the reasoning style of these systems with the learning structure of neural networks. The absence of interpretability in neural networks, as well as the limited learning capabilities of fuzzy systems, are analogous issues that restrict the practical use of these techniques. Neuro-fuzzy systems are hybrid systems that aim to address this issue by integrating the learning capacity of connectionist models with the interpretability characteristic of fuzzy systems. When dealing with a constantly changing work environment, it becomes essential to have an automatic correction system in place for the knowledge base in fuzzy systems. Conversely, artificial neural networks are effectively utilized in difficulties related to acquiring knowledge by learning by examples with the necessary level of accuracy.

The essence of the neuro-fuzzy approach lies in the integration of neural networks into fuzzy systems. This integration involves utilizing neural networks for tasks such as constructing fuzzy rules, fuzzification, implementing fuzzy reasoning, optimizing and adapting the fuzzy knowledge base, and performing defuzzification. Consequently, neural networks get more transparency, while fuzzy systems acquire the ability to learn. The neuro-fuzzy system can be enhanced by integrating expert knowledge into its structure, and simultaneously, the connectionist structure circumvents fuzzy inference, which involves a significant computing load.

Type-1 neuro-fuzzy systems, which are considered ordinary, have proven to be effective in several applications. In order to create conventional neuro-fuzzy systems, it is necessary to have access to the expertise of human professionals and experimental data. This information is used to develop fuzzy rules and membership functions, which are in accord with linguistic or numeric data that is already available. Nevertheless, in numerous instances, the accessible information or facts are linked to diverse forms of uncertainty that must be considered.

The uncertainty can be effectively represented by employing higher order fuzzy sets. Therefore, a highly efficient approach is to utilize type-2 fuzzy sets, which enhance fuzzy models by providing them with greater expressive capability, enabling the development of models that effectively represent the element of uncertainty. Fuzzy type-2 neuro-fuzzy systems are more effective than fuzzy type-1 neuro-fuzzy systems in representing and handling uncertain information. They also add to the robustness and the inference stability. Type-2 fuzzy sets provide increased degrees of freedom when combined with neural networks, which are considered parallel computational models having adaptive characteristics. This combination allows for a more direct and effective consideration of uncertainties in the model that arise from many sources. Professor L. Zadeh originally developed the notion of type-2 fuzzy sets as an expansion of regular fuzzy sets. Mendel and Karnik have formulated a theory pertaining to type-2 fuzzy systems.

Uncertainty impacts the process of making decisions and manifests in several forms. The notion of information is intrinsically linked to the notion of uncertainty.[1,2] The primary factor in this relation is that the uncertainty arises from an information shortage in any problem-solving scenario, which might be due to incompleteness, imprecision, fragmentation, unreliability, vagueness, contradictions, or other deficiencies. Uncertainty is a characteristic of information.[3] The overarching structure of fuzzy reasoning enables the management of a significant portion of this uncertainty. If fuzzy systems utilize type-1 fuzzy sets, these sets aim to depict uncertainty

---

[1] Melin, P., Castillo, O.: Modelling, Simulation and Control of Non-Linear Dynamical Systems. Taylor and Francis, London, Great Britain (2002).

[2] Mendel, J.M.: Uncertainty, fuzzy logic, and signal processing. Sig. Process. J. 80, 913–933 (2000).

[3] Zadeh, L.A.: The concept of a linguistic variable and its application to approximate reasoning. Inf. Sci. 8, 43–80 (1975).

by employing numbers inside the range of [0, 1]. When there is uncertainty regarding an entity, such as a measurement, it becomes challenging to determine its precise value. In such cases, a type-1 fuzzy set is more appropriate than a standard set.

Although there has been significant progress in the theory and design methodologies of type-2 neuro-fuzzy systems, the notion of type-2 neuro-fuzzy system is still in its early phases of development and clarification. There is little advancement in the field of extracting type-2 fuzzy rules and integrating type-2 fuzzy logic systems with other components of soft computing, such as evolutionary computing, etc.

Type-2 fuzzy systems have recently become a potent method for addressing intricate challenges across several application domains. This book provides an overview of the fundamental principles of type-2 fuzzy systems. It then presents the suggested definitions for interval type-3 fuzzy sets and relations, as well as interval type-3 inference and systems.

Furthermore, we provide an explanation of techniques for creating interval type-3 fuzzy systems and demonstrate the use of these techniques through examples and simulations. In addition, we offer a comparison of the outcomes obtained from the type-3 approach in relation to type-2 fuzzy systems.

The fundamental objective of this book is to explore the possibilities of interval type-3 fuzzy models in managing uncertainty in control utilizations and other fields, as they have been shown to provide improved performance in this regard.[4]Recent data suggests that type-3 fuzzy systems can enhance results compared to type-2 in certain situations. Therefore, this book is important since it offers fundamental concepts for developing these type-3 fuzzy systems. One approach is to begin with a well-designed type-2 fuzzy model and then expand it into a type-3 fuzzy model. Alternatively, it is also feasible to directly develop a type-3 fuzzy model from data, although this may be more challenging.

In this book, we present a type-2 fuzzy systems comparison of the type-3 fuzzy method. For scientists and engineers interested in using interval type-3 fuzzy logic approaches to solve problems in a variety of application areas, this book is meant to be a valuable resource. Graduate courses in subjects like fuzzy logic, hybrid intelligent systems, soft computing, and others can benefit from using this book as a reference.

[4] Mohammadzadeh, et al.: A novel fractional-order multiple-model type-3 fuzzy control for nonlinear systems with unmodeled dynamics. Int. J. Fuzzy Syst. 23, 1633–1651 (2021).

Contents

Chapter 1
Basic Theory of Type-3 Fuzzy Logic Systems

1.1 Proposed Type-3 Fuzzy System

1.1.1 Primary Definitions

This section presents the type-3 fuzzy system that is proposed. Figure 1.1 displays the proposed interval type-3 membership function (IT3MF) [1–3] featuring horizontal and vertical slices. Figure 1.2 displays the general type-2 membership function (GT2MF) and Fig. 1.3 shows the interval type-2 membership function (IT2MF) for comparison. The secondary membership function in the IT2MF is constant and in the GT2MF is type-1. The secondary membership function in the suggested IT3MF is an interval type-2 membership function.

The following determines the general type-2 MF [4]:

$$\tilde{A} = \sum\nolimits_{x \in X}, \sum\nolimits_{u \in J_x} \frac{\mu_{\tilde{A}}(x, u)}{(x, u)} \; J_x \subseteq [0, 1] \tag{1.1}$$

The symbol $\Sigma\Sigma$ denotes the union over x and u. The general type-2 MF is akin to the interval type-2 MF, but with the distinction that the secondary MF remains constant:

$$\tilde{A} = \sum\nolimits_{x \in X}, \sum\nolimits_{u \in J_x} \frac{1}{(x, u)} \; J_x \subseteq [0, 1] \tag{1.2}$$

From Eq. (1.2), it is evident that in the interval type-2 MF, for all values of x and u, $\mu_{\tilde{A}}(x, u) = 1$. The interval type-3 MF is defined as below:

$$\tilde{A} = \sum_{x \in X} \sum_{u \in J_x} \frac{\left[\frac{1}{\mu_{\tilde{A}}(x,u)}\right]}{(x, u)} \; J_x \subseteq [0, 1] \tag{1.3}$$

R. Sabetahd et al., *Applied Type-3 Fuzzy Logic Systems and Controllers*,
Studies in Fuzziness and Soft Computing 417,
https://doi.org/10.1007/978-3-031-75988-8_1

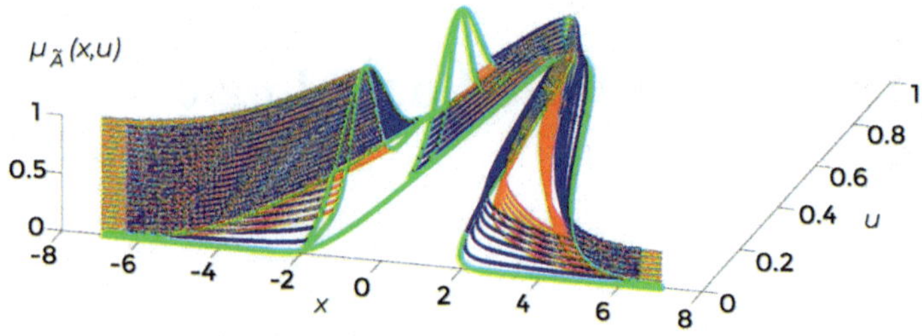

Fig. 1.1 The suggested interval type-3 Gaussian MF

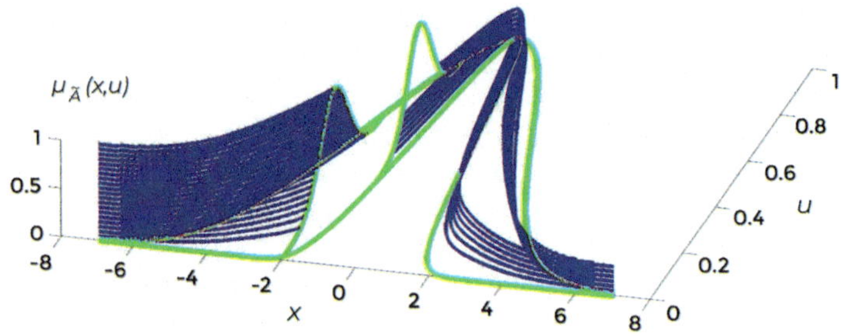

Fig. 1.2 The general form of type-2 Gaussian MF

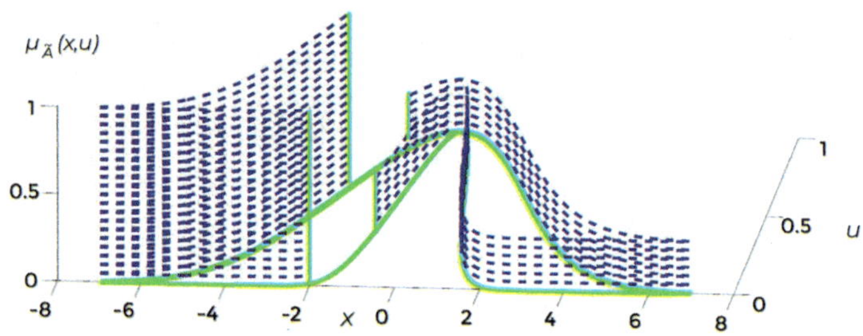

Fig. 1.3 The interval form of type-2 Gaussian MF

where, $1/\mu_{\tilde{A}}(x, u)$ depicts the interval type-2 fuzzy MF.

1.1.2 Vertical Slice Representation

The IT2MF can be expressed as the combination of many MFs by examining its vertical slices. Divide the domain of x into N points $(x_1, x_2, \dots, x_N)$ and discretize the associated u_i into M_i points for each x_i, $i = 1, \dots, N$. The interval type-2 MF is expressed as follows [4]:

$$\tilde{A} = \frac{\sum_{k=1}^{M_1} \frac{1}{u_{1k}}}{x_1} + \cdots + \frac{\sum_{k=1}^{M_N} \frac{1}{u_{Nk}}}{x_N} \tag{1.4}$$

where the symbol " + " represents the union operation. Now, in the case of a general type-2 MF, the secondary grade associated with each u_i is not equal to unity, but it is represented as α_i. The GT2MF can then be rewritten as below:

$$\tilde{A} = \frac{\sum_{k=1}^{M_1} \frac{\alpha_k}{u_{1k}}}{x_1} + \cdots + \frac{\sum_{k=1}^{M_N} \frac{\alpha_k}{u_{Nk}}}{x_N} \tag{1.5}$$

The vertical slice representation of the considered interval type-3 MF is defined similarly to Eq. (1.5). The domain of x is divided into N points ($x_1, x_2, ..., x_N$). At each x_i, the domain of primary membership u_i is divided into M_i points (u_{ik}, $k = 1, ..., M_i$), and the domain of corresponding secondary grade to each u_{ik} is divided into P_k points α_{ikp}, $p = 1, ..., P_k$. The suggested interval type-3 MF is demonstrated as follows:

$$\tilde{A} = \frac{\sum_{k=1}^{M_1} \sum_{p=1}^{P_1} \frac{\alpha_{kp}}{u_{1k}}}{x_1} + \cdots + \frac{\sum_{k=1}^{M_N} \sum_{p=1}^{P_N} \frac{\alpha_{kp}}{u_{Nk}}}{x_N} \tag{1.6}$$

The definitions provided are visually demonstrated by an example. Figure 1.4 illustrates an interval type-2 membership function. The domain of x is divided into two points, x_1 and x_2. At each x_i, $i = 1, 2$, the domain of u_i is divided into 6 points, and the secondary grade equals unity for each u_{ik}. Figure 1.5 illustrates a general type-2 MF. The x domain is divided into two points, x_1 and x_2. At each x_i, $i = 1, 2$, the u_i domain is divided into 6 points, u_{ik}, $k = 1, ..., 6$, with each u_{ik} corresponding to a secondary grade α_{ik}, $k = 1, ..., 6$. Figure 1.6 shows an interval type-3 membership function. The x domain is divided into two points, x_1 and x_2. At each x_i, $i = 1, 2$, the u_i domain is divided into two points, u_{i1} and u_{i2}. The secondary grade relevant to each u_{ik}, $k = 1, 2$ is divided into three points $\alpha_{ik1}, \alpha_{ik2}, \alpha_{ik3}$.

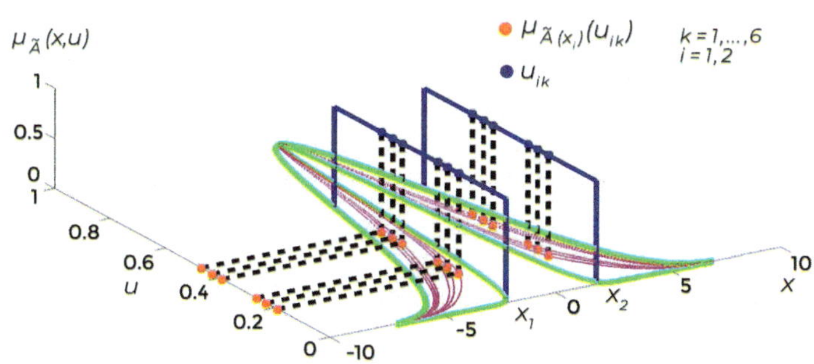

Fig. 1.4 A vertical slice example for the interval type-2 Gaussian MF

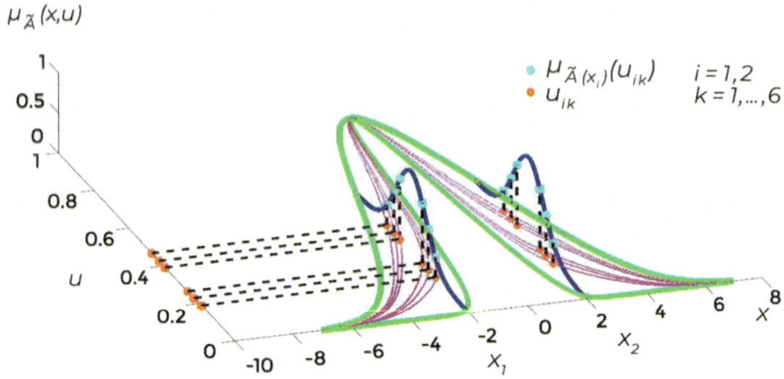

Fig. 1.5 A vertical slice example for the general type-2 Gaussian MF

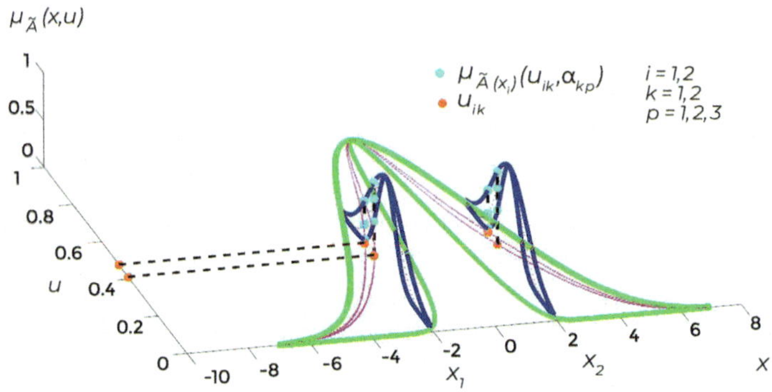

Fig. 1.6 A vertical slice example for the suggested interval type-3 Gaussian MF

1.1.3 Horizontal Slice Representation

Taking into account a Gaussian type-2 MF denoted as $\tilde{A}$, the horizontal slice representation of $\tilde{A}$ is: (see Fig. 1.7)

$$\tilde{A} = \frac{\alpha_1}{\sum_{x \in X} \frac{\tilde{A}_{\alpha_1}}{x}} + \cdots + \frac{\alpha_K}{\sum_{x \in X} \frac{\tilde{A}_{\alpha_K}}{x}}, \alpha_k \subseteq [0, 1] \tag{1.7}$$

where, $\alpha_k, k = 1, ..., K$ show the secondary grade (the horizontal slice level). At each level α_k, $\tilde{A}_{\alpha_k}$ can be expressed as its upper and lower MFs. Then Eq. (1.7) is reformulated as:

$$\tilde{A} = \frac{\alpha_1}{\sum_{x \in X} \frac{\left[\underline{A}_{\alpha_1}, \overline{A}_{\alpha_1}\right]}{x}} + \cdots + \frac{\alpha_K}{\sum_{x \in X} \frac{\left[\underline{A}_{\alpha_K}, \overline{A}_{\alpha_K}\right]}{x}}, \alpha_k \subseteq [0, 1] \tag{1.8}$$

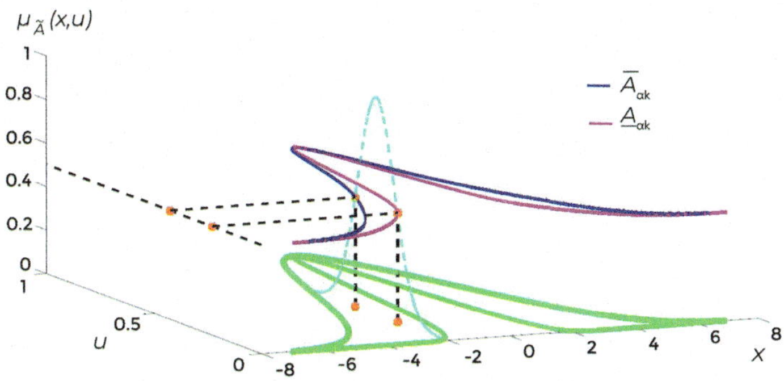

Fig. 1.7 A horizontal slice example for the general type-2 Gaussian MF

It is worth mentioning that, as shown in Fig. 1.8, the secondary grade at all slices is equal to one ($\alpha_k = 1, k = 1, ..., K$) and also the upper and the lower MF at all slice equal $\underline{A}_{\alpha_k} = \underline{A}, \underline{A}_{\alpha_k} = \underline{A}, k = 1, ..., K$ if $\tilde{A}$ is an interval type-2 MF. Now, contemplate an interval type-3 membership function. Unlike the general type-2 MF, the secondary grade is not equivalent to α_k, but rather an interval type-1 MF. The horizontal representation of $\tilde{A}$ is shown below:

$$\tilde{A} = \frac{1/\tilde{\alpha}_1}{\sum_{x \in X} \frac{\tilde{A}_{\tilde{\alpha}_1}}{x}} + \cdots + \frac{1/\tilde{\alpha}_K}{\sum_{x \in X} \frac{\tilde{A}_{\tilde{\alpha}_K}}{x}}, \alpha_k \subseteq \left[\underline{\alpha}_k, \underline{\alpha}_k\right] \tag{1.9}$$

where we have $\alpha_k \subseteq \left[\underline{\alpha}_k, \underline{\alpha}_k\right]$. Associated with $\underline{\alpha}_k$ and $\underline{\alpha}_k$, we have $\tilde{A}_{\tilde{\alpha}_k} \subseteq \left[\underline{A}_{\underline{\alpha}_k}, \underline{A}_{\underline{\alpha}_k}\right]$ and $\tilde{A}_{\tilde{\alpha}_k} \subseteq \left[\underline{A}_{\underline{\alpha}_k}, \underline{A}_{\underline{\alpha}_k}\right]$, respectively. Then Eq. (1.9) is restated as follows (see Fig. 1.9):

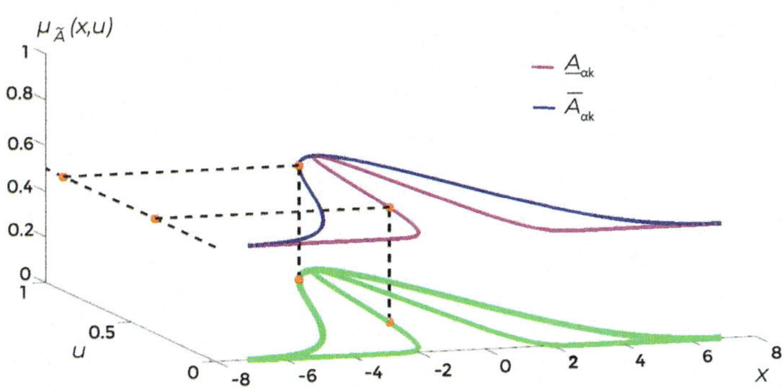

Fig. 1.8 A horizontal slice example for the interval type-2 Gaussian MF

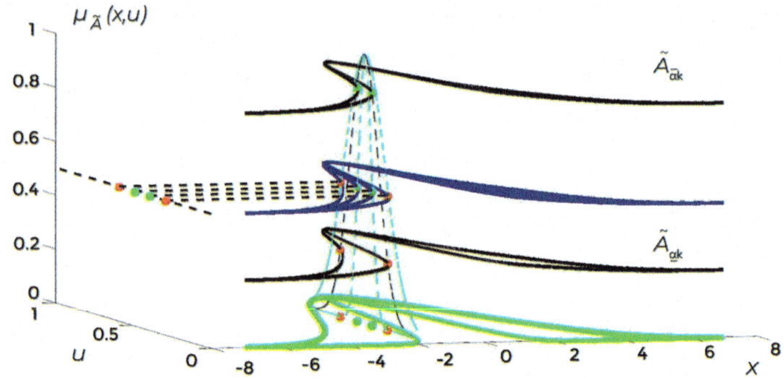

Fig. 1.9 A horizontal slice example for the proposed interval type-3 Gaussian MF

$$\tilde{A} = \left[\frac{1/\underline{\alpha}_1}{\sum_{x\in X} \frac{\left[\underline{A}_{\underline{\alpha}_1}, \underline{A}_{\underline{\alpha}_1}\right]}{x}}, \frac{1/\underline{\alpha}_1}{\sum_{x\in X} \frac{\left[\underline{A}_{\underline{\alpha}_1}, \underline{A}_{\underline{\alpha}_1}\right]}{x}} \right] + \tag{1.10}$$

$$\cdots + \left[\frac{1/\underline{\alpha}_K}{\sum_{x\in X} \frac{\left[\underline{A}_{\underline{\alpha}_K}, \underline{A}_{\underline{\alpha}_K}\right]}{x}}, \frac{1/\underline{\alpha}_K}{\sum_{x\in X} \frac{\left[\underline{A}_{\underline{\alpha}_K}, \underline{A}_{\underline{\alpha}_K}\right]}{x}} \right] \tag{1.11}$$

Note 1 *The suggested fuzzy system differs from the general type-2 fuzzy system. That is, in the proposed MF, the secondary domain is a fuzzy number for all items in the primary domain, rather than a crisp value between [0,1] as in the general type-2 fuzzy system. The secondary membership function is an interval type-2 MF. The secondary domain elements have a third domain value of 1 (refer to Fig. 1.10).*

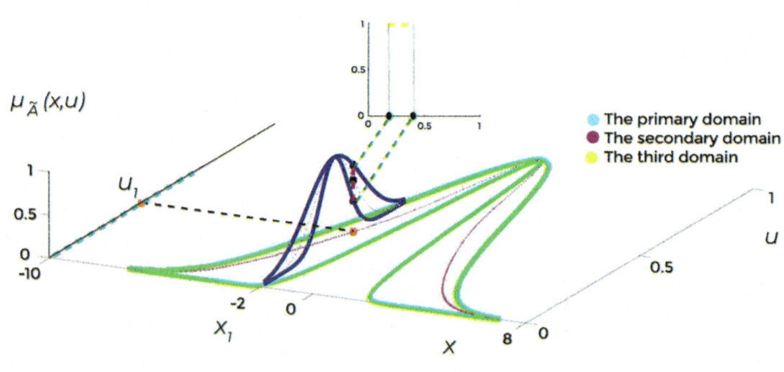

Fig. 1.10 The 3 domains of the suggested MF

1.1.4 Interval Type-3 Fuzzy System

When $x = x'$ in the interval type-2 MF shown in Fig. 1.8, the upper and lower initial memberships (u) at all horizontal slice levels are constant and have specific values. In type-2 MF, demonstrated in Fig. 1.7, the upper and lower initial memberships shift at various horizontal slice levels to specific values. Now, examine the interval type-3 MF depicted in Fig. 1.9. The upper and lower initial memberships are altered at various horizontal slice levels, and they possess unknown values. Put simply, the upper and lower memberships are interval type-1 MFs.

Figure 1.9 illustrates that a horizontal section of an interval type-3 MF is equivalent to a horizontal slice of a general type-2 MF at two specific levels $\underline{\alpha}_k, \underline{\alpha}_k$. The interval type-3 MF can be formed by combining all horizontal slices $\tilde{\alpha}_1, \tilde{\alpha}_2, \ldots, \tilde{\alpha}_K$. An IT3MF at level $\tilde{\alpha}_k$ can be represented by a horizontal slice, indicating a type-2 MF with a secondary grade falling within the range of $\left[\underline{\alpha}_k, \underline{\alpha}_k\right]$. Figure 1.11 displays the comprehensive perspective of the IT3FS. The output of interval type-3 MF is acquired sequentially through the following steps:

(1) Obtain inputs $x_1, \ldots, x_n$.. This step has a time complexity of O(1) in big-O notation.
(2) Take into account the membership functions $\tilde{A}_i^j, j = 1, \ldots, M$ for each input. $\tilde{A}_i^j$ represents the jth membership function for the ith input. This step has a complexity of O(1).

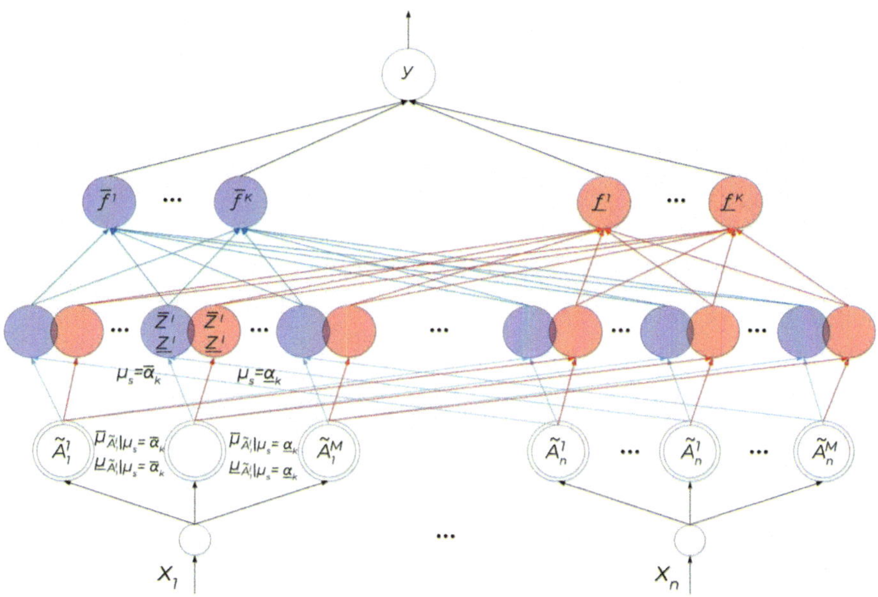

Fig. 1.11 A general view of the Mamdani type IT3FS

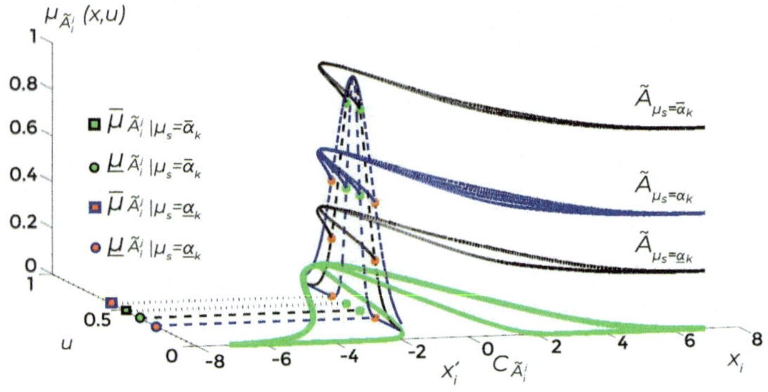

Fig. 1.12 A horizontal slice example for the suggested interval type-3 Gaussian MF

(3) Calculate the memberships of $\tilde{A}_i^j, j = 1, ..., n_{A_i}$ at various horizontal slice levels $\mu_s = \alpha_k, k = 1, ..., K$. Given x_i equals x_i' and $\mu_s = \tilde{\alpha}_k$, then we have: (see Fig. 1.12)

$$\underline{\mu}_{\tilde{A}_i^j|\mu_s=\underline{\alpha}_k}(x_i') = m_{\tilde{A}_i^j|\mu_s=\alpha_0}(x_i') + \sqrt{ln(1/\underline{\alpha}_k)\sigma_{V_i}^2} \tag{1.12}$$

$$\underline{\mu}_{\tilde{A}_i^j|\mu_s=\underline{\alpha}_k}(x_i') = m_{\tilde{A}_i^j|\mu_s=\alpha_0}(x_i') - \sqrt{ln(1/\underline{\alpha}_k)\sigma_{V_i}^2} \tag{1.13}$$

$$\underline{\mu}_{\tilde{A}_i^j|\mu_s=\underline{\alpha}_k}(x_i') = m_{\tilde{A}_i^j|\mu_s=\alpha_0}(x_i') + \sqrt{ln(1/\underline{\alpha}_k)\sigma_{V_i}^2} \tag{1.14}$$

$$\underline{\mu}_{\tilde{A}_i^j|\mu_s=\underline{\alpha}_k}(x_i') = m_{\tilde{A}_i^j|\mu_s=\alpha_0}(x_i') - \sqrt{ln(1/\underline{\alpha}_k)\sigma_{V_i}^2} \tag{1.15}$$

where, $\underline{\mu}_{\tilde{A}_i^j|\mu_s=\underline{\alpha}_k}(x_i')/\underline{\mu}_{\tilde{A}_i^j|\mu_s=\underline{\alpha}_k}(x_i')$ and $\underline{\mu}_{\tilde{A}_i^j|\mu_s=\underline{\alpha}_k}(x_i')/\underline{\mu}_{\tilde{A}_i^j|\mu_s=\underline{\alpha}_k}(x_i')$ are respectively the upper and the lower memberships of $\tilde{A}_i^j$, for when the horizontal slice level is $\underline{\alpha}_k/\underline{\alpha}_k$. The $\left[\underline{\alpha}_k, \underline{\alpha}_k\right]$ shows the secondary grades uncertainty.

$$\underline{\alpha}_k = \alpha_k^{1/\Delta} \underline{\alpha}_k = \alpha_k^{\Delta} \tag{1.16}$$

where $\Delta \geq 1$ is a constant representing the standard division uncertainty of Gaussian vertical slice at $x_i = x_i'$. In other words, if $\tilde{A}_i^j$ represents the general type-2 MF, then the vertical slice at $x_i = x_i'$ is a type-1 MF with standard division $\sigma_{V_i}^2(x_i')$ (see Fig. 1.7), however, as $\tilde{A}_i^j$ is the interval type-3 MF, the vertical slice is an interval type-2 MF with standard division $\left[1/\Delta \cdot \sigma_{V_i}^2(x_i'), \Delta \cdot \sigma_{V_i}^2(x_i')\right]$ at $x_i = x_i'$ (see Fig. 1.9). $\sigma_{V_i}^2(x_i')$ is calculated as follows:

$$\sigma_{V_i}^2(x_i') = \left(\underline{\mu}_{\tilde{A}_i^j|\mu_s=\alpha_0}(x_i') - m_{\tilde{A}_i^j|\mu_s=\alpha_0}(x_i')\right)^2 / ln\left(\frac{1}{\varepsilon}\right) \tag{1.17}$$

where,

$$m_{\tilde{A}_i^j|\mu_s=\alpha_0}(x'_i) = \left(\underline{\mu}_{\tilde{A}_i^j|\mu_s=\alpha_0}(x'_i) + \underline{\mu}_{\tilde{A}_i^j|\mu_s=\alpha_0}(x'_i)\right)/2 \tag{1.18}$$

$$\underline{\mu}_{\tilde{A}_i^j|\mu_s=\alpha_0}(x'_i) = exp\left(-\frac{\left(x'_i - C_{\tilde{A}_i^j}\right)^2}{\underline{\sigma}^2_{\tilde{A}_i^j}}\right) \tag{1.19}$$

$$\underline{\mu}_{\tilde{A}_i^j|\mu_s=\alpha_0}(x'_i) = exp\left(-\frac{\left(x'_i - C_{\tilde{A}_i^j}\right)^2}{\underline{\sigma}^2_{\tilde{A}_i^j}}\right) \tag{1.20}$$

where, $C_{\tilde{A}_i^j}$, $\underline{\sigma}_{\tilde{A}_i^j}$ and $\underline{\sigma}_{\tilde{A}_i^j}$ are, respectively, center, upper width and lower width of MF $\tilde{A}_i^j$ at horizontal slice level $\mu_s = \alpha_0$. This step's complexity is calculated to be O($n \times M \times K$).

(D) Calculate the rule firings. Take into account the $l - th$ rule as follows:

$$if\ x_1\ is\ \tilde{A}_1^l\ and \cdots and\ x_n\ is\ \tilde{A}_n^l \qquad then\ y \subset \left[\underline{w}_l, \underline{w}_l\right],\ l = 1, ..., M \tag{1.21}$$

where, M represents the number of rules, while $\underline{w}_l$, $\underline{w}_l$ are the consequent parameters for the $l - th$ rule. The rules are regarded fixed since the subsequent parameters are adjusted online. The ranges of input variables are segmented into M portions. A Gaussian MF is proposed for each segment. The number of rules is equivalent to the number of membership functions (MFs). The $l - th$ rule can be understood in the following manner. If the input vector $[x_1, ..., x_n]^T$ belongs to the cluster $\left[C_{\tilde{A}_1^l}, ..., C_{\tilde{A}_n^l}\right]^T$, then $y \subset \left[\underline{w}_l, \underline{w}_l\right]$. The rule firing degrees are calculated as:

$$\underline{z}^l_{\mu_s=\underline{\alpha}_k} = T_i^n \underline{\mu}_{\tilde{A}_i^l|\mu_s=\underline{\alpha}_k} \tag{1.22}$$

$$\underline{z}^l_{\mu_s=\underline{\alpha}_k} = T_i^n \underline{\mu}_{\tilde{A}_i^l|\mu_s=\underline{\alpha}_k} \tag{1.23}$$

$$\underline{z}^l_{\mu_s=\underline{\alpha}_k} = T_i^n \underline{\mu}_{\tilde{A}_i^l|\mu_s=\underline{\alpha}_k} \tag{1.24}$$

$$\underline{z}^l_{\mu_s=\underline{\alpha}_k} = T_i^n \underline{\mu}_{\tilde{A}_i^l|\mu_s=\underline{\alpha}_k} \tag{1.25}$$

where, T shows the T-norm. The step's complexity is O(M).

(E) Calculate the output of the suggested fuzzy system as follows:

$$y = \frac{\sum_{k=1}^{K}\left(\underline{\alpha}_k \underline{f}_k + \underline{\alpha}_k \underline{f}_k\right)}{\sum_{k=1}^{K}\left(\underline{\alpha}_k + \underline{\alpha}_k\right)} \tag{1.26}$$

where,

$$\underline{f}_k = \frac{\sum_{l=1}^{M}\left(\underline{z}^l_{\mu_s=\underline{\alpha}_k}\underline{w}_l + \underline{z}^l_{\mu_s=\underline{\alpha}_k}\underline{w}_l\right)}{\sum_{l=1}^{M}\left(\underline{z}^l_{\mu_s=\underline{\alpha}_k} + \underline{z}^l_{\mu_s=\underline{\alpha}_k}\right)} \tag{1.27}$$

$$\underline{f}_k = \frac{\sum_{l=1}^{M}\left(\underline{z}^l_{\mu_s=\underline{\alpha}_k}\underline{w}_l + \underline{z}^l_{\mu_s=\underline{\alpha}_k}\underline{w}_l\right)}{\sum_{l=1}^{M}\left(\underline{z}^l_{\mu_s=\underline{\alpha}_k} + \underline{z}^l_{\mu_s=\underline{\alpha}_k}\right)} \tag{1.28}$$

This step has a complexity of $O(M \times K)$.

1.2 Stable Fractional-Order Learning Algorithm

This section introduces a new algorithm designed to enhance the parameters of the suggested fuzzy system. The adaptation laws are determined by a stability analysis to ensure that the specified sliding surface converges to zero:

$$s = D_t^{1-q} e + \lambda D_t^{-q} e \tag{1.29}$$

where, $0 < q < 1$, e is the estimation error, and λ is a positive constant, which is defined as below:

$$e = y(\theta|u) - y_d \tag{1.30}$$

The output of the proposed fuzzy system is represented by y, the desired output is represented by y_d, the input vector of the fuzzy system is represented by u, and the vector of succesive parameters is denoted by θ:

$$\theta = \left[\underline{w}_1, \ldots, \underline{w}_M, \underline{w}_1, \ldots, \underline{w}_M\right]^T \tag{1.31}$$

where, M is the number of rules. $D_t^q g(t)$ represents the Caputo derivative of the function g of the order q, which is defined as follows:

$$D_t^q g(t) = \frac{1}{\Gamma(1-q)} \int_0^t (t-\tau)^{-q} \frac{d}{dt} g(\tau) d\tau \tag{1.32}$$

Here, $\Gamma(\cdot)$ denotes the Gamma function.

Theorem 1 *If the rules of adaptation are selected in a specific manner shown below for the suggested fuzzy system, then the estimate error $e(t)$ will converge to zero as t approaches infinity.*

$$\dot{\theta} = \left(-\lambda e - \left(\left|\underline{\dot{y}}_d\right| + \left|\varsigma^T\right|\left|\underline{\dot{u}}\right|\right) sign(s)\right) \frac{\xi}{\xi^T \xi} \tag{1.33}$$

where, ξ and ς are designated as $\xi = \frac{\partial y(\theta|u)}{\partial \theta}$, $\varsigma = \frac{\partial y(\theta|u)}{\partial u}$. Moreover, $\left|\underline{\dot{y}}_d\right|$ and $\left|\underline{\dot{u}}\right|$ are the upper bounds of $\dot{y}_d$ and $\dot{u}$, respectively.

Proof Take into account the provided Lyapunov function:

$$V = \frac{1}{2} s^2 \tag{1.34}$$

where, s has previously been specified in Eq. (1.29). The fractional time derivative of V gives:

$$D_t^q V \leq s D_t^q s \tag{1.35}$$

From Eq. (1.29), $D_t^q s$ is acquired in the following manner:

$$D_t^q s = \dot{e} + \lambda e \tag{1.36}$$

From Eq. (1.30), $\dot{e}$ can be obtained as:

$$\dot{e} = \frac{\partial y(\theta|u)}{\partial \theta}^T \dot{\theta} + \frac{\partial y(\theta|u)}{\partial u}^T \dot{u} - \dot{y}_d \tag{1.37}$$

Moreover, $\frac{\partial y}{\partial \theta}$ is obtained using the following equation:

$$\frac{\partial y(\theta|u)}{\partial \theta} = \left[\frac{\partial y}{\underline{w}_1}, \ldots, \frac{\partial y}{\underline{w}_M}, \frac{\partial y}{\overline{w}_1}, \ldots, \frac{\partial y}{\overline{w}_M}\right]^T \tag{1.38}$$

where,

$$\frac{\partial y}{\underline{w}_l} = \frac{\sum_{k=1}^{K} \underline{\alpha}_k \frac{\underline{z}^l_{\mu_s=\underline{\alpha}_k}}{\sum_{l=1}^{M}\left(\underline{z}^l_{\mu_s=\underline{\alpha}_k} + \underline{z}^l_{\mu_s=\underline{\alpha}_k}\right)}}{\sum_{k=1}^{K}\left(\underline{\alpha}_k + \underline{\alpha}_k\right)} + \frac{\sum_{k=1}^{K} \underline{\alpha}_k \frac{\underline{z}^l_{\mu_s=\underline{\alpha}_k}}{\sum_{l=1}^{M}\left(\underline{z}^l_{\mu_s=\underline{\alpha}_k} + \underline{z}^l_{\mu_s=\underline{\alpha}_k}\right)}}{\sum_{k=1}^{K}\left(\underline{\alpha}_k + \underline{\alpha}_k\right)} \qquad l = 1, \cdots, M \tag{1.39}$$

$$\frac{\partial y}{\underline{w}_l} = \frac{\sum_{k=1}^{K} \underline{\alpha}_k \frac{\underline{z}^l_{\mu_s=\underline{\alpha}_k}}{\sum_{l=1}^{M}\left(\underline{z}^l_{\mu_s=\underline{\alpha}_k} + \underline{z}^l_{\mu_s=\underline{\alpha}_k}\right)}}{\sum_{k=1}^{K}\left(\underline{\alpha}_k + \underline{\alpha}_k\right)} + \frac{\sum_{K=1}^{K} \underline{\alpha}_k \frac{\underline{z}^l_{\mu_s=\underline{\alpha}_k}}{\sum_{l=1}^{M}\left(\underline{z}^l_{\mu_s=\underline{\alpha}_k} + \underline{z}^l_{\mu_s=\underline{\alpha}_k}\right)}}{\sum_{k=1}^{K}\left(\underline{\alpha}_k + \underline{\alpha}_k\right)} \qquad l = 1, \ldots, M \tag{1.40}$$

The fractional term $\frac{\partial y}{\partial u_i}$ in Eq. (1.37), is computed to be:

$$\frac{\partial y}{\partial u_i} = \frac{\sum_{k=1}^{K} \underline{\alpha}_k \frac{\partial \underline{f}_k}{\partial x_i} + \underline{\alpha}_k \frac{\partial \underline{f}_k}{\partial x_i}}{\sum_{k=1}^{K}\left(\underline{\alpha}_k + \underline{\alpha}_k\right)} \tag{1.41}$$

where,

$$\begin{aligned}\frac{\partial \underline{f}_k}{\partial x_i} = {} & \frac{\sum_{l=1}^{M}\left(\frac{-\left(x_i - C_{\tilde{A}_i^l}\right)}{\underline{\sigma}^2_{\tilde{A}_i^l}} \underline{z}^l_{\mu_s=\underline{\alpha}_k} \underline{w}_l + \frac{-\left(x_i - C_{\tilde{A}_i^l}\right)}{\underline{\sigma}^2_{\tilde{A}_i^l}} \underline{z}^l_{\mu_s=\underline{\alpha}_k} \underline{w}_l\right)}{\sum_{l=1}^{M}\left(\underline{z}^l_{\mu_s=\underline{\alpha}_k} + \underline{z}^l_{\mu_s=\underline{\alpha}_k}\right)} \\ & - \frac{\underline{f}_k \sum_{l=1}^{M}\left(\frac{-\left(x_i - C_{\tilde{A}_i^l}\right)}{\underline{\sigma}^2_{\tilde{A}_i^l}} \underline{z}^l_{\mu_s=\underline{\alpha}_k} + \frac{-\left(x_i - C_{\tilde{A}_i^l}\right)}{\underline{\sigma}^2_{\tilde{A}_i^l}} \underline{z}^l_{\mu_s=\underline{\alpha}_k}\right)}{\sum_{l=1}^{M}\left(\underline{z}^l_{\mu_s=\underline{\alpha}_k} + \underline{z}^l_{\mu_s=\underline{\alpha}_k}\right)}\end{aligned} \tag{1.42}$$

$$\begin{aligned}\frac{\partial \underline{f}_k}{\partial x_i} = {} & \frac{\sum_{l=1}^{M}\left(\frac{-\left(x_i - C_{\tilde{A}_i^l}\right)}{\underline{\sigma}^2_{\tilde{A}_i^l}} \underline{z}^l_{\mu_s=\underline{\alpha}_k} \underline{w}_l + \frac{-\left(x_i - C_{\tilde{A}_i^l}\right)}{\underline{\sigma}^2_{\tilde{A}_i^l}} \underline{z}^l_{\mu_s=\underline{\alpha}_k} \underline{w}_l\right)}{\sum_{l=1}^{M}\left(\underline{z}^l_{\mu_s=\underline{\alpha}_k} + \underline{z}^l_{\mu_s=\underline{\alpha}_k}\right)} \\ & - \frac{\underline{f}_k \sum_{l=1}^{M}\left(\frac{-\left(x_i - C_{\tilde{A}_i^l}\right)}{\underline{\sigma}^2_{\tilde{A}_i^l}} \underline{z}^l_{\mu_s=\underline{\alpha}_k} + \frac{-\left(x_i - C_{\tilde{A}_i^l}\right)}{\underline{\sigma}^2_{\tilde{A}_i^l}} \underline{z}^l_{\mu_s=\underline{\alpha}_k}\right)}{\sum_{l=1}^{M}\left(\underline{z}^l_{\mu_s=\underline{\alpha}_k} + \underline{z}^l_{\mu_s=\underline{\alpha}_k}\right)}\end{aligned} \tag{1.43}$$

From Eqs. (1.36) and (1.37), we get:

$$D_t^q V \leq s \frac{\partial y(\theta|u)}{\partial \theta}^T \dot{\theta} + s \frac{\partial y(\theta|u)}{\partial u}^T \dot{u} - s\dot{y}_d + s\lambda e \tag{1.44}$$

Considering $\xi = \frac{\partial y(\theta|u)}{\partial \theta}$, $\varsigma = \frac{\partial y(\theta|u)}{\partial u}$, the inequality shown in Eq. (1.44) can be reformulated as follows:

$$D_t^q V \leq s\xi^T \dot{\theta} + s\varsigma^T \dot{u} - s\dot{y}_d + s\lambda e \tag{1.45}$$

By replacing the adaptation law of Eq. (1.33) into Eq. (1.44), we obtain:

$$D_t^q V \leq s\xi^T \left\{ -\lambda e - \left(\left| \dot{\underline{y}}_d \right| + |\varsigma^T| |\underline{\dot{u}}| \right) sign(s) \right\} \frac{\xi}{\xi^T \xi} + s\varsigma^T \dot{u} - s\dot{y}_d + s\lambda e \tag{1.46}$$

Simplification of Eq. (1.46) gives:

$$D_t^q V \leq -|s| \left| \dot{\underline{y}}_d \right| - |s| |\varsigma^T| |\underline{\dot{u}}| + s\varsigma^T \dot{u} - s\dot{y}_d < 0 \tag{1.47}$$

This completes the proof.

1.3 Simulations and Empirical Confirmation

This section evaluates the efficacy of the introduced fuzzy system and its corresponding learning algorithm.

Example 1 *The IT3FS is utilised in this example for Mackey Glass time series online prediction, the dynamic of which is shown as follows:*

$$\dot{x} = \frac{0.2x(t-\tau)}{0.2x^{10}(t-\tau)} - 0.1x(t) \tag{1.48}$$

where $\tau = 30$ and $x(0) = 1.2$. The fuzzy system's inputs consist of $x(t-6)$, $x(t-12)$, $x(t-18)$ and $x(t-24)$. The goal is to approximate $x(t)$ using the suggested fuzzy system. 700/300 out of 1000 data samples are applied for training and testing [5]. Table 1.1 contains the simulation parameters. Figures 1.13 and 1.14 display the trajectories of the actual and predicted outputs along with the estimation error in scenarios with noisy and noise-free conditions. A white noise with zero mean and variance of 0.1 is studied in a noisy scenario. The proposed approach demonstrates good performance. The proposed method's performance is evaluated against various fuzzy models, including the hierarchical type-2 fuzzy system (HT2FS), interval type-2 fuzzy neural network (IT2FNN) [6], evolving type-2 neural fuzzy system (eT2FS) [7], self-constructing neural fuzzy inference network (SONFIN) [8], and self-evolving interval type-2 fuzzy neural network (SEIT2FNN) [9]. Table 1.2 displays the values of root mean square error (RMSE). The suggested method demonstrates superior performance compared to other methods, particularly in noisy conditions.

Note 2 *The proposed learning algorithm is implemented online. Contrary to previous methods, the simulation is run for only one epoch.*

Table 1.1 The simulation parameters in Ex. 1

Brown parameter	Value	See
M	3	Equation (1.21)
q	0.9	Equation (1.29)
K	4	Equation (1.26)
λ	10	Equation (1.29)
ε	0.001	Equation (1.17)
Δ	2	Equation (1.16)
$[\alpha_1, \ldots, \alpha_4]$	[0.2, 0.4, 0.6, 0.8]	Equation (1.16)
$\left[C_{A_i^1}, C_{\tilde{A}_i^2}, C_{\tilde{A}_i^3}\right]$	$[-1, 0, 1]$	Equations (1.19, 1.20), Fig. 1.12
$\underline{\sigma}_{A_i^j}$	1	Equation (1.19)
$\underline{\sigma}_{A_i^j}$	0.5	Equation (1.20)

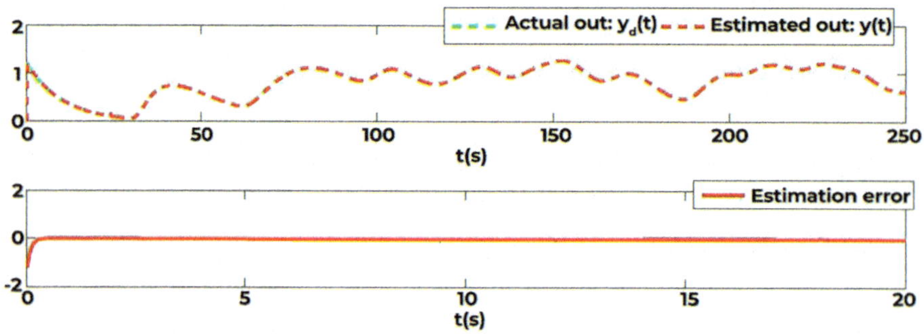

Fig. 1.13 The prediction performance, free noise case in Ex. 1

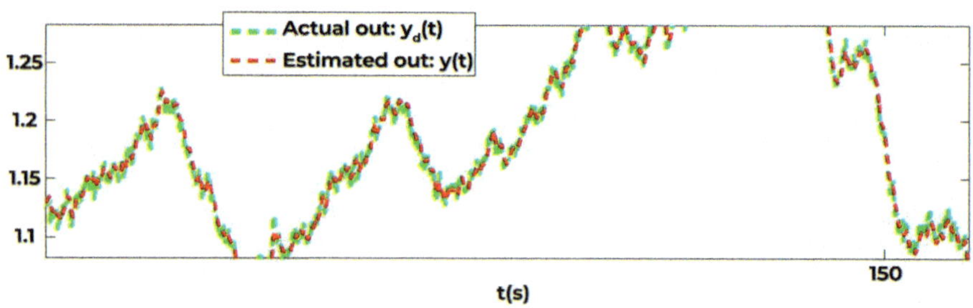

Fig. 1.14 The performance of the prediction in a noisy case of Ex. 1

Table 1.2 The comparison of RMSE for various models in Ex. 1

Brown models	SONFIN [8]	SEIT2FNN [9]	IT2FNN [6]	eT2FS [7]	HT2FS [5]	Proposed method
No. of optimized parameters	130	110	103	—	80	6
Test RMSE (free noise case)	0.054	0.49	0.046	0.059	0.023	5.5256e-4
Test RMSE (noisy case)	0.108	0.097	0.088	0.107	0.070	0.0048

Example 2 *This example demonstrates the utilisation of the proposed method for online dynamic estimate of the Chuas chaotic system:*

$$\dot{y}_1 = y_2, \dot{y}_2 = y_3, \dot{y}_3 = f(y_1, y_2, y_3)$$

$$f(y_1, y_2, y_3) = \frac{14}{1805}y_1 - \frac{168}{9025}y_2 + \frac{1}{38}y_3 - \frac{2}{45}\left(\frac{28}{361}y_1 + \frac{7}{95}y_2 + y_3\right)^3 \quad (1.49)$$

The initial conditions are $y_1(0) = 0$, $y_2(0) = 0$ and $y_3(0) = 0.3$. Figure 1.15 displays the trajectories of the real and predicted states along with the errors in estimation. The calculated trajectories closely align with the real trajectories, indicating convergence. The proposed method's performance is evaluated against the hierarchical type-2 fuzzy system (HT2FS) [5]. All parameters of HT2FS are trained using the square-root cubature Kalman filter in reference [5]. RMSE values are compared in Table 1.3. The results in reference [5] were achieved after 50 epochs. The findings of the proposed method in this research are provided after a single period. The suggested method outperforms existing methods greatly, as seen in Table 1.3.

Example 3 *The suggested fuzzy system's effectiveness is verified by analysing the modelling problem of two real-world data sets. The first model predicts the fuel consumption of automobiles in miles per gallon (MPG)* [10], *while the second model estimates the maintenance expenses of medium-voltage electrical lines (MCV)* [11]. *For MPG and MCV problems, 320/72 and 396/99 samples of entire data sets are utilized for testing. A comparison is made between the performance of the proposed IT3FS and other well-known fuzzy neural systems, like: self-constructing neural fuzzy system (SNFS)* [8], *interval type-2 neural fuzzy system for system identification and feature elimination (SEIT2FNN)* [10], *self-evolving interval type-2 fuzzy neural network (SEIT2FNN)* [9], *radial-basis function neural-network according to supervised fuzzy clustering (RBFNN-SFC)* [16], *general type-2 fuzzy system with sampling defuzzification (GT2FS-SD), general type-2 fuzzy system with vertical slice centroid type-reducer (GT2FS-VSCTR) and interval type-2 fuzzy logic system (IT2FLS)* [11–15]. *Tables* 1.4 *and* 1.5 *provide the RMSE comparison for several models. The suggested fuzzy system outperforms conventional models by achieving much greater performance with fewer training iterations and parameters. The suggested IT3FS*

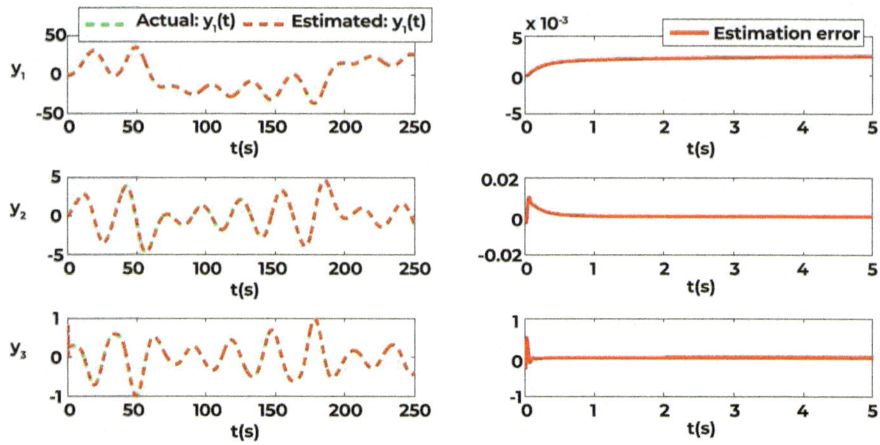

Fig. 1.15 The comparison of RMSE for various models in Ex. 1

Table 1.3 The comparison of RMSE for various models in Ex. 2

Brown models	HT2FS [5]	Proposed method
No. of optimized parameters	80	6
Test RMSE	0.0654	2.2891×10^{-4}

subsequent parameters are learned using the extended Kalman filter in this example, but the simulation parameters of IT3FS remain the same as in Example 1. The inputs of IT3FS are separated into 200 and 5000 in the MCV and MPG issues [17].

Example 4 *The usefulness of the suggested fuzzy system and online learning algorithm is confirmed through experimentation in this example. Figure 1.16 displays the temperature control using an on–off control mechanism, while the system's dynamics are approximated online by the proposed IT3FS. The parameters of Interval Type-3 Fuzzy Sets (IT3FS) are identical to those in Example 1. The estimation performance is displayed in Figs. 1.17 and 1.18. The anticipated output trajectory has closely converged to the actual output trajectory. The real-world application demonstrates*

Table 1.4 The comparison of RMSE for various models in Ex. 3, MPG issue

Brown models	RBFNN-SFC [16]	SNFSSNFS [8]	SEIT2FNN [9]	SEIT2FNN [10]	Proposed IT3FS
No. of rules	6	5	3	3	3
No. of epoches	–	100	100	100	10
Test RMSE	6.77	4.66	4.45	4.23	4.01

Table 1.5 The comparison of RMSE for various models in Ex. 3, maintenance cost issue

Brown models	IT2FLS [11]	GT2FS-VSCTR [11]	GT2FS-SD [11]	Proposed IT3FS
No. of optimized parameters	60	240	240	6
No. of epoches	40	40	40	10
Test RMSE	353.9975	424.3692	317.4332	284.2716

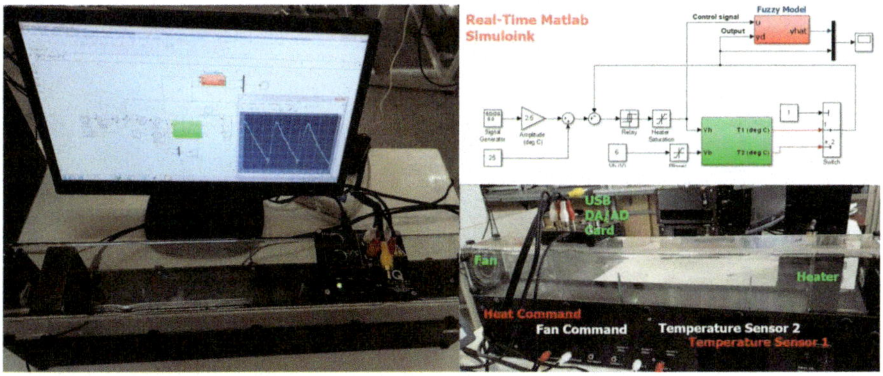

Fig. 1.16 The experimental configuration in Ex. 4

the effectiveness of the proposed strategy for online dynamic approximation problems [17].

Example 5 *This example compares the estimation performance and the time executed forthe planned IT3FS with the GT2FS and IT2FS. The Chuas chaotic system online dynamic estimation involves introducing white noise with zero mean and varying variances as measurement error is added to the state* y_3. *The RMSE values are provided in Table* 1.6. *The simulation situation matches Example 2, and all models have three rules. The suggested IT3FS demonstrates superior performance, particularly in high noise environments. The suggested framework is better equipped to manage uncertainties due to the characteristics of its membership functions. The upper and lower bounds of uncertainty in the general type-2 fuzzy system are represented as crisp values, whereas in the proposed structure, they are depicted as type-1 fuzzy sets (see to Figs. 1.19 and 1.20)* [17].

Note 3 *Only the following parameters are adjusted online in this work. The selection of other parameters is fixed, resulting in the identical values for these parameters throughout five separate examples. The parameters of the membership functions can be established according to the range of fluctuation of the input variables and the degree of uncertainties. The antecedent parameters can also be taught using adaption law (33). To accomplish this, Eqs. (1.31,1.38) need to be taken into account in the following manner:*

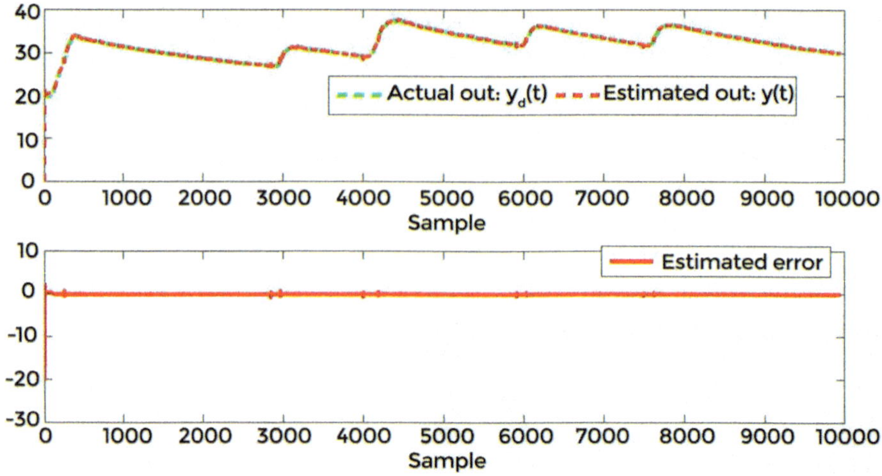

Fig. 1.17 The estimation performance in Ex. 4

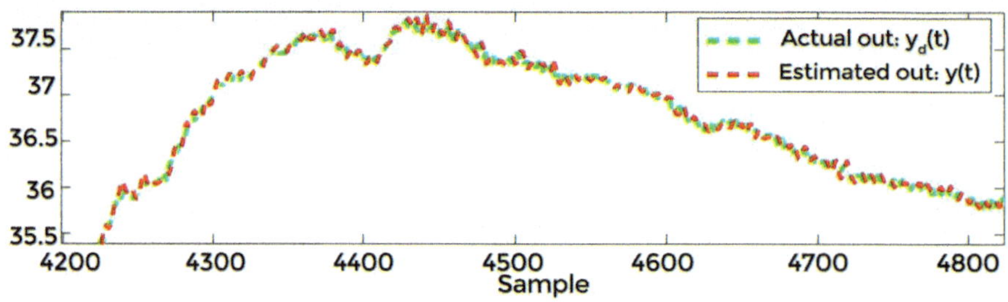

Fig. 1.18 The estimation performance in Ex. 4, zoomed view

Table 1.6 The execution time and RMSE is compared for various models in Ex. 5

Brown models	IT2FS	GT2FS	IT3FS
RMSE (noise variance 0.1)	0.4575	0.4509	0.4350
RMSE (noise variance 1)	302.0601	265.9153	263.3428
Time (s)	13.065	19.735	21.011

$$\theta = \begin{bmatrix} C_{\tilde{A}_1^1}, \cdots C_{\tilde{A}_1^M}, C_{\tilde{A}_n^1}, \cdots C_{\tilde{A}_n^M} \underline{\sigma}_{\tilde{A}_1^1}, \cdots \underline{\sigma}_{\tilde{A}_1^M}, \underline{\sigma}_{\tilde{A}_n^1}, \cdots \\ \underline{\sigma}_{\tilde{A}_n^M} \underline{\sigma}_{\tilde{A}_1^1}, \cdots \underline{\sigma}_{\tilde{A}_1^M}, \underline{\sigma}_{\tilde{A}_n^1}, \cdots \underline{\sigma}_{\tilde{A}_n^M} \underline{w}_1, \ldots, \underline{w}_M, \underline{w}_1, \ldots, \underline{w}_M \end{bmatrix}^T \tag{1.50}$$

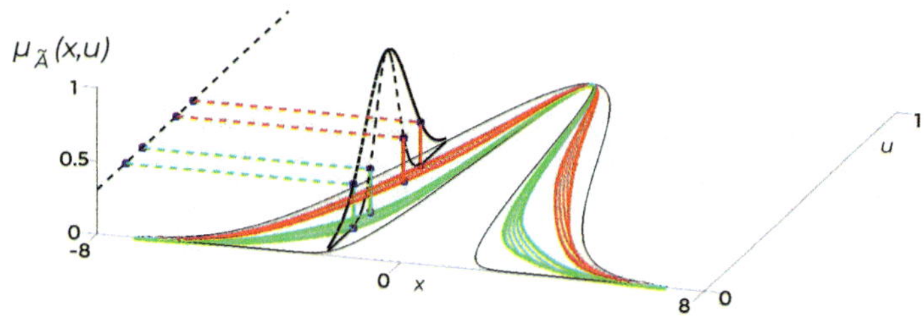

Fig. 1.19 The uncertainty foot print at a horizontal slice as to the interval type-3 fuzzy set in Ex. 5

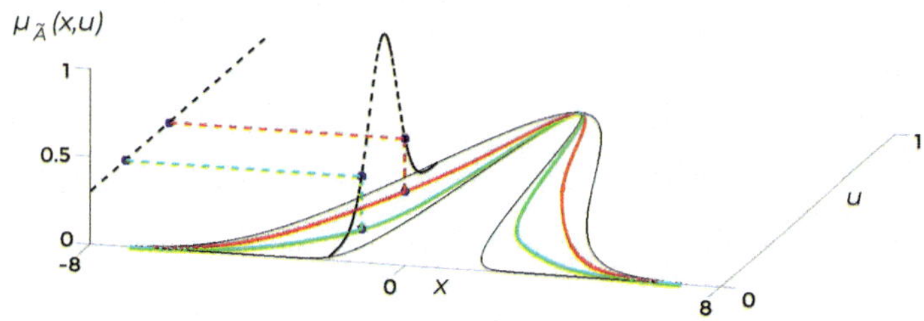

Fig. 1.20 The uncertainty foot print at a horizontal slice regarding the general type-2 fuzzy set in Ex. 5

$$\frac{\partial y(\theta|u)}{\partial \theta} = \begin{bmatrix} \frac{\partial y}{C_{\tilde{A}_1^1}}, \ldots, \frac{\partial y}{C_{\tilde{A}_1^M}}, \cdots \frac{\partial y}{C_{\tilde{A}_n^1}}, \ldots, \frac{\partial y}{C_{\tilde{A}_n^M}} \frac{\partial y}{\underline{\sigma}_{\tilde{A}_1^1}}, \ldots, \frac{\partial y}{\underline{\sigma}_{\tilde{A}_1^M}}, \cdots \\ \frac{\partial y}{\underline{\sigma}_{\tilde{A}_n^1}}, \ldots, \frac{\partial y}{\underline{\sigma}_{\tilde{A}_n^M}} \frac{\partial y}{\underline{\sigma}_{\tilde{A}_1^1}}, \ldots, \frac{\partial y}{\underline{\sigma}_{\tilde{A}_1^M}}, \cdots \frac{\partial y}{\underline{\sigma}_{\tilde{A}_n^1}}, \ldots, \frac{\partial y}{\underline{\sigma}_{\tilde{A}_n^M}} \frac{\partial y}{\underline{w}_1}, \ldots, \\ \frac{\partial y}{\underline{w}_M}, \frac{\partial y}{\underline{w}_1}, \ldots, \frac{\partial y}{\underline{w}_M} \end{bmatrix}^T \tag{1.51}$$

References

1. O. Castillo, J.R. Castro, P. Melin, Interval type-3 fuzzy systems: a natural evolution from type-1 and type-2 fuzzy systems. In *Fuzzy logic and neural networks for hybrid intelligent system design* (pp. 209-221). Cham: Springer International Publishing (2023)
2. O. Castillo, P. Melin, *New horizons for fuzzy logic* (Springer International Publishing AG, Neural Networks and Metaheuristics, 2024)
3. O. Castillo, P. Melin (eds), Hybrid intelligent systems based on extensions of fuzzy logic, neural networks and metaheuristics (Vol. 1096). Springer Nature (2023)
4. J.M. Mendel, R.I. John, F. Liu, Interval type-2 fuzzy logic systems made simple. IEEE Trans. Fuzzy Syst. **14**(6), 808–821 (2006)
5. A. Mohammadzadeh, S. Ghaemi, Synchronization of chaotic systems and identification of nonlinear systems by using recurrent hierarchical type-2 fuzzy neural networks. ISA Trans. **58**, 318–329 (2015)

6. C.-F. Juang, R.-B. Huang, W.-Y. Cheng, An interval type-2 fuzzy neural network with support-vector regression for noisy regression problems. Fuzzy Syst. IEEE Trans. **18**(4), 686–699 (2010)
7. S.W. Tung, C. Quek, C. Guan, Et2fis: an evolving type-2 neural fuzzy inference system. Inf. Sci. **220**, 124–148 (2013)
8. C.-F. Juang, C.-T. Lin, An online self-constructing neural fuzzy inference network and its applications. Fuzzy Syst. IEEE Trans. **6**(1), 12–32 (1998)
9. C.-F. Juang, Y.-W. Tsao, A self-evolving interval type-2 fuzzy neural network with online structure and parameter learning. Fuzzy Syst. IEEE Trans. **16**(6), 1411–1424 (2008)
10. C.-T. Lin, N.R. Pal, S.-L. Wu, Y.-T. Liu, Y.-Y. Lin, An interval type-2 neural fuzzy system for online system identification and feature elimination. IEEE Trans. Neural Netw. Learn. Syst. **26**(7), 1442–1455 (2015)
11. M. Almaraashi, R. John, A. Hopgood, S. Ahmadi, Learning of interval and general type-2 fuzzy logic systems using simulated annealing: theory and practice. Inf. Sci. **360**, 21–42 (2016)
12. L. Zadeh, What is fuzzy logic and what does it have to offer. *Ponencia presentada en WCECS, San Francisco. Resumen recuperado de* (2014). http://www.iaeng.org/WCECS2014/doc/WCECS_2014_keynote_speech_I.pdf
13. L.A. Zadeh, R.A. Aliev, *Fuzzy logic theory and applications: part I and part II* (World Scientific Publishing, 2018)
14. L. Zadeh, Some thoughts about appealing directions for the future of fuzzy theory and technologies along the path traced. In *Fuzzy Logic and Applications: 12th International Workshop, WILF 2018, Genoa, Italy, September 6–7, 2018, Revised Selected Papers* (Vol. 11291, p. 240). Springer (2019)
15. S.N. Shahbazova, M. Sugeno, J. Kacprzyk, L. Zadeh, Recent developments in fuzzy logic and fuzzy sets. Recent Developments in Fuzzy Logic and Fuzzy Sets, Berlin, Germany: Springer, 391 (2020)
16. A. Staiano, R. Tagliaferri, W. Pedrycz, Improving rbf networks performance in regression tasks by means of a supervised fuzzy clustering. Neurocomputing **69**(13–15), 1570–1581 (2006)
17. A. Mohammadzadeh, M.H. Sabzalian, W. Zhang, An interval type-3 fuzzy system and a new online fractional-order learning algorithm: theory and practice. IEEE Trans. Fuzzy Syst. **28**(9), 1940–1950 (2019)

Chapter 2
Applied Type-3 Fuzzy Logic System and Matlab Implementation

2.1 General View

Figure 2.1 provides an example that illustrates the distinctions between IT2, GT2, and IT3 fuzzy sets, allowing for a comprehensive understanding of their differences. The secondary-membership for IT2-FS is a crisp value, GT2-FS is a type-1 fuzzy set, and IT3-FS is a type-2 fuzzy set. IT3-FSs have the capability to represent a greater degree of fuzziness and uncertainty compared to IT2-FS and GT2-FSs. Furthermore, when comparing GT2-FSs with IT2-FSs, it is evident that the secondary membership in GT2-FSs is represented by a fuzzy number, whereas in IT2-FSs it is represented by a fixed number. The GT2-FSs have superior skill in modeling uncertainty compared to IT2-FSs. Another significant distinction is that the upper bound (UB) and lower bound (LB) of the footprint-of-uncertainty (FOU) in IT3-FSs are represented by fuzzy numbers, whereas in IT2-FSs and GT2-FSs, they are represented by crisp values. The Fig. 2.2 illustrates the FOU (footprint-of-uncertainty) for three different types of FS (Fuzzy Sets). Estimating the upper and lower limits of uncertainties is a difficult task in numerous circumstances. IT3-FSs can be useful in circumstances where accurately defining the range of uncertainty is a crucial issue.

The membership function (MF) $\tilde{M}$ can be categorized into different types based on IT2, GT2, and IT3, as shown in Table 2.1. In Table 2.1, the symbol $\sum\sum$ represents the union operation on the variables χ and p. Table 2.1 clearly shows that the primary distinction between IT2-FS and GT2-FS lies in the fact that the secondary membership function remains constant. The distinction between interval and general type-2 membership functions is in the constancy of secondary membership in the interval counterpart. Simply said, in the context of IT2-FS, we have $s_{\tilde{M}}(\chi, p) = 1$. The secondary membership $s_{\tilde{M}}(\chi, p)$ is an interval type-2 fuzzy set, represented as $1/s_{\tilde{M}}(\chi, p)$, when compared to type-3 FS.

R. Sabetahd et al., *Applied Type-3 Fuzzy Logic Systems and Controllers*,
Studies in Fuzziness and Soft Computing 417,
https://doi.org/10.1007/978-3-031-75988-8_2

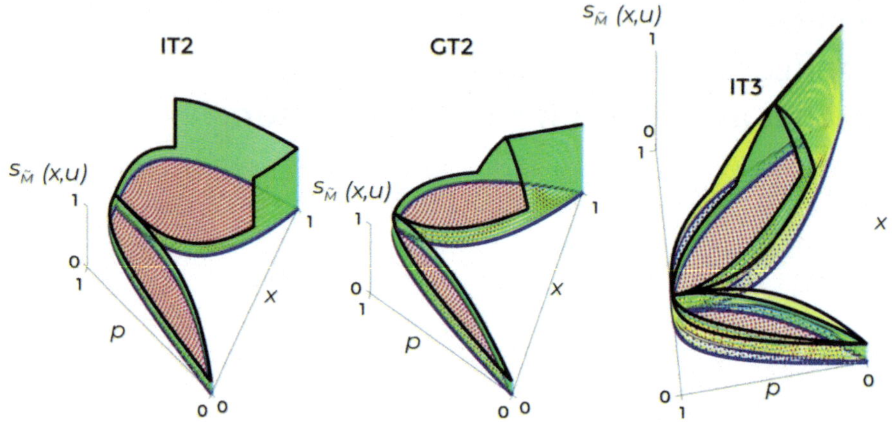

Fig. 2.1 An overview of some horizontal and vertical slices as to IT2, GT2 and IT3 fuzzy sets

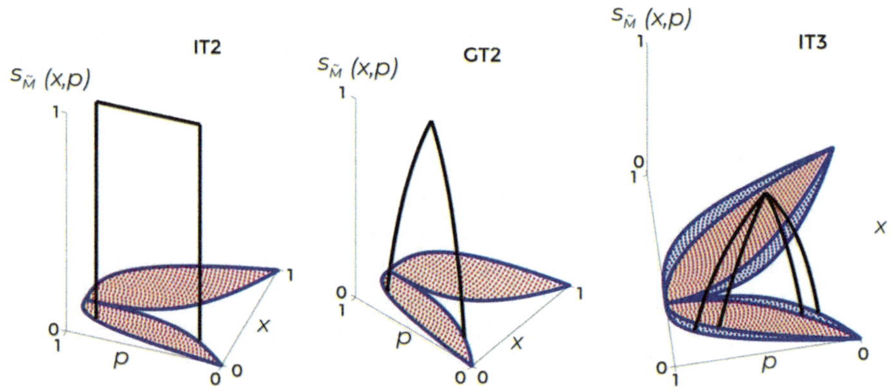

Fig. 2.2 Footprint-of-uncertainty for IT2, GT2 and IT3 fuzzy sets

Table 2.1 Comparison of different type of MFs from the viewpoint of mathematics

Brown	$\tilde{M}$
GT2	$\sum_{\chi\in X}, \sum_{p\in J_\chi} \frac{s_{\tilde M}(\chi,p)}{(\chi,p)}, J_\chi \subseteq [0, 1]$
IT2	$\sum_{\chi\in X}, \sum_{p\in J_\chi} \frac{1}{(\chi,p)}, J_\chi \subseteq [0, 1]$
IT3	$\sum_{\chi\in X}, \sum_{p\in J_\chi} \frac{\left[\frac{1}{s_{\tilde M}(\chi,p)}\right]}{(\chi,p)}, J_\chi \subseteq [0, 1]$

2.2 Proposed Type-3 FLS

Figure 2.3 provides the overall structure. The upper bound (UB) and lower bound (LB) for the footprint-of-uncertainty in type-3 membership functions (MFs) are represented as interval fuzzy sets rather than exact values. Moreover, the SM is classified as a type-2 MF. The calculation is succinctly explained in the following.

(1) Initially, obtain the necessary inputs. The inputs are represented by the symbol χ_i, where i ranges from 1 to n.

(2) According to the information presented in Fig. 2.3, each input χ_i is associated with Q_i membership functions (MFs). The j-th membership function input χ_i is represented as $\tilde{M}_i^j$. The memberships (firing degrees) for $\tilde{M}_i^j$ are calculated. Each horizontal slice α_k has four primary memberships, which are referred to as secondary memberships. The values $pr_{\alpha_{ik}}$ and $\underline{pr}_{\alpha_{ik}}$ indicate the upper and lower limits of the right side, whereas $pl_{\alpha_{ik}}$ and mparison of different type of MFs from $\underline{pl}_{\alpha_{ik}}$ represent the upper and lower limits of the left side. Next, the memberships are calculated for the specified membership function. The memberships for MF, as shown in Fig. 2.4, are calculated as follows:

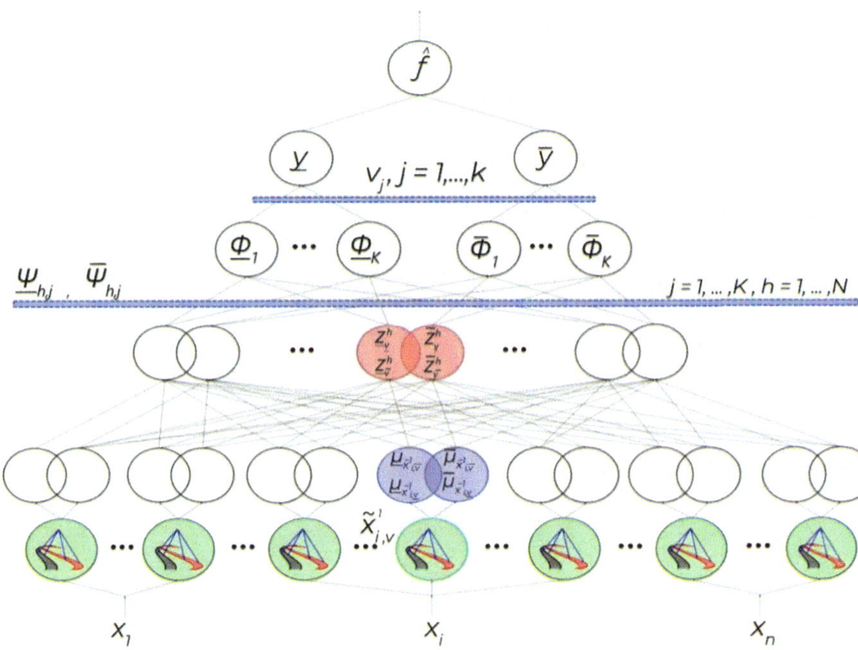

Fig. 2.3 Proposed T3-FLS

$$\underline{p}_{i,r,\alpha_k} = \{ if\ \ c_{\tilde{M}_i^j} - \underline{\vartheta}_{\tilde{M}_i^j} < \chi_i \le c_{\tilde{M}_i^j} 1 - \left(\frac{\left|\chi_i - c_{\tilde{M}_i^j}\right|}{\underline{\vartheta}_{\tilde{M}_i^j}} \right)^{\Delta_r(1-\alpha_k)+1}, \\ if\ \ \tilde{M}_i^j < \chi_i \le c_{\tilde{M}_i^j} + \underline{\vartheta}_{\tilde{M}_i^j} 0 if\ \chi_i > \\ c_{\tilde{M}_i^j} + \underline{\vartheta}_{\tilde{M}_i^j} or \chi_i \le c_{\tilde{M}_i^j} - \underline{\vartheta}_{\tilde{M}_i^j} \tag{2.1}$$

$$\underline{p}_{i,r,\alpha_k} = \{ 1 - \left(\frac{\left|\chi_i - c_{\tilde{M}_i^j}\right|}{\underline{\vartheta}_{\tilde{M}_i^j}} \right)^{1/(\Delta_r(1-\alpha_k)+1)} \quad if\ \ c_{\tilde{M}_i^j} - \underline{\vartheta}_{\tilde{M}_i^j} < \chi_i \le c_{\tilde{M}_i^j}, \\ 1 - \left(\frac{\left|\chi_i - c_{\tilde{M}_i^j}\right|}{\underline{\vartheta}_{\tilde{M}_i^j}} \right)^{1/(\Delta_r(1-\alpha_k)+1)} \quad if_{\tilde{M}_i^j} \left\langle \chi_i \le c_{\tilde{M}_i^j} + \underline{\vartheta}_{\tilde{M}_i^j} 0 if\ \chi_i \right\rangle \\ c_{\tilde{M}_i^j} + \underline{\vartheta}_{\tilde{M}_i^j} or \chi_i \le c_{\tilde{M}_i^j} - \underline{\vartheta}_{\tilde{M}_i^j} \tag{2.2}$$

$$\underline{p}_{i,l,\alpha_k} = \{ 1 - \left(\frac{\left|\chi_i - c_{\tilde{M}_i^j}\right|}{\underline{\vartheta}_{\tilde{M}_i^j}} \right)^{\Delta_l(1-\alpha_k)+1} \\ if\ \ c_{\tilde{M}_i^j} - \underline{\vartheta}_{\tilde{M}_i^j} < \chi_i \le c_{\tilde{M}_i^j} 1 - \left(\frac{\left|\chi_i - c_{\tilde{M}_i^j}\right|}{\underline{\vartheta}_{\tilde{M}_i^j}} \right)^{\Delta_l(1-\alpha_k)+1} \\ if\ \ \tilde{M}_i^j \left\langle \chi_i \le c_{\tilde{M}_i^j} + \underline{\vartheta}_{\tilde{M}_i^j} 0 if\ \chi_i \right\rangle \\ c_{\tilde{M}_i^j} + \underline{\vartheta}_{\tilde{M}_i^j} or \chi_i \le c_{\tilde{M}_i^j} - \underline{\vartheta}_{\tilde{M}_i^j} \tag{2.3}$$

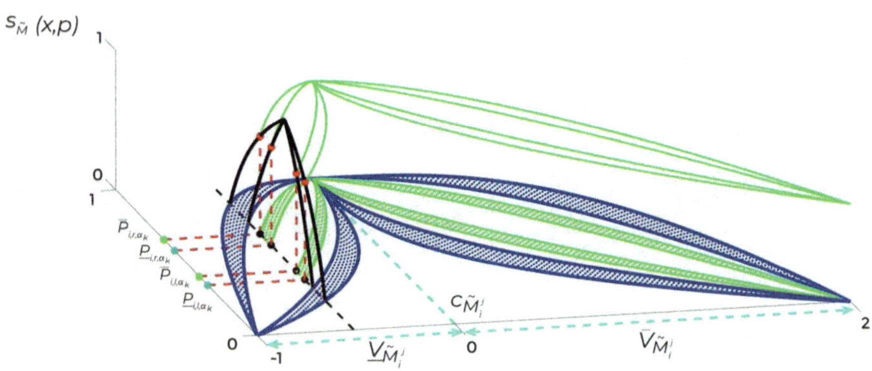

Fig. 2.4 Type-3 MF

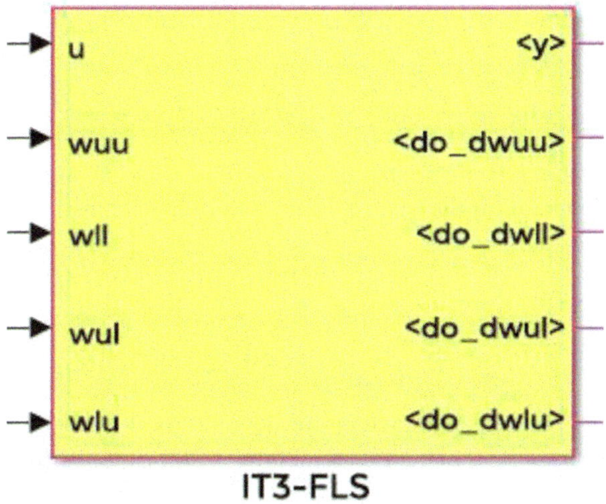

Fig. 2.5 Simulink block

$$\underline{p}_{i,l,\alpha_k} =\{1 - \left(\frac{\left|\chi_i - c_{\tilde{M}_i^j}\right|}{\underline{\vartheta}_{\tilde{M}_i^j}}\right)^{1/(\Delta_l(1-\alpha_k)+1)}$$

$$if\ c_{\tilde{M}_i^j} - \underline{\vartheta}_{\tilde{M}_i^j} < \chi_i \le c_{\tilde{M}_i^j} 1 - \left(\frac{\left|\chi_i - c_{\tilde{M}_i^j}\right|}{\underline{\vartheta}_{\tilde{M}_i^j}}\right)^{1/(\Delta_l(1-\alpha_k)+1)}$$

$$if\ \tilde{M}_i^j$$

$$\left\langle \chi_i \le c_{\tilde{M}_i^j} + \underline{\vartheta}_{\tilde{M}_i^j} 0\, if\ \chi_i \right\rangle c_{\tilde{M}_i^j} + \underline{\vartheta}_{\tilde{M}_i^j}\ or\ \chi_i \le c_{\tilde{M}_i^j} - \underline{\vartheta}_{\tilde{M}_i^j} \tag{2.4}$$

In this context, $c_{\tilde{M}_i^j}$ represents the center of $\tilde{M}_i^j$, whereas $\underline{\vartheta}_{\tilde{M}_i^j}$ and $\underline{\vartheta}_{\tilde{M}_i^j}$ refer to the widths of the left and right sides of $\tilde{M}_i^j$ respectively (see Fig. 2.4). The values of the horizontal slices are represented by $\alpha_k k = 1, ..., K$. Δ_r and Δ_l denote the magnitude of the width of membership functions (MFs). Reduced values of Δ_r and Δ_l lead to the formation of slender shapes of membership functions.

(C) By taking into account the memberships of the upper bound (UB) and lower bound (LB) of the membership functions (MFs) $\tilde{M}_i^j$, the rule firings can be given as follows:

$$\underline{z}_{r,\alpha_k}^h = \underline{s}_{\tilde{M}_1^{h_1}|r,\alpha_k}(\chi_1)\underline{s}_{\tilde{M}_2^{h_2}|r,\alpha_k}(\chi_2)\ldots\underline{s}_{\tilde{M}_n^{h_n}|r,\alpha_k}(\chi_n) \tag{2.5}$$

$$\underline{z}_{r,\alpha_k}^h = \underline{s}_{\tilde{M}_1^{h_1}|r,\alpha_k}(\chi_1)\underline{s}_{\tilde{M}_2^{h_2}|r,\alpha_k}(\chi_2)\ldots\underline{s}_{\tilde{M}_n^{h_n}|r,\alpha_k}(\chi_n) \tag{2.6}$$

$$\underline{z}_{l,\alpha_k}^{h} = \underline{s}_{\tilde{M}_1^{h_1}|l,\alpha_k}(\chi_1)\underline{s}_{\tilde{M}_2^{h_2}|l,\alpha_k}(\chi_2)\ldots.\underline{s}_{\tilde{M}_n^{h_n}|l,\alpha_k}(\chi_n) \tag{2.7}$$

$$\underline{z}_{l,\alpha_k}^{h} = \underline{s}_{\tilde{M}_1^{h_1}|l,\alpha_k}(\chi_1)\underline{s}_{\tilde{M}_2^{h_2}|l,\alpha_k}(\chi_2)\ldots.\underline{s}_{\tilde{M}_n^{h_n}|l,\alpha_k}(\chi_n) \tag{2.8}$$

Let k be a variable that ranges from 1 to K. $\tilde{M}_i^{h_i}$ represents the h_i-th membership function for the input χ_i. The h-th rule is defined as follows:

$$If x_1 is \tilde{M}_1^{h_1}, \ldots, and \ldots x_n is \tilde{M}_n^{h_n} Then$$

$$y_h \in \left[\left[\underline{\Psi}_{l,h}, \underline{\Psi}_{r,h}\right]\left[\underline{\Psi}_{l,h}, \underline{\Psi}_{r,h}\right]\right] \tag{2.9}$$

The rule parameters are denoted as $\underline{\Psi}_{l,h}$, $\underline{\Psi}_{r,h}$, $\underline{\Psi}_{l,h}$, and $\underline{\Psi}_{r,h}$. The output is considered to be an interval type-3 fuzzy number at each rule, where both its lower bound (LB) and upper bound (UB) are likewise represented as fuzzy numbers. To clarify, the lower bound (LB) is a part of the set $\left[\underline{\Psi}_{l,h}, \underline{\Psi}_{r,h}\right]$, while the upper bound (UB) is a part of the set $\left[\underline{\Psi}_{l,h}, \underline{\Psi}_{r,h}\right]$.

(D) The most basic approximation of the output is proposed as:

$$y = \frac{\sum_{k=1}^{K}\left(\underline{\phi}_k + \underline{\phi}_k\right)\alpha_k}{\sum_{k=1}^{K}\alpha_k} \tag{2.10}$$

where,

$$\underline{\phi}_k = \frac{\sum_{h=1}^{N}\left(\underline{z}_{r,\alpha_k}^{h}\underline{\Psi}_{r,k}^{h} + \underline{z}_{l,\alpha_k}^{h}\underline{\Psi}_{l,k}^{h}\right)}{\sum_{h=1}^{N}\underline{z}_{r,\alpha_k}^{h} + \underline{z}_{l,\alpha_k}^{h}} \tag{2.11}$$

$$\underline{\phi}_k = \frac{\sum_{h=1}^{N}\left(\underline{z}_{r,\alpha_k}^{h}\underline{\Psi}_{r,k}^{h} + \underline{z}_{l,\alpha_k}^{h}\underline{\Psi}_{l,k}^{h}\right)}{\sum_{h=1}^{N}\underline{z}_{r,\alpha_k}^{h} + \underline{z}_{l,\alpha_k}^{h}} \tag{2.12}$$

2.3 Learning Approach

Derivative-based optimization approaches can be employed to tune the rule parameters in numerous issues optimally. The calculation of the derivative of the output of FLS with respect to the rule variables is necessary in these efficient algorithms. This section presents the following computations. The derivatives are acquired through the process of differentiation as follows:

$$\frac{\partial y}{\partial \underline{\Psi}_{r,k}^{h}} = \frac{\alpha_k}{\sum_{k=1}^{K}\alpha_k}\frac{\underline{z}_{r,\alpha_k}^{h}}{\sum_{h=1}^{N}\underline{z}_{r,\alpha_k}^{h} + \underline{z}_{l,\alpha_k}^{h}} \tag{2.13}$$

$$\frac{\partial y}{\underline{\Psi}^h_{l,k}} = \frac{\alpha_k}{\sum_{k=1}^{K}\alpha_k} \frac{\underline{z}^h_{l,\alpha_k}}{\sum_{h=1}^{N}\underline{z}^h_{r,\alpha_k} + \underline{z}^h_{l,\alpha_k}} \tag{2.14}$$

$$\frac{\partial y}{\underline{\Psi}^h_{r,k}} = \frac{\alpha_k}{\sum_{k=1}^{K}\alpha_k} \frac{\underline{z}^h_{r,\alpha_k}}{\sum_{h=1}^{N}\underline{z}^h_{r,\alpha_k} + \underline{z}^h_{l,\alpha_k}} \tag{2.15}$$

$$\frac{\partial y}{\underline{\Psi}^h_{l,k}} = \frac{\alpha_k}{\sum_{k=1}^{K}\alpha_k} \frac{\underline{z}^h_{l,\alpha_k}}{\sum_{h=1}^{N}\underline{z}^h_{r,\alpha_k} + \underline{z}^h_{l,\alpha_k}} \tag{2.16}$$

2.4 Performing T3-FLSs in Matlab Simulink

This section proposes a pragmatic implementation method for Matlab. The recommended Toolbox is highly versatile and may be effortlessly utilized for a wide range of applications. Figure 2.4 displays the proposed scheme. The output signals consist of the output of T3-FLS, as well as the derivatives of the output of T3-FLS with respect to rule variables. These derivatives can be utilized in tuning rules.

In order to demonstrate the ease of use of the intended toolbox, a straightforward example is provided below. Consider a system that consists of two inputs, denoted as χ_1 and χ_2. For the given input χ_1, we consider two membership functions, denoted as $\tilde{M}_1^1$ and $\tilde{M}_1^2$. For the input χ_2, three membership functions are taken into account: $\tilde{M}_2^1, \tilde{M}_2^2$, and $\tilde{M}_2^3$. The horizontal slices are denoted by $\alpha_1 = 0, \alpha_2 = 0.5$, and $\alpha_3 = 1$. The widths Δ_r and Δ_l for all MFs are defined as $\Delta_r = 3$ and $\Delta_l = 2$, respectively, as shown in Eqs. 2.1–2.4.

Let us now take into account the following rules:

If χ_1 is $\tilde{M}_1^1$ and χ_2 is $\tilde{M}_2^1$ then y is $y_1 \in \left[\left[\underline{\Psi}_{l,1}, \underline{\Psi}_{r,1}\right]\left[\underline{\Psi}_{l,1}, \underline{\Psi}_{r,1}\right]\right]$.
If χ_1 is $\tilde{M}_1^1$ and χ_2 is $\tilde{M}_2^2$ then y is $y_2 \in \left[\left[\underline{\Psi}_{l,2}, \underline{\Psi}_{r,2}\right]\left[\underline{\Psi}_{l,2}, \underline{\Psi}_{r,2}\right]\right]$.
If χ_1 is $\tilde{M}_1^2$ and χ_2 is $\tilde{M}_2^2$ then y is $y_3 \in \left[\left[\underline{\Psi}_{l,3}, \underline{\Psi}_{r,3}\right]\left[\underline{\Psi}_{l,3}, \underline{\Psi}_{r,3}\right]\right]$.
If χ_1 is $\tilde{M}_1^2$ and χ_2 is $\tilde{M}_2^3$ then y is $y_4 \in \left[\left[\underline{\Psi}_{l,4}, \underline{\Psi}_{r,4}\right]\left[\underline{\Psi}_{l,4}, \underline{\Psi}_{r,4}\right]\right]$.

The variable $\tilde{M}_i^j$ represents the $j-th$ membership function for the $i-th$ input. Therefore, the information in the planned toolbox should be completed in the manner indicated in Figs. 2.6 and 2.7. Let's consider the rules as follows:

- Number of inputs: 2
- MFs centers: {{−1 1},{−1 0 1}}; which means, $c_{\tilde{M}_1^1} = -1$, $c_{\tilde{M}_1^2} = 1$, $c_{\tilde{M}_2^1} = -1$, $c_{\tilde{M}_2^2} = 0$, and, $c_{\tilde{M}_2^3} = 1$.
- Upper MFs Width: {{3 3},{3 3 3}} meaning $\Delta_r = 3$, (see Eqs. 2.1 and 2.2)).
- Lower MFs Width: {{2 2},{2 2 2}} meaning $\Delta_l = 3$, (see Eqs. 2.3–2.4).
- Right distance: {{2 2},{1 1 1}} meaning $\underline{\vartheta}_{\tilde{M}_1^j} = 2, j = 1, 2$, and $\underline{\vartheta}_{\tilde{M}_2^j} = 1, j = 1, 2, 3$ (see Eqs. 2.1–2.4).
- Left distance: {{2 2},{1 1 1}} meaning $\underline{\vartheta}_{\tilde{M}_1^j} = 2, j = 1, 2$, and $\underline{\vartheta}_{\tilde{M}_2^j} = 1, j = 1, 2, 3$ (see Eqs. 2.1–2.4.

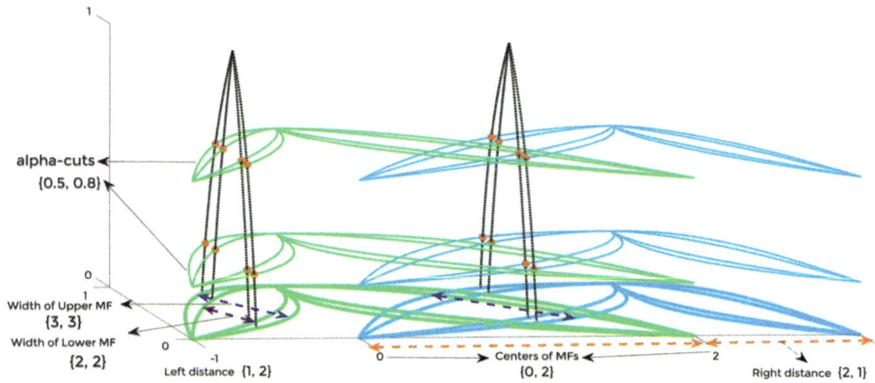

Fig. 2.6 A modal for designed MFs

- The rules are as follows: [1 1;1 2;2 2;2 3]. This indicates that in the first rule, $\tilde{M}_1^1$ and $\tilde{M}_2^1$, for the second rule, rule $\tilde{M}_1^1$ and $\tilde{M}_2^2$, and in the third rule, $\tilde{M}_1^2$ and $\tilde{M}_2^2$ are all fired. Finally, in the last rule, $\tilde{M}_1^2$ and $\tilde{M}_2^3$ are fired.
- The alpha-cuts are defined as: [0 0.5 1], indicating that there are three horizontal slices, with alpha values of $\alpha_1 = 0$, $\alpha_1 = 0.5$, and $\alpha_1 = 1$.

2.5 Performing T3-FLSs in Matlab M-file

A straightforward method to apply the T3-FLSs to different problems is by utilizing a concise M-file code. An exemplification is provided to demonstrate more effectively the use of the designed code. The issue at hand involves the representation of $y(t)$ in Eq. (2.17). The input variables are denoted as as $y(t-1)$, $y(t-2)$ and $u(t-1)$. The function $u(t)$ is defined as $sin(2\pi n/25)$. The training and testing data sets are generated by $t \in [0, 4999]$ and $t \in [5000, 5299]$, respectively.

$$y(t) = y(t-1)(y(t-1) - 0.5)y(t-2)/\left[1 + y^2(t-1) + y^2(t-2)\right] + u(t-1) \tag{2.17}$$

The presented function in Matlab is denoted as "out = Type3_FLS(u)". When the input is "u", we should provide the following information:

- The vector of inputs is denoted as u.x. For this problem, there are 3 inputs. At each epoch, u.x is a vector consisting of 3 elements.
- The vector u.c represents the centers of the membership functions. Refer to Eqs. 2.1–2.4 for more details. The centers should be entered as shown in the preceding section. This problem involves three inputs and we consider three membership functions (MFs) for each input. The value of u.c is determined as u.c = {{−1 0 1}, {−1 0 1}, {−1 0 1}}.

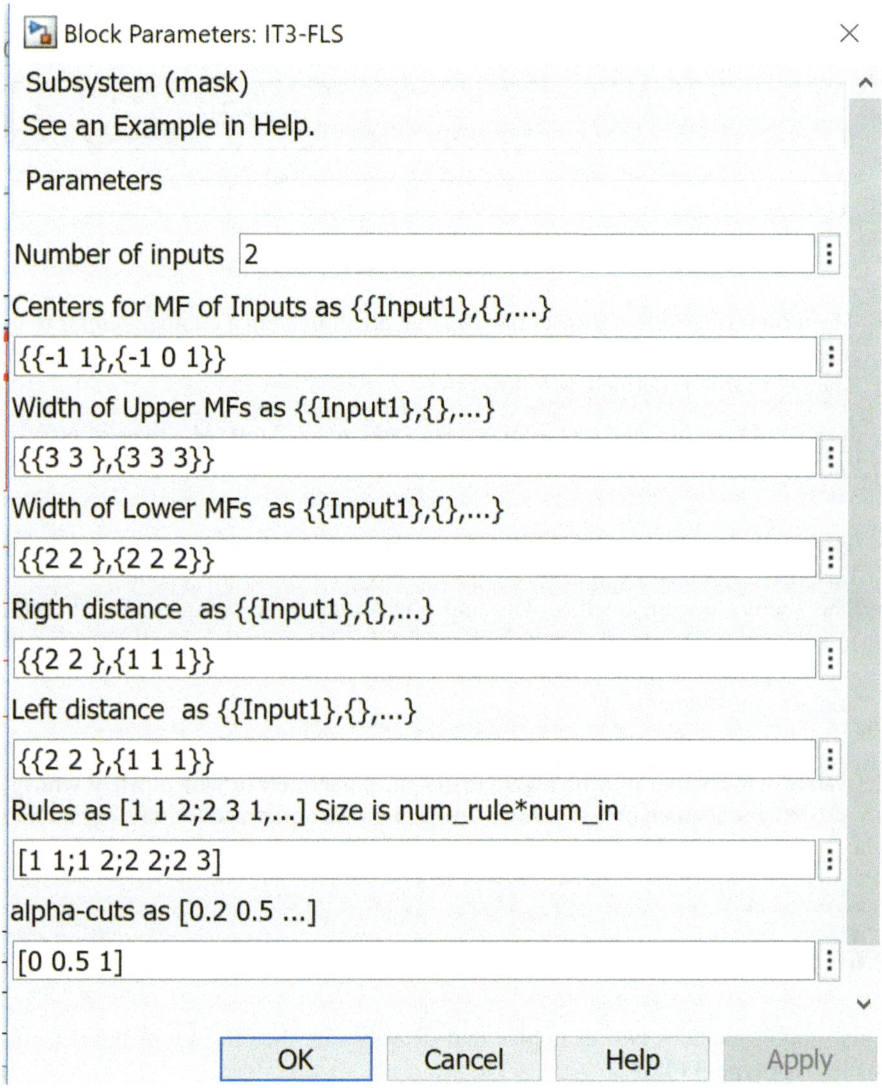

Fig. 2.7 IT3-FLS information

- The symbols u.au and u.al correspond to the values of Δ_r and Δ_l, as indicated in Eqs. 2.1–2.4.
- The variables u.dr and u.dl denote the distances of the center of MFs from the right and left points, respectively. These distances can be seen as $\underline{\vartheta}$ and $\underline{\vartheta}$ in Eqs. 2.1–2.4.

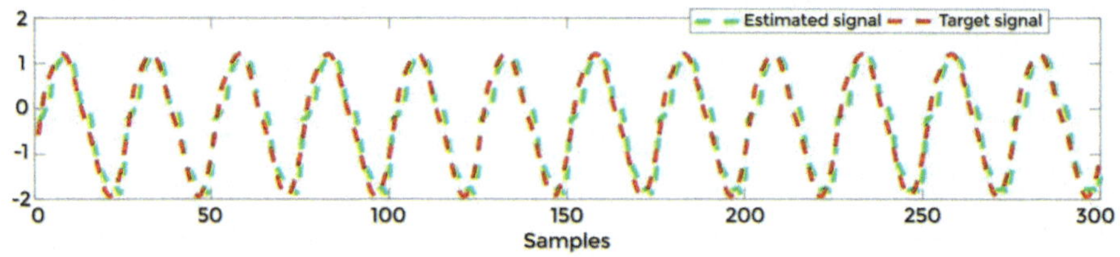

Fig. 2.8 The real and estimated signals

- The term u.rules refers to a set of rules that are defined in a manner similar to the rules mentioned in the previous section. Three rules are defined for this situation. The $i-th$ rule is defined as follows:

  If $y(t-1)$ is $\tilde{M}_1^i$ and $y(t-2)$ is $\tilde{M}_2^i$, and $u(t-1)$ is $\tilde{M}_3^i$ then $\hat{y}$ is $\hat{y}_i \in [[\underline{\Psi}_{l,i}, \underline{\Psi}_{r,i}][\underline{\Psi}_{l,i}, \underline{\Psi}_{r,i}]]$; $i = 1, 2, 3$. Then u.rules is defined as u.rules = [1 1 1;2 2 2; 3 3 3].

- The symbol u.alpha denotes the value of a horizontal slice, as shown by α in Eqs. 2.1–2.4. Only a single slice is taken into account for this problem.
- The vectors u.wuu, u.wll, u.wlu, and u.wul reflect the parameters that follow in sequence. u.wuu contains the elements $\underline{\Psi}_{r,i}$, $i = 1, 2, 3$. u.wll contains the elements $\underline{\Psi}_{l,i}$, $i = 1, 2, 3$. u.wul contains the elements $\underline{\Psi}_{l,i}$, $i = 1, 2, 3$, and u.wlu contains the elements $\underline{\Psi}_{r,i}$, $i = 1, 2, 3$.

The function "out = Type3_FLS(u)" returns the estimated signal "out" and the derivative of the output ($\hat{y}$) with regard to the rule parameters (u.wuu, u.wll, u.wlu, and u.wul). When the main program is executed, then we optimized the rule parameters. The real and estimated signals are displayed in Fig. 2.8.

2.6 Simulation

This section includes two examples that demonstrate the efficacy of the designed method for type-3 FLSs.

Example 1 *This section presents a straightforward simulation to demonstrate the practical application of the built type-3 fuzzy logic system (FLS). The proposed IT3-FLS is utilized to analyze and identify a chaotic system. The mathematical representation of the chaotic system under investigation is as follows:*

$$\begin{aligned}\dot{\chi}_1 &= -10\chi_1 + 10\chi_2 \\ \dot{\chi}_2 &= -\chi_2(28 - \chi_3)\chi_1 \\ \dot{\chi}_3 &= -8\chi_3/3 + \chi_2\chi_1\end{aligned} \tag{2.18}$$

The inputs of the Fuzzy Logic System (FLS) are denoted as $\chi_1/20$, $\chi_2/20$ and $\chi_3/40$. Three membership functions are considered for each input. The centroids of membership functions for the first and second inputs are located at –1, 0, and 1, whereas for the third input they are located at 0, 0.5, and 1. The first and second inputs of MFs width is 2, whereas the width of the third MF is 0.5. There are three alpha-cuts, with values of 0, 0.5, and 1. There are three rules, followed by 12 consequent parameters.

Figures 2.2 and 2.9 illustrates the simulation diagram. The signals χ_i, $i = 1, 2, 3$, are estimated using IT3-FLS approximation. The trajectory estimate $\hat{\chi}_1$ and actual signal χ_1 are displayed in Fig. 2.10. The trajectory $\hat{\chi}_2$ and real signal χ_2 are displayed in Fig. 2.11. Figure 2.12 displays the anticipated trajectory $\hat{\chi}_3$ and real signal χ_3. It is observed that the recommended Simulink functions effectively in online applications.

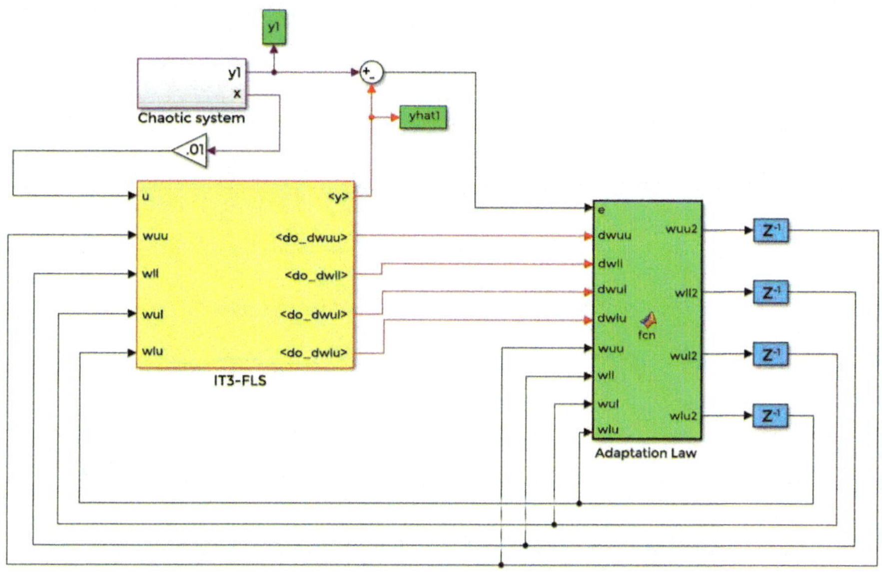

Fig. 2.9 Example 1: Simulation scheme

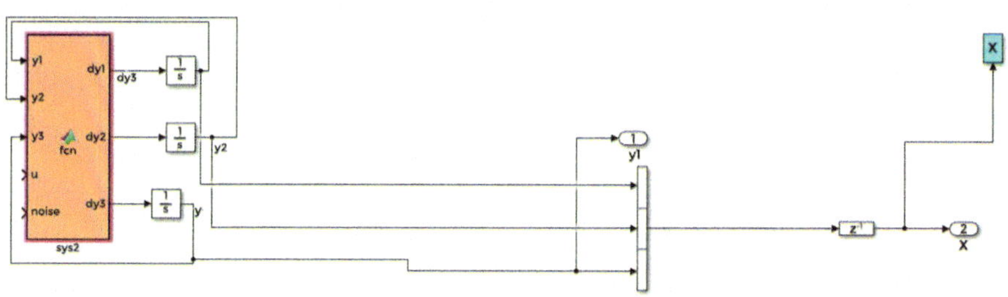

Fig. 2.10 Example 1: Chaotic system

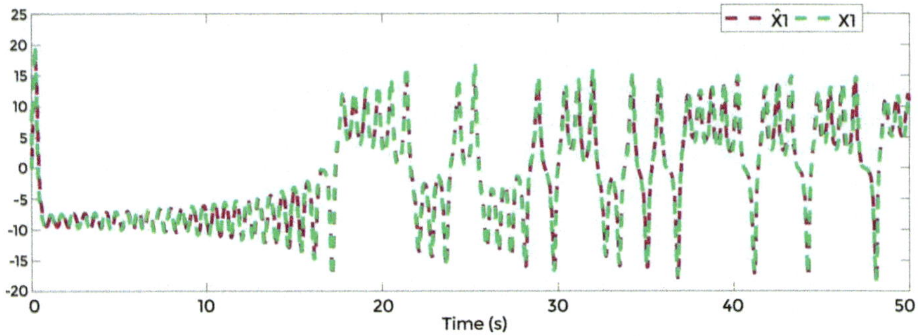

Fig. 2.11 The estimation performance of χ_1 in Ex. 1

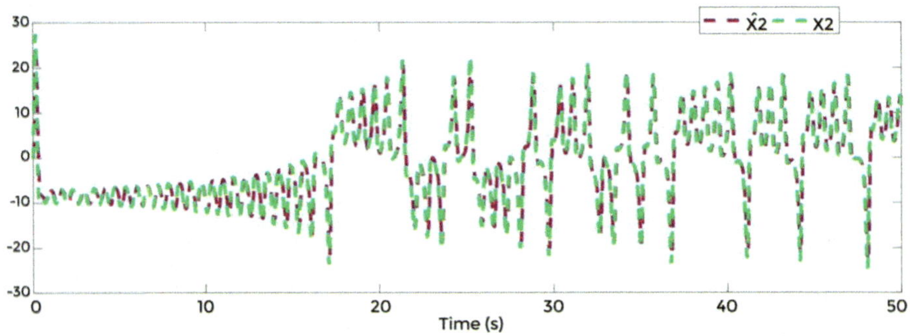

Fig. 2.12 The estimation performance of χ_2 in Ex. 1

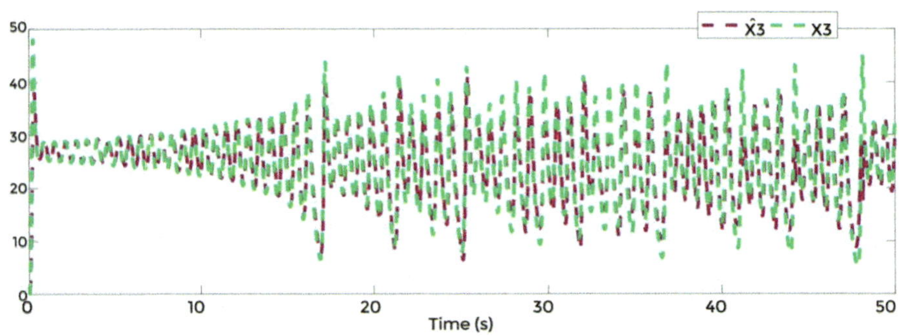

Fig. 2.13 The estimation performance of χ_3 in Ex. 1

Related Matlab Codes for Fig. 2.9:

The corresponding codes related to the orange Adaptation Law box shown in Fig. 2.9 are as follows:

MATLAB script for adaption law

```
function [wuu2,wll2,wul2,wlu2] = fcn(e,dwuu,dwll,dwul,dwlu,wuu,wll,wul,wlu)
%#eml
eta = 1;
wuu2 = wuu + eta*e*dwuu;
wll2 = wll + eta*e*dwll;
wul2 = wul + eta*e*dwul;
wlu2 = wlu + eta*e*dwlu;
```

The corresponding codes related to the dlue chaotic system box shown in Fig. 2.10 are as follows:

MATLAB script for chaotic system

```
function [dy1,dy2,dy3] = fcn(y1,y2,y3,u,noise)
%#eml
p = 28; sigma = 10;beta = 8/3;
dy1 = sigma*(y2-y1);
dy2 = y1*(p-y3)-y2;
dy3 = y1*y2-beta*y3;
Then the following Simulink codes are available:
function out = initialization(u)
pp = Simulink.Mask.get(gcb);%'AIT2FLS/IT2FLS');
[n c au al dr dl rules alpha] = pp.Parameters.Value;
n = eval(n); c = eval(c); au = eval(au);
al = eval(al); rules = eval(rules); alpha = eval(alpha);
dr = eval(dr);dl = eval(dl);
num_rule = size(rules,1);
k = numel(alpha);
% if numel(UU) < 2*num_rule + n
% u = zeros(2*num_rule + n,1);
% else
% u = UU;
% end
%%
x(:,1) = u(1:n);
wuu = reshape(u(n + 1:n + num_rule*k),num_rule,k);
wll = reshape(u(n + num_rule*k + 1:n + 2*num_rule*k),num_rule,k);
wul = reshape(u(n + 2*num_rule*k + 1:n + 3*num_rule*k),num_rule,k);
wlu = reshape(u(n + 3*num_rule*k + 1:n + 4*num_rule*k),num_rule,k);
%%
% u is n*1 where n is number of inputs
% c is n*n_MF where n_MF is the number of MFs for this input
% the dimension of sigma is the same as c
% the variable rule represent the rule database
% rule is n_rule*n, where n_rule is the number of rules
%%
% clear all
% clc
% u = [1 2]';
% c = [-1 0 1; -1 0 1];
```

(continued)

(continued)

MATLAB script for chaotic system

```
% sigma = ones(size(c));
% rule = [1 3; 2 3];
% w = rand(2,1);–2 2
%c = [–2 –1 0 1 2;3 4 0 5 6]
%%
%au = zeros(num_rule,n);al = zeros(num_rule,n);
yuu = {};yll = {};yul = {};ylu = {};dwuu = zeros(num_rule,k);dwll = zeros(num_
rule,k);dwul = zeros(num_rule,k);dwlu = zeros(num_rule,1);
%% MF
y = 0;
for k = 1:length(alpha)
for ii = 1:n
for jj = 1:length(c{ii})
if x(ii) > c{ii}{jj} && x(ii) < = (c{ii}{jj} + dr{ii}{jj})
yuu{ii}(jj) = (1-abs((x(ii)-c{ii}{jj})./dr{ii}{jj}).^au{ii}{jj});
yll{ii}(jj) = (1-abs((x(ii)-c{ii}{jj})./dr{ii}{jj}).^(1/au{ii}{jj}));
yul{ii}(jj) = (1-abs((x(ii)-c{ii}{jj})./dr{ii}{jj}).^al{ii}{jj});
ylu{ii}(jj) = (1-abs((x(ii)-c{ii}{jj})./dr{ii}{jj}).^(1/al{ii}{jj}));
elseif x(ii) < c{ii}{jj} && x(ii) > = (c{ii}{jj}-dl{ii}{jj})
yuu{ii}(jj) = (1-abs((x(ii)-c{ii}{jj})./dl{ii}{jj}).^au{ii}{jj});
yll{ii}(jj) = (1-abs((x(ii)-c{ii}{jj})./dl{ii}{jj}).^(1/au{ii}{jj}));
yul{ii}(jj) = (1-abs((x(ii)-c{ii}{jj})./dl{ii}{jj}).^al{ii}{jj});
ylu{ii}(jj) = (1-abs((x(ii)-c{ii}{jj})./dl{ii}{jj}).^(1/al{ii}{jj}));
elseif x(ii) = = c{ii}{jj}
yuu{ii}(jj) = 1;
yll{ii}(jj) = 1;
yul{ii}(jj) = 1;
ylu{ii}(jj) = 1;
else
yuu{ii}(jj) = 0;
yll{ii}(jj) = 0;
yul{ii}(jj) = 0;
ylu{ii}(jj) = 0;
end
end
end
%% Rule fiering
for ii = 1:n
Zuu(:,ii) = yuu{ii}(rules(:,ii));
Zll(:,ii) = yll{ii}(rules(:,ii));
Zul(:,ii) = yul{ii}(rules(:,ii));
Zlu(:,ii) = ylu{ii}(rules(:,ii));
end
zuu(:,k) = prod(Zuu,2);%min
zll(:,k) = prod(Zll,2);%min
zul(:,k) = prod(Zul,2);%min
zlu(:,k) = prod(Zlu,2);%min
%% Type reduction
sum_zu = (sum(zuu(:,k) + zll(:,k)) + 0.001);
```

(continued)

(continued)

MATLAB script for chaotic system

```
sum_zl = (sum(zlu(:,k) + zul(:,k)) + 0.001);
Pu(k,1) = (zuu(:,k)'*wuu(:,k) + zll(:,k)'*wll(:,k))/sum_zu;
Pl(k,1) = (zul(:,k)'*wul(:,k) + zlu(:,k)'*wlu(:,k))/sum_zl;
dwuu(:,k) = alpha(k)*zuu(:,k)/sum_zu;
dwll(:,k) = alpha(k)*zll(:,k)/sum_zu;
dwul(:,k) = alpha(k)*zul(:,k)/sum_zl;
dwlu(:,k) = alpha(k)*zlu(:,k)/sum_zl;
end%aa
fhat = alpha*(Pu + Pl)/sum(alpha);
dwuu = dwuu/sum(alpha);
dwll = dwll/sum(alpha);
dwul = dwul/sum(alpha);
dwlu = dwlu/sum(alpha);
out = [fhat;reshape(dwuu,num_rule*k,1);reshape(dwll,num_rule*k,1);reshape(dwul,num_
rule*k,1);reshape(dwlu,num_rule*k,1)];
end
```

Initialization codes for u can be considered as follows:

MATLAB script for u initialization

```
function out = initialization(u)
pp = Simulink.Mask.get(gcb);%'AIT2FLS/IT2FLS');
[n c au al dr dl rules alpha] = pp.Parameters.Value;
n = eval(n); c = eval(c); au = eval(au);
al = eval(al); rules = eval(rules); alpha = eval(alpha);
dr = eval(dr);dl = eval(dl);
num_rule = size(rules,1);
k = numel(alpha);
% if numel(UU) < 2*num_rule + n
% u = zeros(2*num_rule + n,1);
% else
% u = UU;
% end
%%
x(:,1) = u(1:n);
wuu = reshape(u(n + 1:n + num_rule*k),num_rule,k);
wll = reshape(u(n + num_rule*k + 1:n + 2*num_rule*k),num_rule,k);
wul = reshape(u(n + 2*num_rule*k + 1:n + 3*num_rule*k),num_rule,k);
wlu = reshape(u(n + 3*num_rule*k + 1:n + 4*num_rule*k),num_rule,k);
%%
% u is n*1 where n is number of inputs
% c is n*n_MF where n_MF is the number of MFs for this input
% the dimension of sigma is the same as c
% the variable rule represent the rule database
% rule is n_rule*n, where n_rule is the number of rules
%%
% clear all
```

(continued)

(continued)

MATLAB script for *u* initialization

```
% clc
% u = [1 2]';
% c = [–1 0 1;–1 0 1];
% sigma = ones(size(c));
% rule = [1 3; 2 3];
% w = rand(2,1);–2 2
%c = [–2 –1 0 1 2;3 4 0 5 6]
%%
%au = zeros(num_rule,n);al = zeros(num_rule,n);
yuu = {};yll = {};yul = {};ylu = {};dwuu = zeros(num_rule,k);dwll = zeros(num_
rule,k);dwul = zeros(num_rule,k);dwlu = zeros(num_rule,1);
%% MF
y = 0;
for k = 1:length(alpha)
for ii = 1:n
for jj = 1:length(c{ii})
if x(ii) > c{ii}{jj} && x(ii) < = (c{ii}{jj} + dr{ii}{jj})
yuu{ii}(jj) = (1-abs((x(ii)-c{ii}{jj})./dr{ii}{jj}).^au{ii}{jj});
yll{ii}(jj) = (1-abs((x(ii)-c{ii}{jj})./dr{ii}{jj}).^(1/au{ii}{jj}));
yul{ii}(jj) = (1-abs((x(ii)-c{ii}{jj})./dr{ii}{jj}).^al{ii}{jj});
ylu{ii}(jj) = (1-abs((x(ii)-c{ii}{jj})./dr{ii}{jj}).^(1/al{ii}{jj}));
elseif x(ii) < c{ii}{jj} && x(ii) > = (c{ii}{jj}-dl{ii}{jj})
yuu{ii}(jj) = (1-abs((x(ii)-c{ii}{jj})./dl{ii}{jj}).^au{ii}{jj});
yll{ii}(jj) = (1-abs((x(ii)-c{ii}{jj})./dl{ii}{jj}).^(1/au{ii}{jj}));
yul{ii}(jj) = (1-abs((x(ii)-c{ii}{jj})./dl{ii}{jj}).^al{ii}{jj});
ylu{ii}(jj) = (1-abs((x(ii)-c{ii}{jj})./dl{ii}{jj}).^(1/al{ii}{jj}));
elseif x(ii) = = c{ii}{jj}
yuu{ii}(jj) = 1;
yll{ii}(jj) = 1;
yul{ii}(jj) = 1;
ylu{ii}(jj) = 1;
else
yuu{ii}(jj) = 0;
yll{ii}(jj) = 0;
yul{ii}(jj) = 0;
ylu{ii}(jj) = 0;
end
end
end
%% Rule fiering
for ii = 1:n
Zuu(:,ii) = yuu{ii}(rules(:,ii));
Zll(:,ii) = yll{ii}(rules(:,ii));
Zul(:,ii) = yul{ii}(rules(:,ii));
Zlu(:,ii) = ylu{ii}(rules(:,ii));
end
zuu(:,k) = prod(Zuu,2);%min
zll(:,k) = prod(Zll,2);%min
zul(:,k) = prod(Zul,2);%min
```

(continued)

(continued)

```
MATLAB script for u initialization
zlu(:,k) = prod(Zlu,2);%min
%% Type reduction
sum_zu = (sum(zuu(:,k) + zll(:,k)) + 0.001);
sum_zl = (sum(zlu(:,k) + zul(:,k)) + 0.001);
Pu(k,1) = (zuu(:,k)'*wuu(:,k) + zll(:,k)'*wll(:,k))/sum_zu;
Pl(k,1) = (zul(:,k)'*wul(:,k) + zlu(:,k)'*wlu(:,k))/sum_zl;
dwuu(:,k) = alpha(k)*zuu(:,k)/sum_zu;
dwll(:,k) = alpha(k)*zll(:,k)/sum_zu;
dwul(:,k) = alpha(k)*zul(:,k)/sum_zl;
dwlu(:,k) = alpha(k)*zlu(:,k)/sum_zl;
end
%aa
fhat = alpha*(Pu + Pl)/sum(alpha);
dwuu = dwuu/sum(alpha);
dwll = dwll/sum(alpha);
dwul = dwul/sum(alpha);
dwlu = dwlu/sum(alpha);
out = [fhat;reshape(dwuu,num_rule*k,1);reshape(dwll,num_rule*k,1);reshape(dwul,num_
rule*k,1);reshape(dwlu,num_rule*k,1)];
end
```

Example 2 *The provided example utilizes the recommended T3-FLS in a control application* [1]. *The proposed T3-FLS is utilized for online regulation of a robotic manipulator* [2]. *The dynamics are expressed in the form of equations, as shown below: (see Fig. 2.14).*

$$[H_{11}H_{12}H_{21}H_{22}][\ddot{q}_1\ddot{q}_2] + [-h\dot{q}_2 - h(\dot{q}_2 + \dot{q}_2)h\dot{q}_1 0][\dot{q}_1\dot{q}_2] = [\tau_1\tau_2] \quad (2.19)$$

$$\begin{aligned} H_{11} &= \alpha_1 + 2\alpha_3 cosq_2 + 2\alpha_4 sinq_2 a H_{12} = H_{21} = \alpha_2 + \alpha_3 cosq_2 + \alpha_4 sinq_2 H_{22} \\ &= \alpha_2 h = \alpha_3 sinq_2 - \alpha_4 cosq_2 \alpha_1 = I_1 + m_1 l_{c1}^2 + I_e + m_e l_{ce}^2 + m_e l_1^2 \alpha_2 \\ &= I_e + m_e l_{ce}^2 \alpha_3 = m_e l_1 l_{ce} cos\delta_e \alpha_4 = m_e l_1 l_{ce} sin\delta_e \end{aligned} \quad (2.20)$$

The dynamics are reformulated as:

$$\begin{aligned} \dot{x}_{11} &= x_{12}\dot{x}_{12} = f_1(x_{11}, x_{12}, x_{21}, x_{22}, u_2) + g_1(x_{11}, x_{12}, x_{21}, x_{22})u_1\dot{x}_{21} = x_{22}\dot{x}_{22} \\ &= f_2(x_{11}, x_{12}, x_{21}, x_{22}, u_1) + g_2(x_{11}, x_{12}, x_{21}, x_{22})u_2 y_1 = x_{12}, y_2 = x_{22} \end{aligned} \quad (2.21)$$

where,

$$x_{11} = q_1, x_{12} = \dot{q}_1, x_{21} = q_2, x_{22} = \dot{q}_2, u_1 = \tau_1, u_2 = \tau_2 \quad (2.22)$$

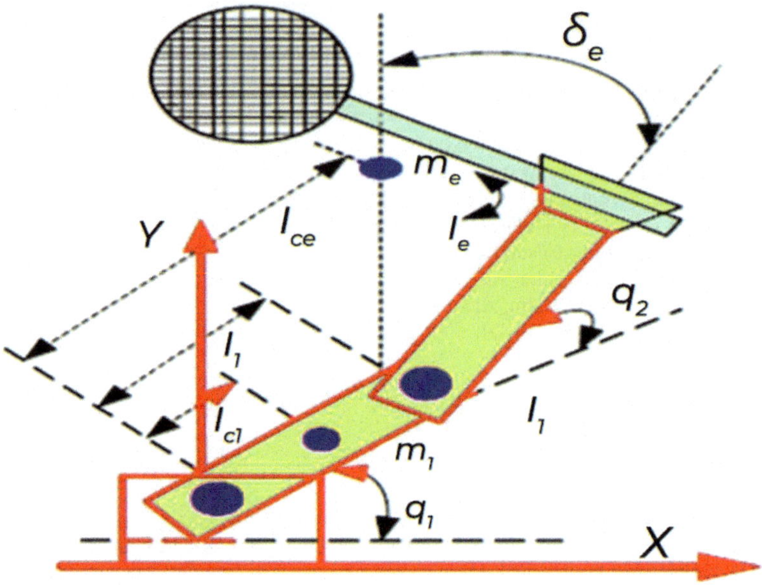

Fig. 2.14 The robotic arm of the case-study in Ex. 2

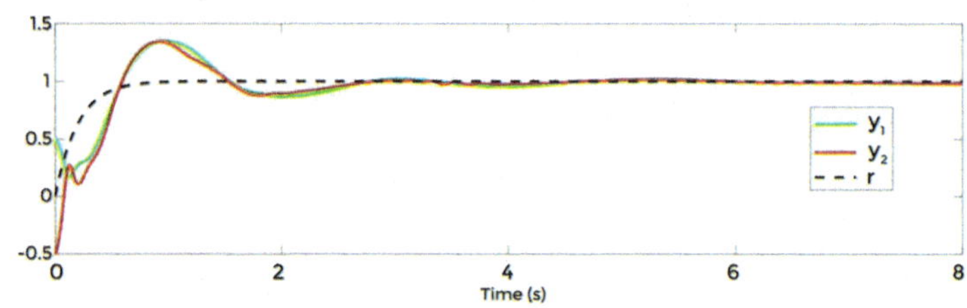

Fig. 2.15 Tracking performance in Ex. 2

$$[f_1 f_2] = [H_{11} H_{12} H_{21} H_{22}]^{-1}[-hx_{22} - h(x_{22} + x_{12})hx_{12}0][x_{12}x_{22}] \tag{2.23}$$

$$g_1 = \frac{H_{22}}{H_{22}H_{11} - H_{12}H_{21}}, g_2 = \frac{H_{11}}{H_{22}H_{11} - H_{12}H_{21}} \tag{2.24}$$

This example aims to demonstrate the suitability of T3-FLSs in an online control system. The recommended controller is provided as:

$$u = \hat{g}^{-1}(\chi_2)\left[-\hat{f}(\chi_1) + y_r^{(n)} - K^T e\right] \tag{2.25}$$

where, $\hat{f}(\chi_1)$ and $\hat{g}(\chi_2)$ denote T3-FLS, K is a positive vector, and e shows the tracking error. T3-FLSs are applied to estimate uncertainties. The T3-FLSs parameters, are tuned according to the below laws:

Here, $\hat{f}(\chi_1)$ and $\hat{g}(\chi_2)$ refer to T3-FLS, K represents a positive vector, and e represents the tracking error. T3-FLSs are utilized for the purpose of estimating uncertainty. The parameters of.

T3-FLSs are adjusted according to the following principles:

$$\underline{\Psi}^h_{r,k}(t+1) = \underline{\Psi}^h_{r,k}(t) + \gamma \frac{\alpha_k}{\sum_{k=1}^K \alpha_k} \frac{\underline{z}^h_{r,\alpha_k}}{\sum_{h=1}^N \underline{z}^h_{r,\alpha_k} + \underline{z}^h_{l,\alpha_k}} e(t) \tag{2.26}$$

$$\underline{\Psi}^h_{l,k}(t+1) = \underline{\Psi}^h_{l,k}(t) + \gamma \frac{\alpha_k}{\sum_{k=1}^K \alpha_k} \frac{\underline{z}^h_{l,\alpha_k}}{\sum_{h=1}^N \underline{z}^h_{r,\alpha_k} + \underline{z}^h_{l,\alpha_k}} e(t) \tag{2.27}$$

$$\underline{\Psi}^h_{r,k}(t+1) = \underline{\Psi}^h_{r,k} + \gamma \frac{\alpha_k}{\sum_{k=1}^K \alpha_k} \frac{\underline{z}^h_{r,\alpha_k}}{\sum_{h=1}^N \underline{z}^h_{r,\alpha_k} + \underline{z}^h_{l,\alpha_k}} e(t) \tag{2.28}$$

$$\underline{\Psi}^h_{l,k}(t+1) = \underline{\Psi}^h_{l,k} + \gamma \frac{\alpha_k}{\sum_{k=1}^K \alpha_k} \frac{\underline{z}^h_{l,\alpha_k}}{\sum_{h=1}^N \underline{z}^h_{r,\alpha_k} + \underline{z}^h_{l,\alpha_k}} e(t) \tag{2.29}$$

The simulation scheme is identical to that of Example 1. The paths of the outputs y_1 and y_2 are shown in Fig. 2.14. The output signals quickly converge to the reference signal r. The signals y_1 and y_2 are both received by r within 4 s. Figures 2.16, 2.17, 2.18 and 2.19 display the control signals and faults. The signals u_1 and u_2 possess highly feasible forms. Figures 2.16 and 2.17 do not exhibit any significant variation in the frequency of signals u_1 and u_2. The rapid reaching of tracking errors e_1 and e_2, as depicted in Figs. 2.18 and 2.19, to the zero level serves as evidence of the effectiveness of the proposed T3-FLSs.

The created toolbox can be assessed in several applications such as machine learning [3], vibration control systems [4], optimization problems [4, 5], mobile robots [6–9], and other problems in engineering for future study [10].

Matlab scripts for, respectively, the main, Type3_FLS, and EKF files are as follows:

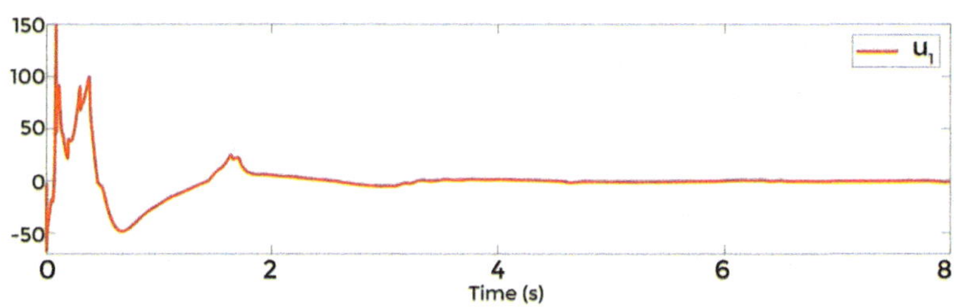

Fig. 2.16 Control signal u_1 in Ex. 2

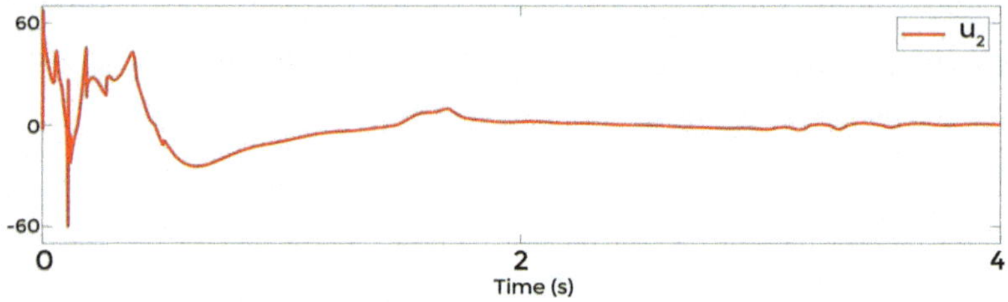

Fig. 2.17 Control signal u_2 in Ex. 2

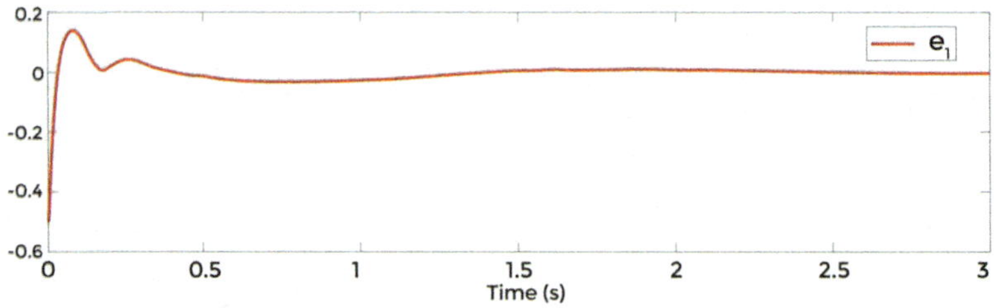

Fig. 2.18 Tracking error e_1 in Ex. 2

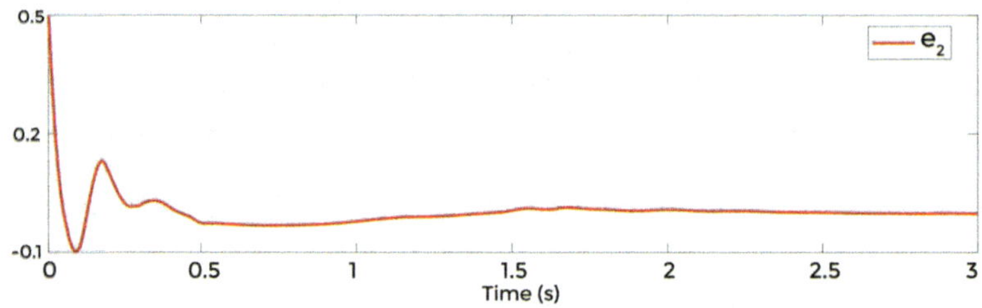

Fig. 2.19 Tracking error e_2 in Ex. 2

MATLAB script for Main file

```
clc
clear all
load ex4_data
%% Normalization
max_tri = 1*(max(max(abs(Train_data))));
max_tsi = 1*(max(max(abs(Test_data))));
max_tr = 1*max(abs(Target_train));
max_ts = 1*max(abs(Target_test));
Train_data = Train_data./max_tri;
Test_data = Test_data./max_tsi;
Target_test = Target_test/max_ts;
```

(continued)

(continued)

MATLAB script for Main file

```
Target_train = Target_train/max_tr;
%% Parameters
num_in = 3; %number of inputs
% MF parameters
c = {{-1 0 1},{-1 0 1},{-1 0 1}};
au = {{3 3 3},{3 3 3},{3 3 3}};
al = {{2 2 2},{2 2 2},{2 2 2}};
dr = {{1 1 1},{1 1 1},{1 1 1}};
dl = {{1 1 1},{1 1 1},{1 1 1}};
rules = [1 1 1;2 2 2; 3 3 3];
alpha = [ 0.5];
% Consewuent parameters
M = size(rules,1);%number of rules
wuu = 0.5*ones(M,1); % 1 equal with the number of alpha-cuts;
wul = 0.25*ones(M,1);
wlu = 0.1*ones(M,1);
wll = 0*ones(M,1);
% Kalman initilization
Puu = 1e-3*eye(M);Plu = Puu;Pll = Puu;Pul = Puu;
u.c = c;
u.au = au;
u.al = al;
u.dr = dr;
u.dl = dl;
u.rules = rules;
u.alpha = alpha;
u.wuu = wuu;
u.wll = wll;
u.wlu = wlu;
u.wul = wul;
for epoch = 1:100
epoch
%% Train
E = 0;
nn = randperm(length(Train_data));
for ii = nn
x(:,1) = Train_data(ii,:);
u.x = x;
out = Type3_FLS(u);
y = out.fhat;
e = Target_train(ii)-y;
[u.wuu,Puu] = EKF(e,u.wuu,Puu,out.dwuu);
[u.wul,Pul] = EKF(e,u.wul,Pul,out.dwul);
[u.wll,Pll] = EKF(e,u.wll,Pll,out.dwll);
[u.wlu,Plu] = EKF(e,u.wlu,Plu,out.dwlu);
E = E + e^2;
end
sqrt(E/size(Train_data,1))
end
```

(continued)

(continued)

MATLAB script for Main file

```
E = 0;
for ii = 1:size(Test_data,1)
x(:,1) = Test_data(ii,:);
u.x = x;
out = Type3_FLS(u);
y(ii,1) = out.fhat;
e = Target_test(ii)-y(ii);
E = E + e^2;
end
Target_test = max_ts*Target_test;
y = max_ts*y;
sqrt(mse(Target_test-y))
%% plot
plot_result
```

MATLAB script for Type3_FLS file

```
function out = Type3_FLS(u)
x = u.x;
c = u.c;
au = u.au;
al = u.al;
dr = u.dr;
dl = u.dl;
rules = u.rules;
alpha = u.alpha;
wuu = u.wuu;
wll = u.wll;
wlu = u.wlu;
wul = u.wul;
num_rule = size(rules,1);
k = length(alpha);
n = length(x);
%%
% u is n*1 where n is number of inputs
% c is n*n_MF where n_MF is the number of MFs for this input
% the dimension of sigma is the same as c
% the variable rule represent the rule database
% rule is n_rule*n, where n_rule is the number of rules
%%
% clear all
% clc
% u = [1 2]';
% c = [-1 0 1;-1 0 1];
% sigma = ones(size(c));
% rule = [1 3; 2 3];
% w = rand(2,1);-2 2
%c = [-2 -1 0 1 2;3 4 0 5 6]
```

(continued)

(continued)

MATLAB script for Type3_FLS file

```
%%
%au = zeros(num_rule,n);al = zeros(num_rule,n);
yuu = c;yll = c;yul = c;ylu = c;dwuu = zeros(num_rule,k);dwll = zeros(num_rule,k);dwul =
zeros(num_rule,k);dwlu = zeros(num_rule,1);
Zuu = zeros(num_rule,n);Zll = zeros(num_rule,n);Zul = zeros(num_rule,n);Zlu = zeros(num_
rule,n);
zuu = zeros(num_rule,k);zll = zeros(num_rule,k);zul = zeros(num_rule,k);zlu = zeros(num_
rule,k);
Pu = zeros(k,1);Pl = zeros(k,1);
%% MF
y = 0;
for k = 1:length(alpha)
for ii = 1:n
for jj = 1:length(c{ii})
if x(ii) > c{ii}{jj} && x(ii) < = (c{ii}{jj} + dr{ii}{jj})
yuu{ii}{jj} = (1-abs((x(ii)-c{ii}{jj})./dr{ii}{jj}).^au{ii}{jj});
yll{ii}{jj} = (1-abs((x(ii)-c{ii}{jj})./dr{ii}{jj}).^(1/au{ii}{jj}));
yul{ii}{jj} = (1-abs((x(ii)-c{ii}{jj})./dr{ii}{jj}).^al{ii}{jj});
ylu{ii}{jj} = (1-abs((x(ii)-c{ii}{jj})./dr{ii}{jj}).^(1/al{ii}{jj}));
elseif x(ii) < c{ii}{jj} && x(ii) > = (c{ii}{jj}-dl{ii}{jj})
yuu{ii}{jj} = (1-abs((x(ii)-c{ii}{jj})./dl{ii}{jj}).^au{ii}{jj});
yll{ii}{jj} = (1-abs((x(ii)-c{ii}{jj})./dl{ii}{jj}).^(1/au{ii}{jj}));
yul{ii}{jj} = (1-abs((x(ii)-c{ii}{jj})./dl{ii}{jj}).^al{ii}{jj});
ylu{ii}{jj} = (1-abs((x(ii)-c{ii}{jj})./dl{ii}{jj}).^(1/al{ii}{jj}));
elseif x(ii) = = c{ii}{jj}
yuu{ii}{jj} = 1;
yll{ii}{jj} = 1;
yul{ii}{jj} = 1;
ylu{ii}{jj} = 1;
else
yuu{ii}{jj} = 0;
yll{ii}{jj} = 0;
yul{ii}{jj} = 0;
ylu{ii}{jj} = 0;
end
end
end
%% Rule fiering
for ii = 1:n
Zuu(:,ii) = [yuu{ii}{rules(:,ii)}]';
Zll(:,ii) = [yll{ii}{rules(:,ii)}]';
Zul(:,ii) = [yul{ii}{rules(:,ii)}]';
Zlu(:,ii) = [ylu{ii}{rules(:,ii)}]';
end
zuu(:,k) = prod(Zuu,2);%min
zll(:,k) = prod(Zll,2);%min
zul(:,k) = prod(Zul,2);%min
zlu(:,k) = prod(Zlu,2);%min
```

(continued)

(continued)

MATLAB script for Type3_FLS file

```
%% Type reduction
sum_zu = (sum(zuu(:,k) + zll(:,k)) + 0.001);
sum_zl = (sum(zlu(:,k) + zul(:,k)) + 0.001);
Pu(k,1) = (zuu(:,k)'*wuu(:,k) + zll(:,k)'*wll(:,k))/sum_zu;
Pl(k,1) = (zul(:,k)'*wul(:,k) + zlu(:,k)'*wlu(:,k))/sum_zl;
dwuu(:,k) = alpha(k)*zuu(:,k)/sum_zu;
dwll(:,k) = alpha(k)*zll(:,k)/sum_zu;
dwul(:,k) = alpha(k)*zul(:,k)/sum_zl;
dwlu(:,k) = alpha(k)*zlu(:,k)/sum_zl;
end%aa
fhat = alpha*(Pu + Pl)/sum(alpha);
dwuu = dwuu/sum(alpha);
dwll = dwll/sum(alpha);
dwul = dwul/sum(alpha);
dwlu = dwlu/sum(alpha);
out.fhat = fhat;
out.dwuu = dwuu;
out.dwll = dwll;
out.dwul = dwul;
out.dwlu = dwlu;
%out.yuu = yuu;
end
```

MATLAB script for EKF file

```
%% UKF
function [w,P] = EKF(e,w,P,H)
% global sat_up
%%
R = 0.001;
Q = 1e-8*eye(size(P));
k = P*H*((R + H'*P*H)\1);
P = (eye(length(w))-k*H')*P + Q;
w = w + k*e;%P*H;
%w(abs(w) > sat_up) = sat_up*sign( w(abs(w) > sat_up));
```

References

1. J. Duan, G. Duan, S. Cheng, S. Cao, G. Wang, Fixed-time time-varying output formation-containment control of heterogeneous general multi-agent systems. ISA Trans (2023)
2. J. Wang, M. Yang, F. Liang, K. Feng, K. Zhang, Q. Wang, An algorithm for painting large objects based on a nine-axis UR5 robotic manipulator. Appl. Sci.**12**, 7219 (2022)
3. H. Leon-Garza, H. Hagras, A. Peña-Rios, A. Conway, G. Owusu, A type-2 fuzzy system-based approach for image data fusion to create building information models. Inform. Fusion **88**, 115–125 (2022)

4. X. Pan, Y. Wang, Evaluation of renewable energy sources in China using an interval type-2 fuzzy large-scale group risk evaluation method. Appl. Soft Comput. **108**, 107458 (2021)
5. S. Sennan, S. Ramasubbareddy, S. Balasubramaniyam, A. Nayyar, M. Abouhawwash, N.A. Hikal, T2FL-PSO: Type-2 fuzzy logic-based particle swarm optimization algorithm used to maximize the lifetime of Internet of Things. IEEE Access **9**, 63966–63979 (2021)
6. C. Peraza, P. Ochoa, O. Castillo, Z.W. Geem, Interval-type 3 fuzzy differential evolution for designing an interval-type 3 fuzzy controller of a unicycle mobile robot. Mathematics **10**, 3533 (2022)
7. O. Castillo, J.R. Castro, P. Melin, Interval type-3 fuzzy systems: theory and design. Springer International Publishing (2022)
8. O. Castillo, P. Melin, *Type-3 fuzzy logic in intelligent control* (Springer, 2023)
9. P. Melin, O. Castillo, New Directions on Hybrid Intelligent Systems Based on Neural Networks, Fuzzy Logic, and Optimization Algorithms (Vol. 1146). Springer Nature (2024)
10. H. Huang, H. Xu, F. Chen, C. Zhang, A. Mohammadzadeh, An applied type-3 fuzzy logic system: practical Matlab Simulink and M-files for robotic, control, and modeling applications. Symmetry **15**(2), 475 (2023)

Chapter 3
Advancements and Applications

3.1 Deep Learned Type-3 Fuzzy Systems for Renewable Energy Prediction

3.1.1 Proposed Type-3 FLS

T3-FLSs are a recently created version of the ordinary type-2 FLSs [1]. In this particular form of FLSs, the boundaries that define the range of uncertainty are not precise numbers, but rather they are represented by fuzzy sets. Moreover, the secondary membership in T3-FLSs is variable. The unique characteristics of T3-FLSs enhance their effectiveness in addressing diverse nonlinear problems. This section introduces a novel recurrent T3-FLSs (RT3-FLS) with a nonlinear consequent portion. The purpose of this model is to accurately represent the behavior of renewable energy generators and make predictions about power output. The diagram illustrating the structure can be found in Fig. 3.1. The process of computing is detailed below.

(1) The inputs consist of the function $u(t)$ and the estimated value $\hat{E}(t-1)$. The variable $u(t)$ denotes the solar irradiation or wind speed, while $\hat{E}(t-1)$ represents the estimated energy feedback.
(2) Two membership functions are considered for each input $u(t)$ and $\hat{E}(t-1)$. The input $u(t)$ is characterized by the membership functions $\underline{M}_{u|h}$ and $\underline{M}_{u|h}$, while the input $\hat{E}(t-1)$ is characterized by the membership functions $\underline{M}_{\hat{E}|h}$ and $\underline{M}_{\hat{E}|h}$. The symbol h repressents the horizontal slice level, as shown in Fig. 3.2.

$$\underline{\psi}_{\underline{M}_{u|\underline{h}}}(u(t)) = exp\left(-\frac{\left(u(t) - c_{\underline{M}_{u|\underline{h}}}\right)^2}{\underline{\sigma}^2_{\underline{M}_{u|\underline{h}}}}\right) \tag{3.1}$$

R. Sabetahd et al., *Applied Type-3 Fuzzy Logic Systems and Controllers*,
Studies in Fuzziness and Soft Computing 417,
https://doi.org/10.1007/978-3-031-75988-8_3

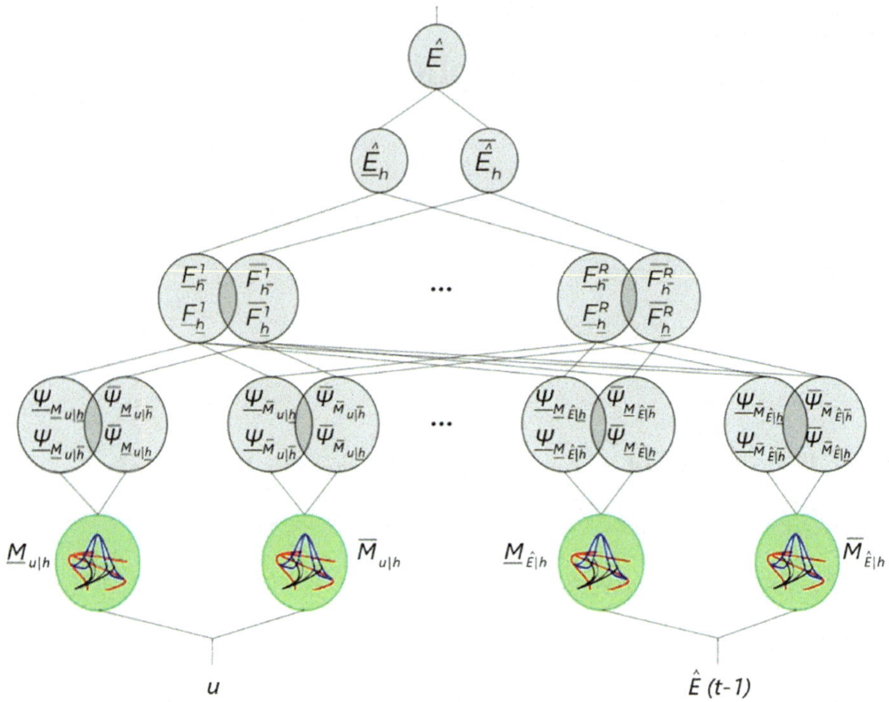

Fig. 3.1 Suggested RT3-FLS

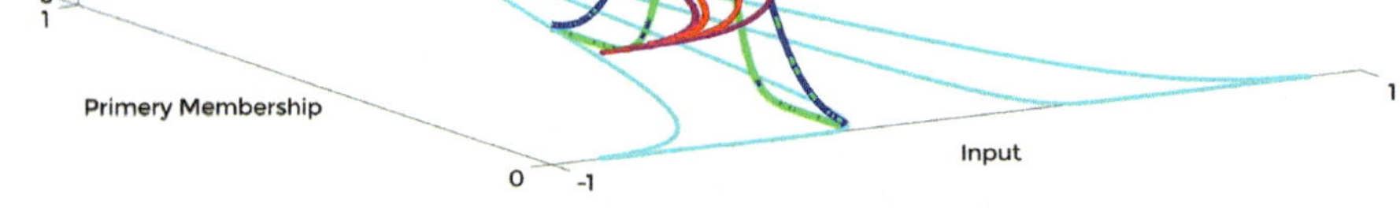

Fig. 3.2 Type-3 MF horizontal slices

$$\underline{\psi}_{\underline{M}_{u|\underline{h}}}(u(t)) = exp\left(-\frac{\left(u(t) - c_{\underline{M}_{u|\underline{h}}}\right)^2}{\underline{\sigma}^2_{\underline{M}_{u|\underline{h}}}}\right) \tag{3.2}$$

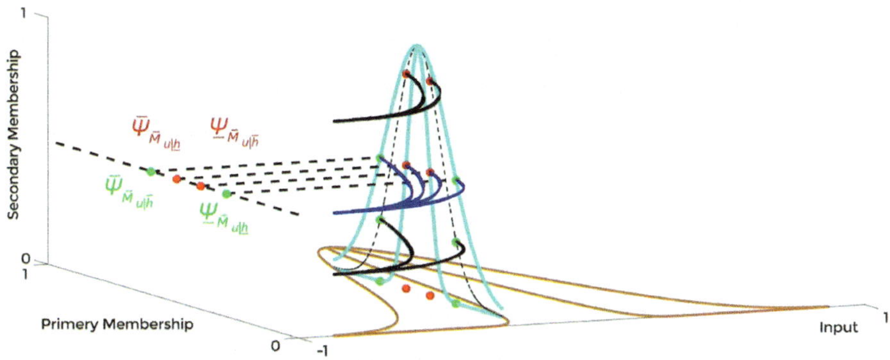

Fig. 3.3 Performance of the upper and lower memberships in two horizontal slices

$$\underline{\psi}_{\underline{M}_{u|\underline{h}}}(u(t)) = exp\left(-\frac{\left(u(t) - c_{\underline{M}_{u|\underline{h}}}\right)^2}{\underline{\sigma}^2_{\underline{M}_{u|\underline{h}}}}\right) \tag{3.3}$$

$$\underline{\psi}_{\underline{M}_{u|\underline{h}}}(u(t)) = exp\left(-\frac{\left(u(t) - c_{\underline{M}_{u|\underline{h}}}\right)^2}{\underline{\sigma}^2_{\underline{M}_{u|\underline{h}}}}\right) \tag{3.4}$$

where, $h = h_1, \ldots, h_n$, the symbol $c_{\underline{M}_{u|\underline{h}}}$ denotes the center of $\underline{M}_{u|\underline{h}}$. The ratios. $\underline{\sigma}_{\underline{M}_{u|\underline{h}}}/\underline{\sigma}_{\underline{M}_{u|\underline{h}}}$ and $\underline{\sigma}_{\underline{M}_{u|\underline{h}}}/\underline{\sigma}_{\underline{M}_{u|\underline{h}}}$ represent the upper and lower widths of the MF $\underline{M}_{u|h}$ for the input $u(t)$. The equations for the MF $\underline{M}_{u|h}$ are analogous to Eqs. (3.5–3.8). The memberships for the input $\hat{E}(t-1)$ are as follows:

$$\underline{\psi}_{\underline{M}_{\hat{E}|\underline{h}}}\left(\hat{E}(t-1)\right) = exp\left(-\frac{\left(\hat{E}(t) - c_{\underline{M}_{\hat{E}|\underline{h}}}\right)^2}{\underline{\sigma}^2_{\underline{M}_{\hat{E}|\underline{h}}}}\right) \tag{3.5}$$

$$\underline{\psi}_{\underline{M}_{\hat{E}|\underline{h}}}\left(\hat{E}(t-1)\right) = exp\left(-\frac{\left(\hat{E}(t) - c_{\underline{M}_{\hat{E}|\underline{h}}}\right)^2}{\underline{\sigma}^2_{\underline{M}_{\hat{E}|\underline{h}}}}\right) \tag{3.6}$$

$$\underline{\psi}_{\underline{M}_{\hat{E}|\underline{h}}}\left(\hat{E}(t-1)\right) = exp\left(-\frac{\left(\hat{E}(t) - c_{\underline{M}_{\hat{E}|\underline{h}}}\right)^2}{\underline{\sigma}^2_{\underline{M}_{\hat{E}|\underline{h}}}}\right) \tag{3.7}$$

$$\underline{\psi}_{\underline{M}_{\hat{E}|\underline{h}}}\left(\hat{E}(t-1)\right) = exp\left(-\frac{\left(\hat{E}(t) - c_{\underline{M}_{\hat{E}|\underline{h}}}\right)^2}{\underline{\sigma}^2_{\underline{M}_{\hat{E}|\underline{h}}}}\right) \tag{3.8}$$

The symbol $c_{\underline{M}_{\hat{E}|h}}$ denotes the center of $\underline{M}_{\hat{E}|h}$. The ratios $\sigma_{\underline{M}_{\hat{E}|h}}/\sigma_{\underline{M}_{\hat{E}|h}}$ and $\sigma_{\underline{M}_{\hat{E}|h}}/\sigma_{\underline{M}_{\hat{E}|h}}$ represent, respectively, the upper and lower widths of the MF $\underline{M}_{\hat{E}|h}$ for the input $\hat{E}(t)$. The equations for MF $\underline{M}_{\hat{E}|h}$ are analogous to Eqs. (3.5–3.8).

(3) The rules can be stated as:

$$\begin{aligned}
&if\ u(t)\ is\ \underline{M}_{u|h}\ and\ \hat{E}(t-1)\ is\ \underline{M}_{\hat{E}|h}\ Then\ \hat{E}(t) \in \left[\underline{\chi}_h^1, \underline{\chi}_h^1\right]\\
&if\ u(t)\ is\ \underline{M}_{u|h}\ and\ \hat{E}(t-1)\ is\ \underline{M}_{\hat{E}|h}\ Then\ \hat{E}(t) \in \left[\underline{\chi}_h^2, \underline{\chi}_h^2\right]\\
&if\ u(t)\ is\ \underline{M}_{u|h}\ and\ \hat{E}(t-1)\ is\ \underline{M}_{\hat{E}|h}\ Then\ \hat{E}(t) \in \left[\underline{\chi}_h^3, \underline{\chi}_h^3\right]\\
&if\ u(t)\ is\ \underline{M}_{u|h}\ and\ \hat{E}(t-1)\ is\ \underline{M}_{\hat{E}|h}\ Then\ \hat{E}(t) \in \left[\underline{\chi}_h^4, \underline{\chi}_h^4\right]
\end{aligned} \tag{3.9}$$

The rule parameters $\underline{\chi}_h^i$ and $\underline{\chi}_h^i$, $i = 1, \ldots, 4$ are formulated as:

$$\underline{\chi}_h^i = \underline{\alpha}_h^{i_0} + \underline{\alpha}_h^{i_1} u(t) + \underline{\alpha}_h^{i_2} \hat{E}(t-1) + \underline{\alpha}_h^{i_3} u(t)\hat{E}(t-1) + \underline{\alpha}_h^{i_4} u^2(t) + \underline{\alpha}_h^{i_5} \hat{E}^2(t-1) \tag{3.10}$$

$$\underline{\chi}_h^i = \underline{\alpha}_h^{i_0} + \underline{\alpha}_h^{i_1} u(t) + \underline{\alpha}_h^{i_2} \hat{E}(t-1) + \underline{\alpha}_h^{i_3} u(t)\hat{E}(t-1) + \underline{\alpha}_h^{i_4} u^2(t) + \underline{\alpha}_h^{i_5} \hat{E}^2(t-1) \tag{3.11}$$

where tuneable parameters are denoted by $\underline{\alpha}_h^{i_0}, \underline{\alpha}_h^{i_1}, \underline{\alpha}_h^{i_2}, \underline{\alpha}_h^{i_3}, \underline{\alpha}_h^{i_4}, \underline{\alpha}_h^{i_5}, \underline{\alpha}_h^{i_0}, \underline{\alpha}_h^{i_1}, \underline{\alpha}_h^{i_2}, \underline{\alpha}_h^{i_3}$, $\underline{\alpha}_h^{i_4}$ and $\underline{\alpha}_h^{i_5}$. From Eq. (3.9), the upper rule firings are stated as:

$$\underline{F}_h^1 = \psi_{\underline{M}_{u|h}} \cdot \psi_{\underline{M}_{\hat{E}|h}} \underline{F}_h^2 = \psi_{\underline{M}_{u|h}} \cdot \psi_{\underline{M}_{\hat{E}|h}} \underline{F}_h^3 = \psi_{\underline{M}_{u|h}} \cdot \psi_{\underline{M}_{\hat{E}|h}} \underline{F}_h^4 = \psi_{\underline{M}_{u|h}} \cdot \psi_{\underline{M}_{\hat{E}|h}} \tag{3.12}$$

$$\underline{F}_h^1 = \psi_{\underline{M}_{u|h}} \cdot \psi_{\underline{M}_{\hat{E}|h}} \underline{F}_h^2 = \psi_{\underline{M}_{u|h}} \cdot \psi_{\underline{M}_{\hat{E}|h}} \underline{F}_h^3 = \psi_{\underline{M}_{u|h}} \cdot \psi_{\underline{M}_{\hat{E}|h}} \underline{F}_h^4 = \psi_{\underline{M}_{u|h}} \cdot \psi_{\underline{M}_{\hat{E}|h}} \tag{3.13}$$

In the same way, the firings of the lower rule can be defined as:

$$\underline{F}_h^1 = \psi_{\underline{M}_{u|h}} \cdot \psi_{\underline{M}_{\hat{E}|h}} \underline{F}_h^2 = \psi_{\underline{M}_{u|h}} \cdot \psi_{\underline{M}_{\hat{E}|h}} \underline{F}_h^3 = \psi_{\underline{M}_{u|h}} \cdot \psi_{\underline{M}_{\hat{E}|h}} \underline{F}_h^4 = \psi_{\underline{M}_{u|h}} \cdot \psi_{\underline{M}_{\hat{E}|h}} \tag{3.14}$$

$$\underline{F}_h^1 = \psi_{\underline{M}_{u|h}} \cdot \psi_{\underline{M}_{\hat{E}|h}} \underline{F}_h^2 = \psi_{\underline{M}_{u|h}} \cdot \psi_{\underline{M}_{\hat{E}|h}} \underline{F}_h^3 = \psi_{\underline{M}_{u|h}} \cdot \psi_{\underline{M}_{\hat{E}|h}} \underline{F}_h^4 = \psi_{\underline{M}_{u|h}} \cdot \psi_{\underline{M}_{\hat{E}|h}} \tag{3.15}$$

(4) The right and left points in consequent part are calculated to be:

$$\underline{\hat{E}}_h = \frac{\sum_{i=1}^{R}\left(\underline{F}_h^i + \underline{F}_h^i\right)\underline{\chi}_h^i}{\sum_{l=1}^{R}\left(\underline{F}_h^i + \underline{F}_h^i\right)} \tag{3.16}$$

$$\underline{\hat{E}}_h = \frac{\sum_{i=1}^{R}\left(\underline{F}_h^i + \underline{F}_h^i\right)\underline{\chi}_h^i}{\sum_{l=1}^{R}\left(\underline{F}_h^i + \underline{F}_h^i\right)} \tag{3.17}$$

where, $R = 4$ shows the number of rules.

(5) Eventually, the output is calculated as:

$$\hat{E} = 0.5\frac{\sum_{\underline{h}\in\{\underline{h}_1,\ldots,\underline{h}_N\}} \underline{h}\underline{\hat{E}}_h}{\sum_{\underline{h}\in\{\underline{h}_1,\ldots,\underline{h}_N\}} \underline{h}} + 0.5\frac{\sum_{\underline{h}\in\{\underline{h}_1,\ldots,\underline{h}_N\}} \underline{h}\underline{\hat{E}}_h}{\sum_{\underline{h}\in\{\underline{h}_1,\ldots,\underline{h}_N\}} \underline{h}} \tag{3.18}$$

3.1.2 Leaning Algorithm

In this part, the centers of membership functions, consequent parameters, and slices level are adjusted.

3.1.2.1 Rule Parameters

The rule parameters $\underline{\alpha}_h^{i_0}$, $\underline{\alpha}_h^{i_1}$, $\underline{\alpha}_h^{i_2}$, $\underline{\alpha}_h^{i_3}$, $\underline{\alpha}_h^{i_4}$, $\underline{\alpha}_h^{i_5}$, $\underline{\alpha}_h^{i_0}$, $\underline{\alpha}_h^{i_1}$, $\underline{\alpha}_h^{i_2}$, $\underline{\alpha}_h^{i_3}$, $\underline{\alpha}_h^{i_4}$, and $\underline{\alpha}_h^{i_5}$ are tuned in order to minimize the cost function shown in Eq. (3.19):

$$J = \frac{1}{2}\left(E_d(t) - \hat{E}(t)\right)^2 \tag{3.19}$$

In this context, E_d represents the desired or intended signal, while $\hat{E}$ represents the signal that has been approximated or predicted (estimated signal). The rules for adjusting $\underline{\alpha}_h^{i_0}$, $\underline{\alpha}_h^{i_1}$, $\underline{\alpha}_h^{i_2}$, $\underline{\alpha}_h^{i_3}$, $\underline{\alpha}_h^{i_4}$, and $\underline{\alpha}_h^{i_5}$ are provided as follows:

$$\underline{\alpha}_h^{i_0}(t+1) = \underline{\alpha}_h^{i_0}(t) + \gamma \cdot \frac{\underline{h}}{\sum_{h\in\{h_1,\ldots,h_N\}} \underline{h}} \cdot \frac{\underline{F}_h^i + \underline{F}_h^i}{\sum_{l=1}^{R}\left(\underline{F}_h^i + \underline{F}_h^i\right)} \cdot \left(E_d(t) - \hat{E}(t)\right) \tag{3.20}$$

$$\underline{\alpha}_h^{i_1}(t+1) = \underline{\alpha}_h^{i_1}(t) + \gamma \cdot \frac{\underline{h}}{\sum_{h\in\{h_1,\ldots,h_N\}} \underline{h}} \cdot \frac{\underline{F}_h^i + \underline{F}_h^i}{\sum_{l=1}^{R}\left(\underline{F}_h^i + \underline{F}_h^i\right)} \cdot \left(E_d(t) - \hat{E}(t)\right)u(t) \tag{3.21}$$

$$\underline{\alpha}_h^{i_2}(t+1)=\underline{\alpha}_h^{i_2}(t)+\gamma\cdot\frac{\underline{h}}{\sum_{h\in\{h_1,v,h_N\}}\underline{h}}\cdot\frac{\underline{F}_{\underline{h}}^i+\underline{F}_{\underline{h}}^i}{\sum_{l=1}^R\left(\underline{F}_{\underline{h}}^i+\underline{F}_{\underline{h}}^i\right)}\cdot\big(E_d(t)-\hat{E}(t)\big)\hat{E}(t-1) \tag{3.22}$$

$$\underline{\alpha}_h^{i_3}(t+1)=\underline{\alpha}_h^{i_3}(t)+\gamma\cdot\frac{\underline{h}}{\sum_{h\in\{h_1,\dots,h_N\}}\underline{h}}\cdot\frac{\underline{F}_{\underline{h}}^i+\underline{F}_{\underline{h}}^i}{\sum_{l=1}^R\left(\underline{F}_{\underline{h}}^i+\underline{F}_{\underline{h}}^i\right)}\cdot\big(E_d(t)-\hat{E}(t)\big)u(t)\hat{E}(t-1) \tag{3.23}$$

$$\underline{\alpha}_h^{i_4}(t+1)=\underline{\alpha}_h^{i_4}(t)+\gamma\cdot\frac{\underline{h}}{\sum_{h\in\{h_1,\dots,h_N\}}\underline{h}}\cdot\frac{\underline{F}_{\underline{h}}^i+\underline{F}_{\underline{h}}^i}{\sum_{l=1}^R\left(\underline{F}_{\underline{h}}^i+\underline{F}_{\underline{h}}^i\right)}\cdot\big(E_d(t)-\hat{E}(t)\big)u^2(t) \tag{3.24}$$

$$\underline{\alpha}_h^{i_5}(t+1)=\underline{\alpha}_h^{i_5}(t)+\gamma\cdot\frac{\underline{h}}{\sum_{h\in\{h_1,\dots,h_N\}}\underline{h}}\cdot\frac{\underline{F}_{\underline{h}}^i+\underline{F}_{\underline{h}}^i}{\sum_{l=1}^R\left(\underline{F}_{\underline{h}}^i+\underline{F}_{\underline{h}}^i\right)}\cdot\big(E_d(t)-\hat{E}(t)\big)\hat{E}^2(t-1) \tag{3.25}$$

In the same way, for the parameters $\underline{\alpha}_h^{i_0}$, $\underline{\alpha}_h^{i_1}$, $\underline{\alpha}_h^{i_2}$, $\underline{\alpha}_h^{i_3}$, $\underline{\alpha}_h^{i_4}$, and $\underline{\alpha}_h^{i_5}$, we have:

$$\underline{\alpha}_h^{i_0}(t+1)=\underline{\alpha}_h^{i_0}(t)+\gamma\cdot\frac{\underline{h}}{\sum_{h\in\{h_1,\dots,h_N\}}\underline{h}}\cdot\frac{\underline{F}_{\underline{h}}^i+\underline{F}_{\underline{h}}^i}{\sum_{l=1}^R\left(\underline{F}_{\underline{h}}^i+\underline{F}_{\underline{h}}^i\right)}\cdot\big(E_d(t)-\hat{E}(t)\big) \tag{3.26}$$

$$\underline{\alpha}_h^{i_1}(t+1)=\underline{\alpha}_h^{i_1}(t)+\gamma\cdot\frac{\underline{h}}{\sum_{h\in\{h_1,\dots,h_N\}}\underline{h}}\cdot\frac{\underline{F}_{\underline{h}}^i+\underline{F}_{\underline{h}}^i}{\sum_{l=1}^R\left(\underline{F}_{\underline{h}}^i+\underline{F}_{\underline{h}}^i\right)}\cdot\big(E_d(t)-\hat{E}(t)\big)u(t) \tag{3.27}$$

$$\underline{\alpha}_h^{i_2}(t+1)=\underline{\alpha}_h^{i_2}(t)+\gamma\cdot\frac{\underline{h}}{\sum_{h\in\{h_1,\dots,h_N\}}\underline{h}}\cdot\frac{\underline{F}_{\underline{h}}^i+\underline{F}_{\underline{h}}^i}{\sum_{l=1}^R\left(\underline{F}_{\underline{h}}^i+\underline{F}_{\underline{h}}^i\right)}\cdot\big(E_d(t)-\hat{E}(t)\big)\hat{E}(t-1) \tag{3.28}$$

$$\underline{\alpha}_h^{i_3}(t+1)=\underline{\alpha}_h^{i_3}(t)+\gamma\cdot\frac{\underline{h}}{\sum_{h\in\{h_1,\dots,h_N\}}\underline{h}}\cdot\frac{\underline{F}_{\underline{h}}^i+\underline{F}_{\underline{h}}^i}{\sum_{l=1}^R\left(\underline{F}_{\underline{h}}^i+\underline{F}_{\underline{h}}^i\right)}\cdot\big(E_d(t)-\hat{E}(t)\big)u(t)\hat{E}(t-1) \tag{3.29}$$

$$\underline{\alpha}_h^{i_4}(t+1) = \underline{\alpha}_h^{i_4}(t) + \gamma \cdot \frac{\underline{h}}{\sum_{h\in\{h_1,\dots,h_N\}} \underline{h}} \cdot \frac{\underline{F}_{\underline{h}}^i + \underline{F}_{\underline{h}}^i}{\sum_{l=1}^{R}\left(\underline{F}_{\underline{h}}^i + \underline{F}_{\underline{h}}^i\right)} \cdot \left(E_d(t) - \hat{E}(t)\right)u^2(t) \tag{3.30}$$

$$\underline{\alpha}_h^{i_5}(t+1) = \underline{\alpha}_h^{i_5}(t) + \gamma \cdot \frac{\underline{h}}{\sum_{h\in\{h_1,\dots,h_N\}} \underline{h}} \cdot \frac{\underline{F}_{\underline{h}}^i + \underline{F}_{\underline{h}}^i}{\sum_{l=1}^{R}\left(\underline{F}_{\underline{h}}^i + \underline{F}_{\underline{h}}^i\right)} \cdot \left(E_d(t) - \hat{E}(t)\right)\hat{E}^2(t-1) \tag{3.31}$$

3.1.2.2 Learning Rules as to MF Centers

This section provides an explanation of the tuning rules for the centers of membership functions ($c_{\underline{M}_{u|h}}$, $c_{\underline{M}_{u|h}}$, $c_{\underline{M}_{\hat{E}|h}}$, and $c_{\underline{M}_{\hat{E}|h}}$). The tuning rules for $c_{\underline{M}_{u|h}}$ are as follows:

$$c_{\underline{M}u|h}(t+1) = c_{\underline{M}u|h}(t) + \frac{\gamma\left(E_d(t) - \hat{E}(t)\right)}{\sum_{h\in\{h_1\dots h_N\}} \underline{h}} \sum_{\underline{h}\in\{\underline{h}_1,\dots,h_N\}} h \frac{\partial \hat{\underline{E}}_h}{\partial c_{\underline{M}u|h}} + \frac{\gamma\left(E_d(t) - \hat{E}(t)\right)}{\sum_{h\in\{h_1\dots,h_N\}} \underline{h}} \sum_{h\in\{\underline{h}_1,\dots,\underline{h}_N\}} h \frac{\partial \hat{\underline{E}}_h}{\partial c_{\underline{M}u|h}} \tag{3.32}$$

Regarding the $c_{\underline{M}_{u|h}}$ mentioned in rules 1 and 2:

$$\frac{\partial \hat{\underline{E}}_h}{\partial c_{\underline{M}_{u|\underline{h}}}} = \frac{\left(u(t) - c_{\underline{M}_{u|\underline{h}}}\right)\left(\underline{\chi}_h^1 \underline{F}_{\underline{h}}^1 + \underline{\chi}_h^2 \underline{F}_{\underline{h}}^2\right)}{B\sigma^2_{\underline{M}_{u|\underline{h}}}} + \frac{\left(u(t) - c_{\underline{M}_{u|\underline{h}}}\right)\left(\underline{\chi}_h^1 \underline{F}_{\underline{h}}^1 + \underline{\chi}_h^2 \underline{F}_{\underline{h}}^2\right)}{B\sigma^2_{\underline{M}_{u|\underline{h}}}} - \left[\frac{\left(u(t) - c_{\underline{M}_{u|\underline{h}}}\right)\left(\underline{F}_{\underline{h}}^1 + \underline{F}_{\underline{h}}^2\right)}{\underline{B}^2\sigma^2_{\underline{M}_{u|\underline{h}}}} + \frac{\left(u(t) - c_{\underline{M}_{u|\underline{h}}}\right)\left(\underline{F}_{\underline{h}}^1 + \underline{F}_{\underline{h}}^2\right)}{\underline{B}^2\sigma^2_{\underline{M}_{u|\underline{h}}}}\right] \cdot \sum_{i=1}^{R}\left(\underline{F}_{\underline{h}}^i + \underline{F}_{\underline{h}}^i\right)\underline{\chi}_h^i \tag{3.33}$$

$$\frac{\partial \hat{\underline{E}}_h}{\partial c_{\underline{M}_{u|\underline{h}}}} = \frac{\left(u(t) - c_{\underline{M}_{u|\underline{h}}}\right)\left(\underline{\chi}_h^1 \underline{F}_{\underline{h}}^1 + \underline{\chi}_h^2 \underline{F}_{\underline{h}}^2\right)}{B\sigma^2_{\underline{M}_{u|\underline{h}}}} + \frac{\left(u(t) - c_{\underline{M}_{u|\underline{h}}}\right)\left(\underline{\chi}_h^1 \underline{F}_{\underline{h}}^1 + \underline{\chi}_h^2 \underline{F}_{\underline{h}}^2\right)}{B\sigma^2_{\underline{M}_{u|\underline{h}}}} - \left[\frac{\left(u(t) - c_{\underline{M}_{u|\underline{h}}}\right)\left(\underline{F}_{\underline{h}}^1 + \underline{F}_{\underline{h}}^2\right)}{\underline{B}^2\sigma^2_{\underline{M}_{u|\underline{h}}}} + \frac{\left(u(t) - c_{\underline{M}_{u|\underline{h}}}\right)\left(\underline{F}_{\underline{h}}^1 + \underline{F}_{\underline{h}}^2\right)}{\underline{B}^2\sigma^2_{\underline{M}_{u|\underline{h}}}}\right] \cdot \sum_{i=1}^{R}\left(\underline{F}_{\underline{h}}^i + \underline{F}_{\underline{h}}^i\right)\underline{\chi}_h^i \tag{3.34}$$

where,

$$\underline{B} = \sum_{l=1}^{R} \left(\underline{F}_{\underline{h}}^{i} + \underline{F}_{\underline{h}}^{i} \right) \tag{3.35}$$

$$\underline{B} = \sum_{l=1}^{R} \left(\underline{F}_{\underline{h}}^{i} + \underline{F}_{\underline{h}}^{i} \right) \tag{3.36}$$

For $c_{\underline{M}_{u|h}}$, the tuning rules are:

$$\begin{aligned} c_{\underline{M}_{u|h}}(t+1) = c_{\underline{M}_{u|h}}(t) &+ \frac{\gamma\left(E_d(t) - \hat{E}(t)\right)}{\sum_{h\in\{h_1,\ldots,h_N\}} \underline{h}} \sum_{h\in\{h_1,\ldots,h_N\}} \underline{h} \frac{\partial \hat{\underline{E}}_h}{\partial c_{\underline{M}_{u|\underline{h}}}} \\ &+ \frac{\gamma\left(E_d(t) - \hat{E}(t)\right)}{\sum_{h\in\{h_1,\ldots,h_N\}} \underline{h}} \sum_{h\in\{h_1,\ldots,h_N\}} \underline{h} \frac{\partial \hat{\underline{E}}_h}{\partial c_{\underline{M}_{u|\underline{h}}}} \end{aligned} \tag{3.37}$$

Given that the term $c_{\underline{M}_{u|h}}$ is present in rules 1 and 2, we can conclude the following:

$$\begin{aligned} \frac{\partial \hat{\underline{E}}_h}{\partial c_{\underline{M}_{u|\underline{h}}}} &= \frac{\left(u(t) - c_{\underline{M}_{u|\underline{h}}}\right)\left(\underline{\chi}_h^3 \underline{F}_{\underline{h}}^3 + \underline{\chi}_h^4 \underline{F}_{\underline{h}}^4\right)}{\underline{B}\sigma_{\underline{M}_{u|\underline{h}}}^2} + \frac{\left(u(t) - c_{\underline{M}_{u|\underline{h}}}\right)\left(\underline{\chi}_h^3 \underline{F}_{\underline{h}}^3 + \underline{\chi}_h^4 \underline{F}_{\underline{h}}^4\right)}{\underline{B}\sigma_{\underline{M}_{u|\underline{h}}}^2} - \\ &\left[\frac{\left(u(t) - c_{\underline{M}_{u|\underline{h}}}\right)\left(\underline{F}_{\underline{h}}^3 + \underline{F}_{\underline{h}}^4\right)}{\underline{B}^2\sigma_{\underline{M}_{u|\underline{h}}}^2} + \frac{\left(u(t) - c_{\underline{M}_{u|\underline{h}}}\right)\left(\underline{F}_{\underline{h}}^3 + \underline{F}_{\underline{h}}^4\right)}{\underline{B}^2\sigma_{\underline{M}_{u|\underline{h}}}^2} \right] \cdot \sum_{i=1}^{R} \left(\underline{F}_{\underline{h}}^{i} + \underline{F}_{\underline{h}}^{i} \right) \underline{\chi}_h^i \end{aligned} \tag{3.38}$$

$$\begin{aligned} \frac{\partial \hat{\underline{E}}_h}{\partial c_{\underline{M}_{u|\underline{h}}}} &= \frac{\left(u(t) - c_{\underline{M}_{u|\underline{h}}}\right)\left(\underline{\chi}_h^3 \underline{F}_{\underline{h}}^3 + \underline{\chi}_h^4 \underline{F}_{\underline{h}}^4\right)}{\underline{B}\sigma_{\underline{M}_{u|\underline{h}}}^2} + \frac{\left(u(t) - c_{\underline{M}_{u|\underline{h}}}\right)\left(\underline{\chi}_h^3 \underline{F}_{\underline{h}}^3 + \underline{\chi}_h^4 \underline{F}_{\underline{h}}^4\right)}{\underline{B}\sigma_{\underline{M}_{u|\underline{h}}}^2} - \\ &\left[\frac{\left(u(t) - c_{\underline{M}_{u|\underline{h}}}\right)\left(\underline{F}_{\underline{h}}^3 + \underline{F}_{\underline{h}}^4\right)}{\underline{B}^2\sigma_{\underline{M}_{u|\underline{h}}}^2} + \frac{\left(u(t) - c_{\underline{M}_{u|\underline{h}}}\right)\left(\underline{F}_{\underline{h}}^3 + \underline{F}_{\underline{h}}^4\right)}{\underline{B}^2\sigma_{\underline{M}_{u|\underline{h}}}^2} \right] \cdot \sum_{i=1}^{R} \left(\underline{F}_{\underline{h}}^{i} + \underline{F}_{\underline{h}}^{i} \right) \underline{\chi}_h^i \end{aligned} \tag{3.39}$$

For $c_{\underline{M}_{\hat{E}|h}}$, the tuning rules are:

$$\begin{aligned} c_{\underline{M}_{\hat{E}|h}}(t+1) = c_{\underline{M}_{\hat{E}|h}}(t) &+ \frac{\gamma\left(E_d(t) - \hat{E}(t)\right)}{\sum_{\underline{h}\in\{\underline{h}_1,\ldots,\underline{h}_N\}} \underline{h}} \sum_{\underline{h}\in\{\underline{h}_1,\ldots,\underline{h}_N\}} \underline{h} \frac{\partial \hat{\underline{E}}_h}{\partial c_{\underline{M}_{\hat{E}|\underline{h}}}} \\ &+ \frac{\gamma\left(E_d(t) - \hat{E}(t)\right)}{\sum_{\underline{h}\in\{\underline{h}_1,\ldots,\underline{h}_N\}} \underline{h}} \sum_{\underline{h}\in\{\underline{h}_1,\ldots,\underline{h}_N\}} \underline{h} \frac{\partial \hat{\underline{E}}_h}{\partial c_{\underline{M}_{\hat{E}|\underline{h}}}} \end{aligned} \tag{3.40}$$

Given that the $c_{\underline{M}_{\hat{E}|h}}$ is present in rules 1 and 3, it follows that:

$$\frac{\partial \underline{\hat{E}}_h}{\partial c_{\underline{M}_{\hat{E}|\underline{h}}}} = \frac{\left(\hat{E}(t-1) - c_{\underline{M}_{\hat{E}|\underline{h}}}\right)\left(\underline{\chi}_h^1 \underline{F}_{\underline{h}}^1 + \underline{\chi}_h^3 \underline{F}_{\underline{h}}^3\right)}{\underline{B\sigma}^2_{\underline{M}_{\hat{E}|\underline{h}}}}$$
$$+ \frac{\left(\hat{E}(t-1) - c_{\underline{M}_{\hat{E}|\underline{h}}}\right)\left(\underline{\chi}_h^1 \underline{F}_{\underline{h}}^1 + \underline{\chi}_h^3 \underline{F}_{\underline{h}}^3\right)}{\underline{B\sigma}^2_{\underline{M}_{\hat{E}|\underline{h}}}} -$$
$$\left[\frac{\left(\hat{E}(t-1) - c_{\underline{M}_{\hat{E}|\underline{h}}}\right)\left(\underline{F}_{\underline{h}}^1 + \underline{F}_{\underline{h}}^3\right)}{\underline{B}^2\underline{\sigma}^2_{\underline{M}_{\hat{E}|\underline{h}}}} + \frac{\left(\hat{E}(t-1) - c_{\underline{M}_{\hat{E}|\underline{h}}}\right)\left(\underline{F}_{\underline{h}}^1 + \underline{F}_{\underline{h}}^3\right)}{\underline{B}^2\underline{\sigma}^2_{\underline{M}_{\hat{E}|\underline{h}}}}\right]$$
$$\cdot \sum_{i=1}^{R} \left(\underline{F}_{\underline{h}}^i + \underline{F}_{\underline{h}}^i\right)\underline{\chi}_h^i \tag{3.41}$$

$$\frac{\partial \underline{\hat{E}}_h}{\partial c_{\underline{M}_{\hat{E}|\underline{h}}}} = \frac{\left(\hat{E}(t-1) - c_{\underline{M}_{\hat{E}|\underline{h}}}\right)\left(\underline{\chi}_h^1 \underline{F}_{\underline{h}}^1 + \underline{\chi}_h^3 \underline{F}_{\underline{h}}^3\right)}{\underline{B\sigma}^2_{\underline{M}_{\hat{E}|\underline{h}}}}$$
$$+ \frac{\left(\hat{E}(t-1) - c_{\underline{M}_{\hat{E}|\underline{h}}}\right)\left(\underline{\chi}_h^1 \underline{F}_{\underline{h}}^1 + \underline{\chi}_h^3 \underline{F}_{\underline{h}}^3\right)}{\underline{B\sigma}^2_{\underline{M}_{\hat{E}|\underline{h}}}} -$$
$$\left[\frac{\left(\hat{E}(t-1) - c_{\underline{M}_{\hat{E}|\underline{h}}}\right)\left(\underline{F}_{\underline{h}}^1 + \underline{F}_{\underline{h}}^3\right)}{\underline{B}^2\underline{\sigma}^2_{\underline{M}_{\hat{E}|\underline{h}}}} + \frac{\left(\hat{E}(t-1) - c_{\underline{M}_{\hat{E}|\underline{h}}}\right)\left(\underline{F}_{\underline{h}}^1 + \underline{F}_{\underline{h}}^3\right)}{\underline{B}^2\underline{\sigma}^2_{\underline{M}_{\hat{E}|\underline{h}}}}\right]$$
$$\cdot \sum_{i=1}^{R} \left(\underline{F}_{\underline{h}}^i + \underline{F}_{\underline{h}}^i\right)\underline{\chi}_h^i \tag{3.42}$$

For $c_{\underline{M}_{\hat{E}|h}}$, the tuning rules are:

$$c_{\underline{M}_{\hat{E}|h}}(t+1) = c_{\underline{M}_{\hat{E}|h}}(t) + \frac{\gamma\left(E_d(t) - \hat{E}(t)\right)}{\sum_{\underline{h}\in\{\underline{h}_1,\ldots,\underline{h}_N\}} \underline{h}} \sum_{\underline{h}\in\{\underline{h}_1,\ldots,\underline{h}_N\}} \underline{h}\frac{\partial \underline{\hat{E}}_h}{\partial c_{\underline{M}_{\hat{E}|\underline{h}}}}$$
$$+ \frac{\gamma\left(E_d(t) - \hat{E}(t)\right)}{\sum_{\underline{h}\in\{\underline{h}_1,\ldots,\underline{h}_N\}} \underline{h}} \sum_{\underline{h}\in\{\underline{h}_1,\ldots,\underline{h}_N\}} \underline{h}\frac{\partial \underline{\hat{E}}_h}{\partial c_{\underline{M}_{\hat{E}|\underline{h}}}} \tag{3.43}$$

Given that the $c_{\underline{M}_{\hat{E}|h}}$ is present in rules 2 and 4, we can conclude:

$$\frac{\partial \underline{\hat{E}}_h}{\partial c_{\underline{M}_{\hat{E}|\underline{h}}}} = \frac{\left(\hat{E}(t-1) - c_{\underline{M}_{\hat{E}|\underline{h}}}\right)\left(\underline{\chi}_h^2 \underline{F}_{\underline{h}}^2 + \underline{\chi}_h^4 \underline{F}_{\underline{h}}^4\right)}{\underline{B\sigma}^2_{\underline{M}_{\hat{E}|\underline{h}}}}$$
$$+ \frac{\left(\hat{E}(t-1) - c_{\underline{M}_{\hat{E}|\underline{h}}}\right)\left(\underline{\chi}_h^2 \underline{F}_{\underline{h}}^2 + \underline{\chi}_h^4 \underline{F}_{\underline{h}}^4\right)}{\underline{B\sigma}^2_{\underline{M}_{\hat{E}|\underline{h}}}} -$$
$$\left[\frac{\left(\hat{E}(t-1) - c_{\underline{M}_{\hat{E}|\underline{h}}}\right)\left(\underline{F}_{\underline{h}}^2 + \underline{F}_{\underline{h}}^4\right)}{\underline{B}^2\underline{\sigma}^2_{\underline{M}_{\hat{E}|\underline{h}}}} + \frac{\left(\hat{E}(t-1) - c_{\underline{M}_{\hat{E}|\underline{h}}}\right)\left(\underline{F}_{\underline{h}}^2 + \underline{F}_{\underline{h}}^4\right)}{\underline{B}^2\underline{\sigma}^2_{\underline{M}_{\hat{E}|\underline{h}}}}\right]$$

$$\cdot \sum_{i=1}^{R} \left(\underline{F}_h^i + \underline{F}_h^i\right) \underline{\chi}_h^i \tag{3.44}$$

$$\frac{\partial \underline{\hat{E}}_h}{\partial c_{\underline{M}_{\hat{E}|\underline{h}}}} = \frac{\left(\hat{E}(t-1) - c_{\underline{M}_{\hat{E}|\underline{h}}}\right)\left(\underline{\chi}_h^2 \underline{F}_h^2 + \underline{\chi}_h^4 \underline{F}_h^4\right)}{\underline{B\sigma}^2_{\underline{M}_{\hat{E}|\underline{h}}}} + \frac{\left(\hat{E}(t-1) - c_{\underline{M}_{\hat{E}|\underline{h}}}\right)\left(\underline{\chi}_h^2 \underline{F}_h^2 + \underline{\chi}_h^4 \underline{F}_h^4\right)}{\underline{B\sigma}^2_{\underline{M}_{\hat{E}|\underline{h}}}} - \left[\frac{\left(\hat{E}(t-1) - c_{\underline{M}_{\hat{E}|\underline{h}}}\right)\left(\underline{F}_h^2 + \underline{F}_h^4\right)}{\underline{B}^2\underline{\sigma}^2_{\underline{M}_{\hat{E}|\underline{h}}}} + \frac{\left(\hat{E}(t-1) - c_{\underline{M}_{\hat{E}|\underline{h}}}\right)\left(\underline{F}_h^2 + \underline{F}_h^4\right)}{\underline{B}^2\underline{\sigma}^2_{\underline{M}_{\hat{E}|\underline{h}}}}\right] \cdot \sum_{i=1}^{R} \left(\underline{F}_h^i + \underline{F}_h^i\right) \underline{\chi}_h^i \tag{3.45}$$

3.1.2.3 How to Optimize the Horizontal Slices

The UKF algorithm optimizes the level of the slices. The state space can be expressed as:

$$h(t+1) = h(t) + \delta(t)$$

$$\hat{E}(t+1) = T3FS(z(t), \theta(t)|h(t)) + \rho(t) \tag{3.46}$$

The variables $z(t) = \left[u(t), \hat{E}(t-1)\right]$, $\delta(t)$, and $\rho(t)$ represent noise with covariance ω_P and ω_m, and $\theta(t)$ is a vector of learnable parameters that contains MF and rule parameters. The sigma points $\tilde{\beta}$ are precisely specified as:

$$\tilde{\beta}_j = \tilde{\beta}_j + \tilde{w}_j, j = 1, ..., 2n \tag{3.47}$$

where,

$$\{\tilde{w}_j = \left(\sqrt{n\Psi(t)}\right)^T, j = 1, ..., n\tilde{w}_{j+n} = -\left(\sqrt{n\Psi(t)}\right)^T, j = 1, ..., n \tag{3.48}$$

Here, n is the number of slices, while $\Psi(t)$ denotes the covariance matrix. The estimates for $\tilde{E}_j(t+1)$ at each point $\tilde{\beta}_j$ are calculated as follows:

$$\hat{E}_j(t+1) = T3FS\left(z(t), \theta(t)|\tilde{\beta}_j(t)\right) \tag{3.49}$$

The mean of $\hat{E}_j$ is:

$$\underline{E} = \sum_{j=1,\ldots,2n} \tilde{E}_j/(2n) \tag{3.50}$$

Cross-covariance, denoted as Ψ_{hE}, is defined as follows:

$$\Psi_{hE} = \frac{1}{2n}\sum_{z=1}^{2n} \left(\tilde{w}_j - \hat{h}\right)\underline{E} \tag{3.51}$$

where, $\hat{h}$ is:

$$\hat{\alpha} = \frac{1}{2n}\sum_{z=1}^{2n} \tilde{\alpha}_z \tag{3.52}$$

The Kalman gain is:

$$K(t) = \Psi_{hE}\Psi^{-1} \tag{3.53}$$

Then the vector h is tunned as:

$$h(t+1) = h(t) + K(t)\left(E_d(t) - \underline{E}(t)\right) \tag{3.54}$$

3.1.3 Stability

In order to analyze stability, it is necessary to consider the Lyapunov function Eq. (3.55).

$$V(t) = \frac{1}{2}e^2 = \left(E_d(t) - \hat{E}(t)\right)^2 \tag{3.55}$$

Time difference of Eq. (3.55), yields:

$$\Delta V(k) = V(t+1) - V(t) = \frac{1}{2}\left(\left(E_d(t+1) - \hat{E}(t+1)\right)^2 - \left(E_d(t) - \hat{E}(t)\right)^2\right) \tag{3.56}$$

Now, consider Eq. (3.57) as below:

$$e(t+1) = e(t) + \Delta e(t) \cong e(t) + \left[\frac{\partial e(t)}{\partial \theta}\right]^T \Delta\theta \tag{3.57}$$

The symbol $\theta(t)$ denotes avector of adjustable parameters, which encompasses both the MF (membership function) parameters and the rule ones. The adaption laws of θ yield the following:

$$\theta(t+1) = \theta(t) + \Delta\theta(t) = \theta(t) + \gamma.\left(-\frac{\partial e(t)}{\partial \theta}\right) \tag{3.58}$$

where,

$$\frac{\partial e(t)}{\partial \theta} = -e(t).\frac{\partial \hat{E}}{\partial \theta} \tag{3.59}$$

From Eq. (3.56), $\Delta V(k)$, we have:

$$\begin{aligned}\Delta V(t) &= \frac{1}{2}\left(e^2(t+1) - e^2(t)\right)\\ &= \frac{1}{2}(e(t+1) - e(t)).(e(t+1) + e(t))\\ &= \frac{1}{2}\Delta e(k)[2e(k) + \Delta e(k)]\end{aligned} \tag{3.60}$$

From Eq. (3.57), $\Delta V(k)$ in Eq. (3.60), changes to:

$$\Delta V(t) = \left[\frac{\partial e(t)}{\partial \theta}\right]^T.\gamma.e(t).\frac{\partial \hat{E}(t)}{\partial \theta}.\left\{e(t) + \frac{1}{2}\left[\frac{\partial e(t)}{\partial \theta}\right]^T.\gamma.e(t).\frac{\partial \hat{E}(t)}{\partial \theta}\right\} \tag{3.61}$$

Replacing $\frac{\partial e(t)}{\partial \theta}$ form Eq. (3.59) yields:

$$\Delta V(t) == -\left[\frac{\partial \hat{E}(t)}{\partial \theta}\right]^T.\gamma.e(t).\frac{\partial \hat{E}(t)}{\partial \theta}.\left\{e(t) - \frac{1}{2}\left[\frac{\partial \hat{E}(t)}{\partial \theta}\right]^T.\gamma.e(t).\frac{\partial \hat{E}(t)}{\partial \theta}\right\} \tag{3.62}$$

From Eq. (3.62), one has:

$$\Delta V(t) == -\gamma.e^2(t)\left|\frac{\partial \hat{E}(t)}{\partial \theta}\right|^2.\left[1 - \frac{1}{2}\gamma.\left|\frac{\partial \hat{E}(t)}{\partial \theta}\right|^2\right] \tag{3.63}$$

Then by:

$$0 < \gamma < 2/max\left|\frac{\partial \hat{E}(t)}{\partial \theta}\right|^2 \tag{3.64}$$

The stability is guaranteed.

3.1.4 Evaluation Indexes and Simulation

In order to assess the effectiveness of the suggested strategy, the following indices are utilized:

$$RMSE = \sqrt{\frac{1}{N}\sum_{i=1}^{N}(y_i - \hat{y}_i)^2} \tag{3.65}$$

$$R^2 = 1 - \frac{\sum_{i=1}^{N}(y_i - \hat{y}_i)^2}{\sum_{i=1}^{N}\left(\frac{1}{N}\sum_{i=1}^{N}y_i - \hat{y}_i\right)^2} \tag{3.66}$$

In this context, N represents the data number, y_i represents the power, and $\hat{y}_i$ represents the estimated power. RMSE stands for root-mean-square-error. This section includes two simulation examples and one comparison with other FLSs to validate their usefulness.

Example 1 *The first example involves considering a 660 kilowatt wind turbine. The data sets are sourced from reference* [2]. *The specifics of WT are outlined in Table 3.1, while the progression of wind speed is illustrated in Fig. 3.4. Table 3.2 provides a description of the T3-FLS parameters.*

Table 3.1 WT description in in Ex. 1

Parameters	Value	Unit
Type	Asynchronous	–
Grid frequency	50	Hz
Tip speed	70.10	m/s
Max of rotor speed	28.50	U/min
Blades number	3	
Swept area	1.735	m^2
Number	1.0	–
Cut-out wind-speed	25	m/s
Material	GFK	–
Type	22.90	–
Grid connection	Thyristor	–
Survival wind speed	60	m/s
Voltage	400	V
Diameter	47	m
Cut-in wind-speed	4	m/s
Speed, max	1.650	U/min
Rated wind-speed	15	m/s

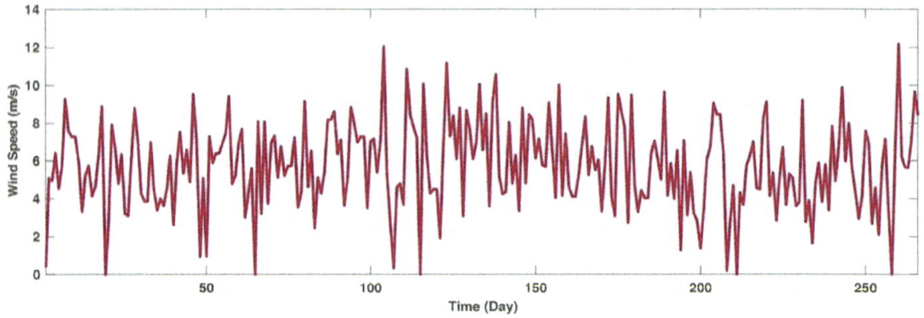

Fig. 3.4 Wind speed in Ex. 1

Table 3.2 T3-FLS parameters in Ex. 1

Parameters	Value
${}^c\underline{M}_{u\|h}$	4
${}^c\underline{M}_{u\|h}$	6
${}^c\underline{M}_{\hat{E}\|h}$	4
${}^c\underline{M}_{\hat{E}\|h}$	6
$\underline{h}$	0, 0.25, 0.5
$\underline{h}$	0.75, 0.9, 1

Figure 3.5 displays the approximated trajectory alongside the real power trajectory. It is evident that the estimated power closely follows the actual power path for both the test and train data sets. Figures 3.6 and 3.7 display the RMSE trajectories over iterations, presenting the Box plot of RMSE for both the testing and training data sets. The Box plot of $R^2$for the testing data set, Histogram plot of R^2, and the value of R^2 with respect to iterations are shown in Fig. 3.8, while the R^2 values for the training data set are displayed in Fig. 3.6. These figures also include the histogram plot and box plot of the R^2 values. The observed R^2 value of approximately 0.95 indicates a very dependable model.

Example 2 *In this instance, the solar radiation produced by a sunlight simulator, as depicted in Fig. 3.10, is supplied to the solar cell illustrated in Fig. 3.11, and the output power is determined. Like Example 1, we assess the effectiveness of the offered method. The simulation situation is outlined in Table 3.3.*

Figure 3.13 illustrates the estimated trajectory alongside the real power trajectory. As with Example 1, we observe that the estimated power closely follows the actual solar power path for both the test and train data sets. Figures 3.14 and 3.15 display the RMSE trajectories for iterations and the Box plot of RMSE for the testing and training data sets. The R^2 values for each iteration, as well as the Histogram and box plot of R^2 for the testing data set, are shown in Fig. 3.16. The values of R^2 for the training data set can be found in Fig. 3.17. The R^2 values of approximately 0.98 indicates a very dependable model [3].

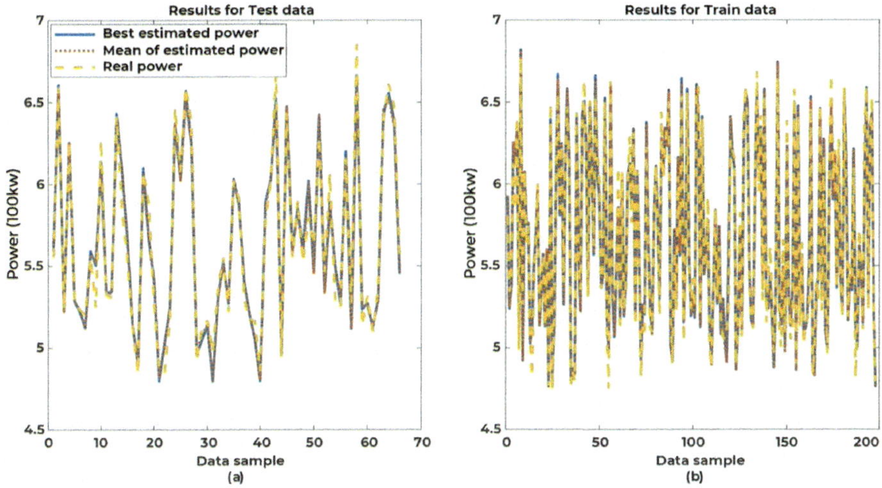

Fig. 3.5 Estimation performance in Ex. 1; **a** Testing data; **b** Training data

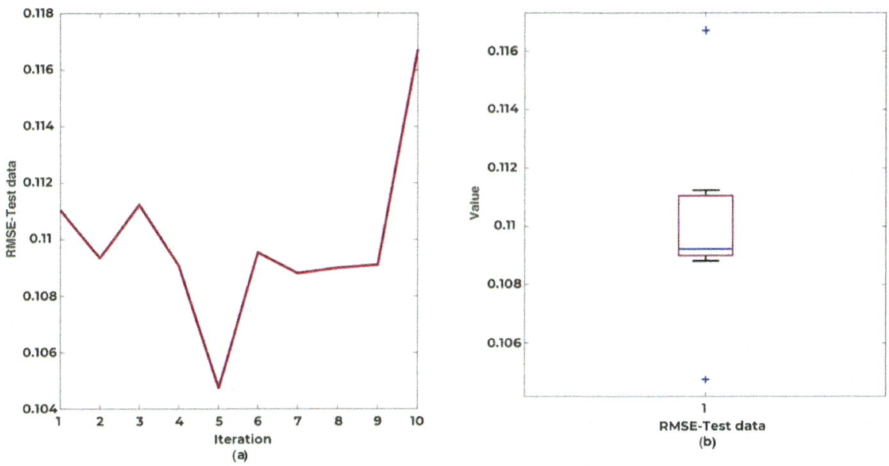

Fig. 3.6 Testing data results in Ex. 1; **a** RMSE at each iterations; **b** RMSE box plot

3.1.4.1 Comparison and Discussion

In order to thoroughly assess the superiority of the suggested method, a comparison is provided with conventional procedures. The root mean square errors (RMSEs) in both of the preceding examples are compared with the multi-layer perceptron (MLP), T1-FLS, T2-FLS, and interval T3-FLS (IT3-FLS). The findings presented in Table 3.4 clearly demonstrate the substantial advantage of the proposed T3-FLS with a deep learning method and a nonlinear consequent part [3].

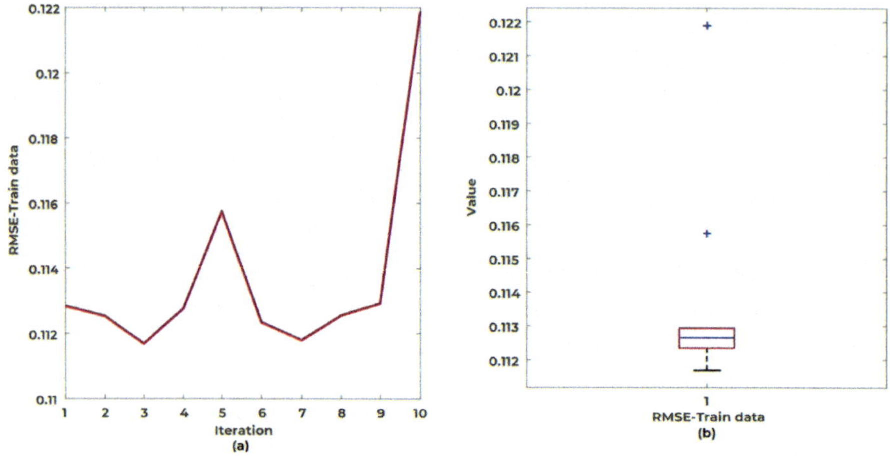

Fig. 3.7 Training data results in Ex. 1; **a** RMSE at each iterations; **a** RMSE box plot

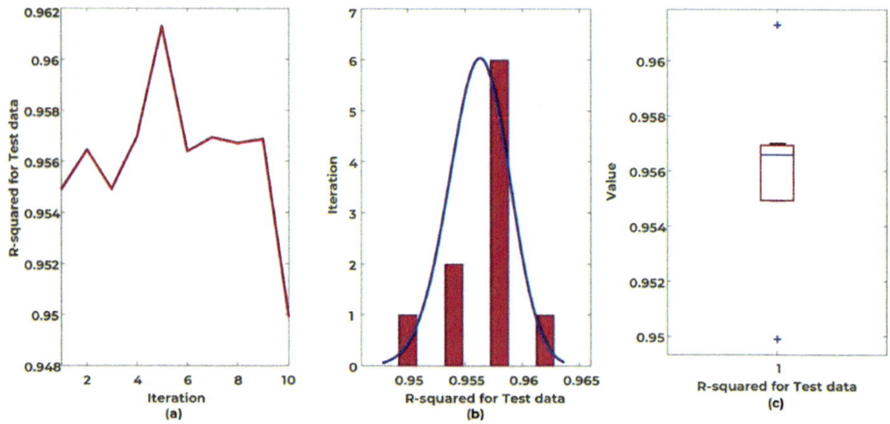

Fig. 3.8 Testing data results in Ex. 1; **a** R^2 value at each iteration; **b** R^2 histogram plot; **c** R^2 box plot

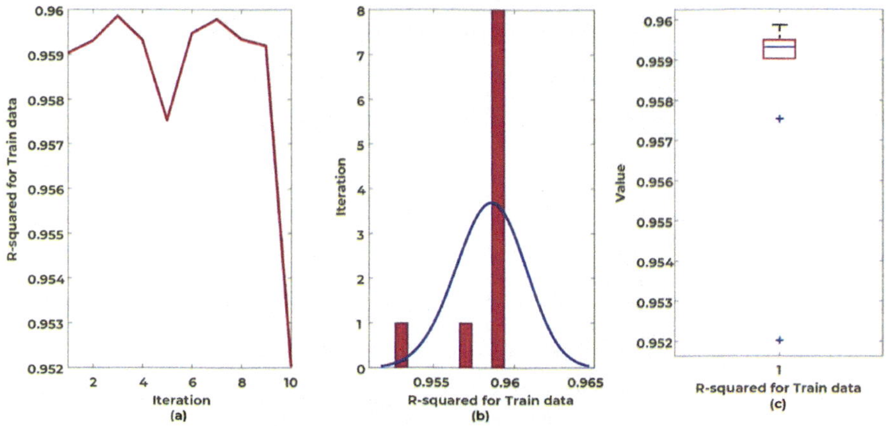

Fig. 3.9 Training data results in Ex. 1; **a** R^2 value at each iteration; **b** R^2 histogram plot; **c** R^2 box plot

Fig. 3.10 Sunlight simulator in Ex. 2

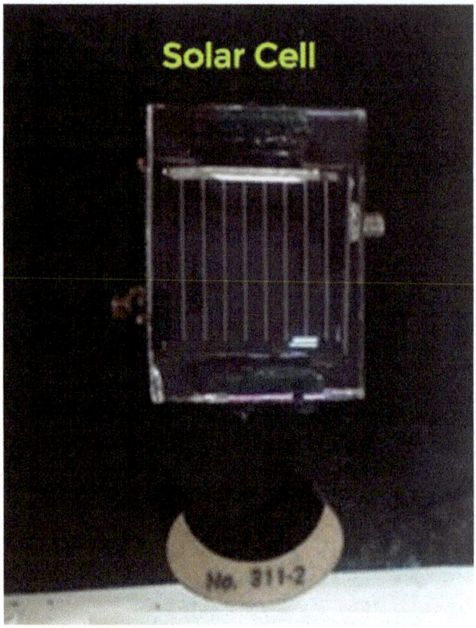

Fig. 3.11 Solar cell in Ex. 2

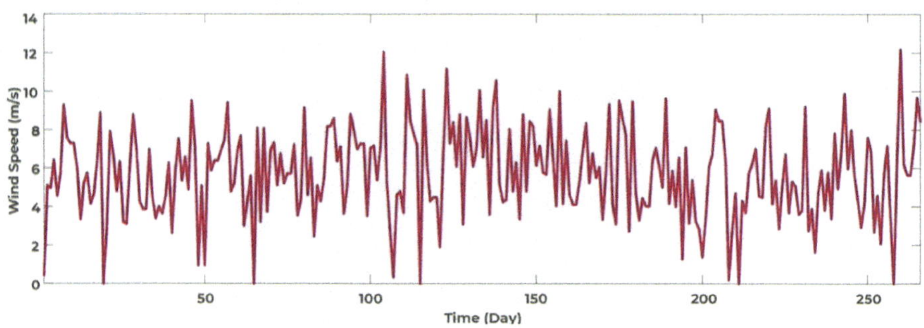

Fig. 3.12 Solar radiation in Ex. 2

Table 3.3 Parameters of T3-FLS in Ex. 2

Parameters	Value
$c_{\underline{M}_{u\|h}}$	500
$c_{\underline{M}_{u\|h}}$	600
$c_{\underline{M}_{\hat{E}\|h}}$	40
$c_{\underline{M}_{\hat{E}\|h}}$	60
$\underline{h}$	0, 0.25, 0.5
$\underline{h}$	0.75, 0.9, 1

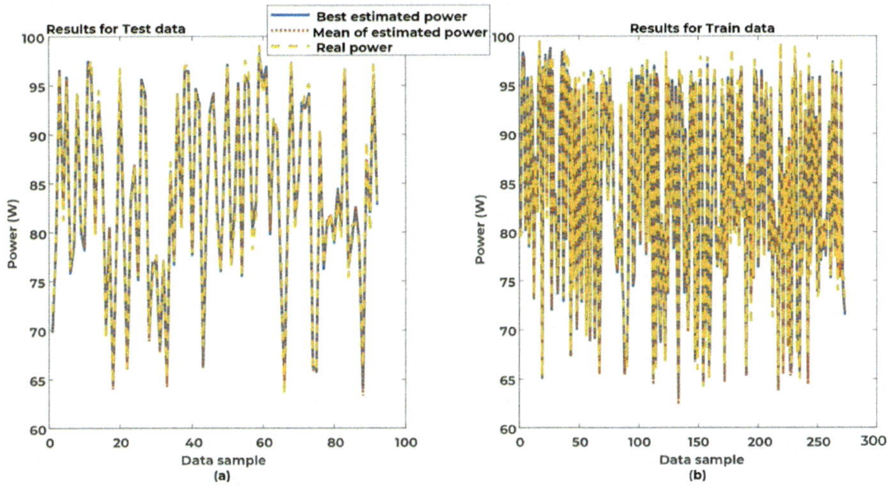

Fig. 3.13 Estimation performance in Ex. 2; **a** Testing data; **b** Training data

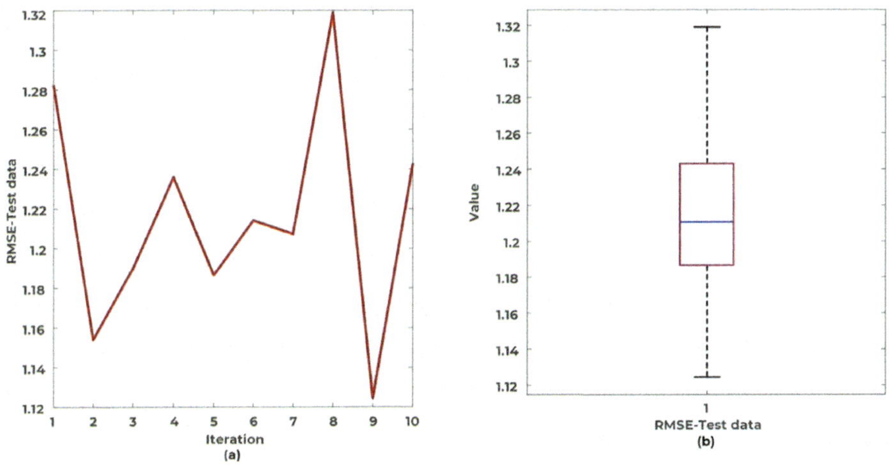

Fig. 3.14 Testing data results in Ex. 2; **a** RMSE at each iterations; **b** RMSE box plot

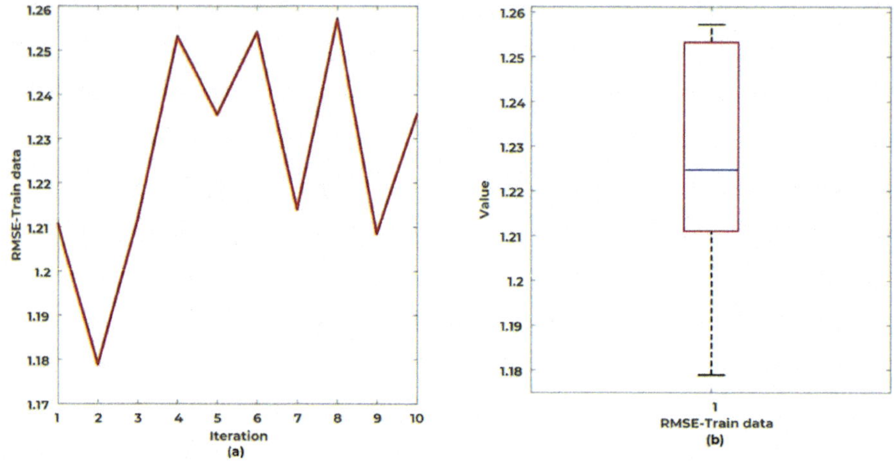

Fig. 3.15 Training data results in Ex. 2; **a** RMSE at each iterations; **b** RMSE box plot

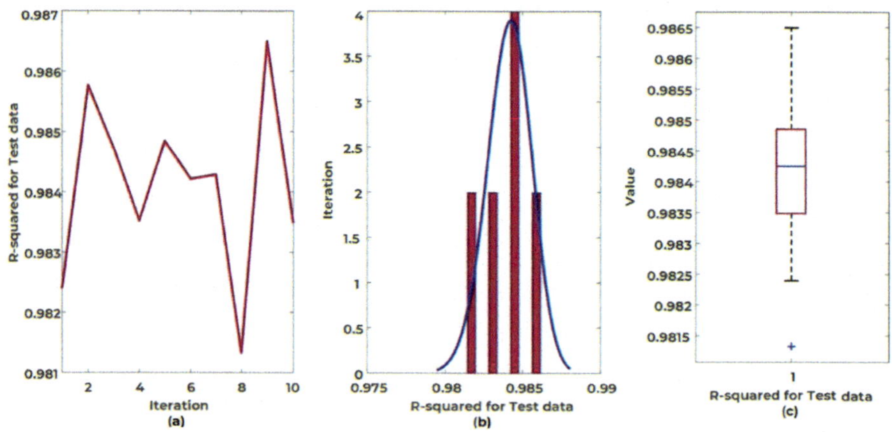

Fig. 3.16 Testing data results in Ex. 2; **a** R^2 value at each iteration; **b** R^2 histogram plot; **c** R^2 box plot

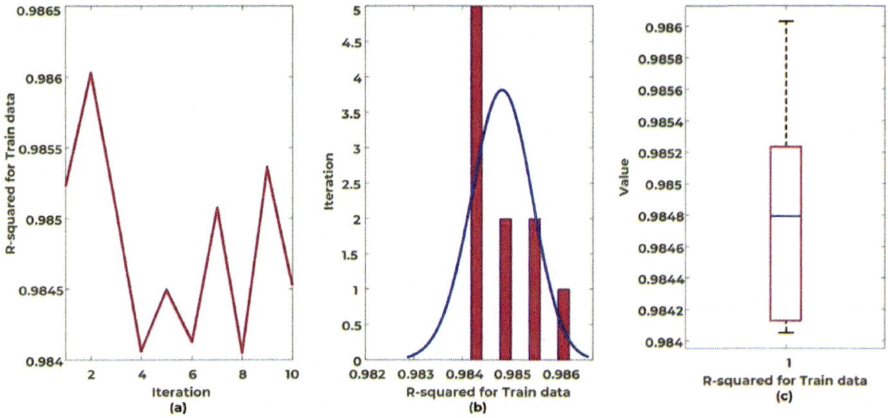

Fig. 3.17 Training data results in Ex. 2; **a** R^2 value at each iteration; **b** R^2 histogram plot; **c** R^2 box plot

Table 3.4 RMSE comparison in Ex. 2

Method	Example 1	Example 2
MLP	2.0214	2.8721
T1-FLS	2.0187	2.2306
T2-FLS	1.8701	1.9701
IT3-FLS	1.2354	1.3517
Proposed FLS	1.1065	1.1245

3.2 Fuzzy Control with Online Learning for Solar Energy Management

3.2.1 Problem Description

3.2.1.1 General Investigation

Figures 3.1 and 3.2 illustrate the block schematics of the primary hybrid system and control system. The system comprises of a photovoltaic (PV) panel, a battery, a boost converter, and a bidirectional boost converter. The control purpose is to devise a suitable charging and/or discharging plan for the battery, as well as determine an appropriate current for the photovoltaic (PV) system, in order to regulate the load voltage at a desired level. The subsequent section demonstrates the mathematical dynamics of all units.

3.2.1.2 Converters

The Boost and bidirectional Boost converters are responsible for constructing the switching unit. The function of these devices is to create a switching mechanism that alternates between the battery and the photovoltaic (PV) system. The switching unit model is represented in state space form by averaging four modes, as shown in Fig. 3.3:

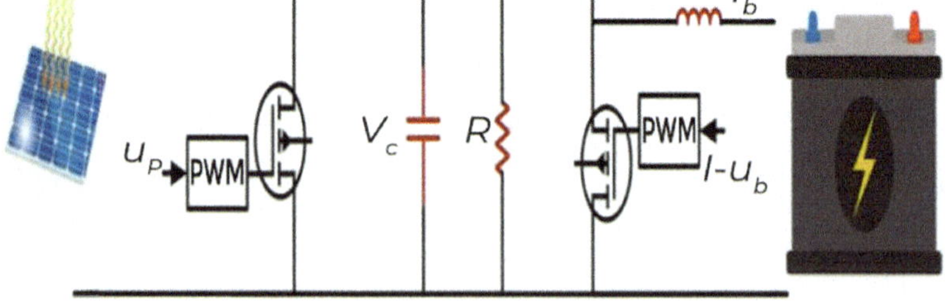

Fig. 3.1 The block diagram of PV/battery hybrid plant

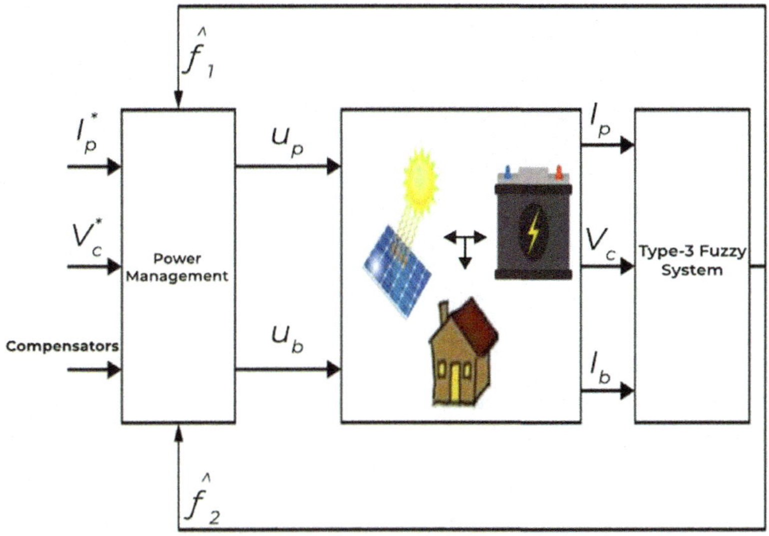

Fig. 3.2 The suggested control system

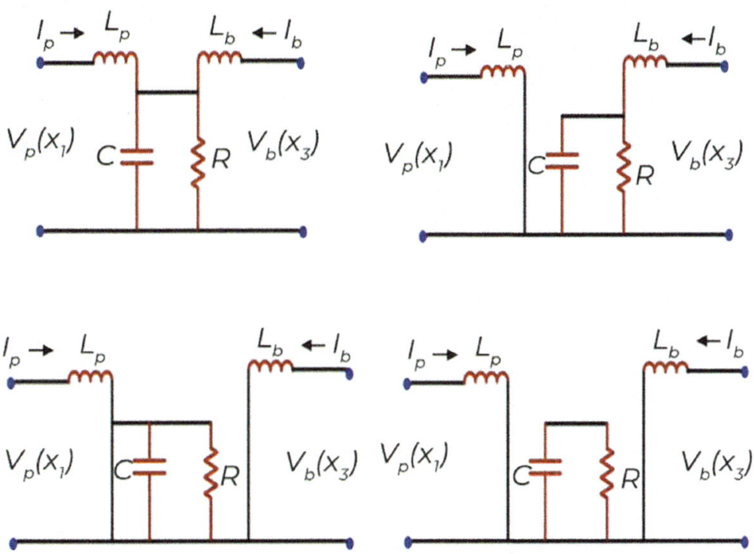

Fig. 3.3 Switching circuit modes

$$\begin{aligned}\dot{y}_1 &= \frac{1}{L_p}\left(-y_2 + y_2 u_p + V_p(y_1)\right)\\ \dot{y}_2 &= \frac{1}{C}\left(y_1 - y_1 u_p - y_2/R + y_3 u_b\right)\\ \dot{y}_3 &= \frac{1}{L_b}(-y_2 u_b + V_b(y_3))\end{aligned} \tag{3.1}$$

where, I_p and I_b are respectively PV and Battery currents, and V_c is the load voltage.

3.2.1.3 PV Modeling

The often cited method for modeling the behavior of a photovoltaic (PV) panel is the single-diode approach [4]:

$$\begin{aligned}i_{ph} = s(k_i(T - T_r) + i_{sc})I_p = -i_o exp\left(\frac{q}{nTk_b}\left(V_p + I_p R_{sg}\right) - 1\right)\\ + GI_{phg} - \left(I_p R_{sg} + V_p\right)/R_{shg} i_0 =\\ exp\left[\frac{E_g q}{k_b A}\left(\frac{1}{T_r + 273} - \frac{1}{T + 273}\right)\right]\left(\frac{T + 273}{T_r + 273}\right)^3 i_r\end{aligned} \tag{3.2}$$

The parameters description can be found in Table 3.1 [5]. The power of photovoltaic is illustrated in Fig. 3.4. It is evident that the highest power is achieved at a specific current. The controller should regulate the current of the photovoltaic system to this specific level.

Table 3.1 PV model parameters and descriptions

Parameter	Description
n	Number of cells
$G(w/m^2)$	Solar radiation
$k_b(J/K)$	Boltzmann's constant
R_{sh} and $R_s(\Omega)$	Equivalent resistances
$T(°c)$	Temperature of PV
q	Electron charge
A	Constant of the diode ideality
$E_g(ev)$	Energy of Band-Gap
$i_r(A)$	Saturation current
$i_{ph}(A)$	Photo generated currents
$T_r(°c)$	Reference temperature

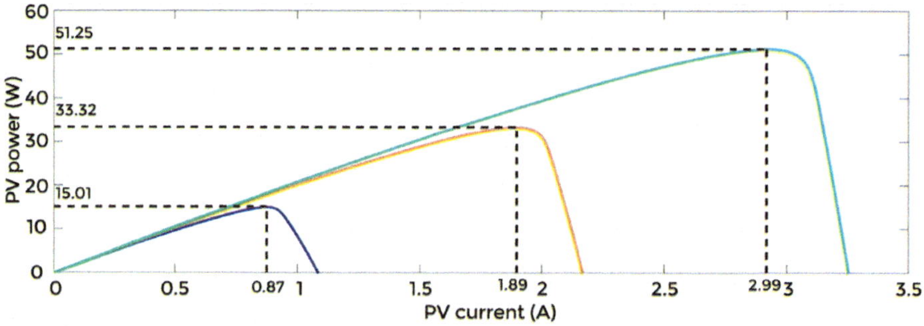

Fig. 3.4 The power of the PV panel as a function of its current

3.2.1.4 Battery Modeling

The mathematical representation of a battery is expressed as equation [4]:

$$E(t) = -\int W_{Loss} + \beta I_b V_{boc} dt, V_b = V_{boc} - r_b I_b$$

$$\beta = \{\beta_1 I_b \geq 0 \beta_2 I_b < 0, SoC(t) = E(T) = E_{max} \tag{3.3}$$

The parameter explanations can be found in Table 3.2.

3.2.2 *Proposed Type-3 FLS*

The interval type-3 fuzzy logic systems (IT3-FLSs) have been designed to effectively manage a greater degree of uncertainty [6, 7]. This study utilizes IT3-FLSs for the purpose of online dynamic identification. The structure of IT3-FLS is depicted in Fig. 3.5, as well as below:

(1) The inputs of IT3-FLSs are $y_1 = I_p(t)$, $y_2 = V_c(t - \tau)$, $y_3 = I_b(t)$, where, I_p represents the current generated by the photovoltaic (PV) system, I_b represents

Table 3.2 Battery model parameters and descriptions

Parameter	Description
$r_b(\Omega)$	Internal resistance
$E_{Max}(J)$	Maximum savable energy
$V_{boc}(v)$	Open circuit voltage
$W_{Loss}(w)$	Power losses
β_1 and β_2	Charge/discharge rates

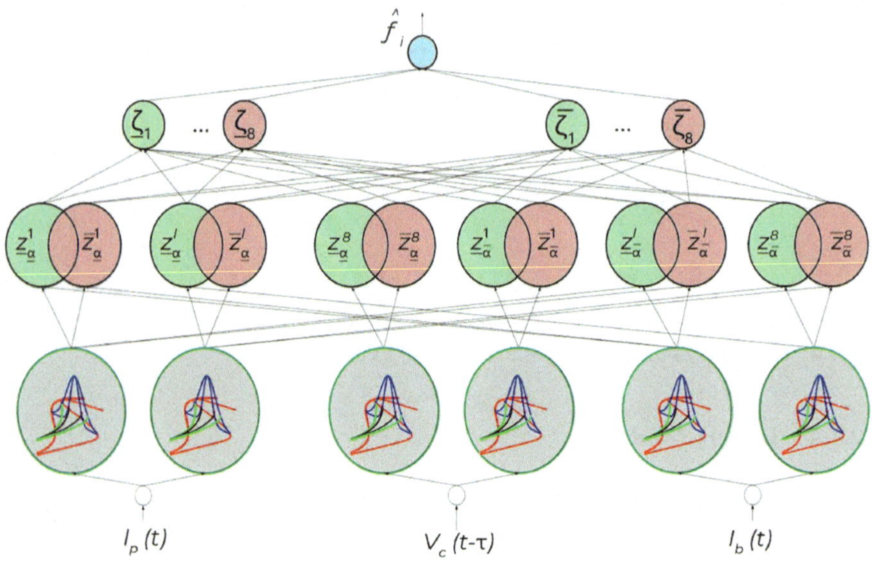

Fig. 3.5 IT3-FLS structure

the current from the battery, and V_c represents the voltage of the load. The symbol τ denotes the time interval at which samples are taken.

(2) Two Gaussian membership functions (MFs) are considered for each input, as shown in Fig. 3.6. The centroids of membership functions are positioned at the upper bound (UB) and lower bound (LB) of each input. The UB and LB memberships for the input I_p are calculated as follows:

$$\underline{\mu}_{\tilde{A}^1_{I_p}|\alpha}(I_p) = exp\left(-\frac{\left(I_p - \underline{c}_{\tilde{A}^1_{I_p}}\right)^2}{\underline{\sigma}^2_{\tilde{A}^1_{I_p}|\alpha}}\right)\underline{\mu}_{\tilde{A}^2_{I_p}|\alpha}(I_p) = exp\left(-\frac{\left(I_p - \underline{c}_{\tilde{A}^2_{I_p}}\right)^2}{\underline{\sigma}^2_{\tilde{A}^2_{I_p}|\alpha}}\right) \tag{3.4}$$

$$\underline{\mu}_{\tilde{A}^1_{I_p}|\alpha}(I_p) = exp\left(-\frac{\left(I_p - \underline{c}_{\tilde{A}^1_{I_p}}\right)^2}{\underline{\sigma}^2_{\tilde{A}^1_{I_p}|\alpha}}\right)\underline{\mu}_{\tilde{A}^2_{I_p}|\alpha}(I_p) = exp\left(-\frac{\left(I_p - \underline{c}_{\tilde{A}^2_{I_p}}\right)^2}{\underline{\sigma}^2_{\tilde{A}^2_{I_p}|\alpha}}\right) \tag{3.5}$$

Here, α represents the horizontal slice level. $\tilde{A}^1_{I_p}$ and $\tilde{A}^2_{I_p}$ represent the initial and subsequent membership functions for the input variable I_p. The centers of $\tilde{A}^1_{I_p}$ and $\tilde{A}^2_{I_p}$ are denoted as $\underline{c}_{\tilde{A}^1_{I_p}}$ and $\underline{c}_{\tilde{A}^2_{I_p}}$ respectively. The ratios. $\underline{\sigma}_{\tilde{A}^1_{Ip}|\alpha}/\underline{\sigma}_{\tilde{A}^1_{Ip}|\alpha}$ and $\underline{\sigma}_{\tilde{A}^2_{Ip}|\alpha}/\underline{\sigma}_{\tilde{A}^2_{Ip}|\alpha}$ represent the standard deviations for the upper and lower bounds of $\tilde{A}^1_{I_p}$ and $\tilde{A}^2_{I_p}$, respectively. Similarly, when considering the input I_b, the same relationship holds:

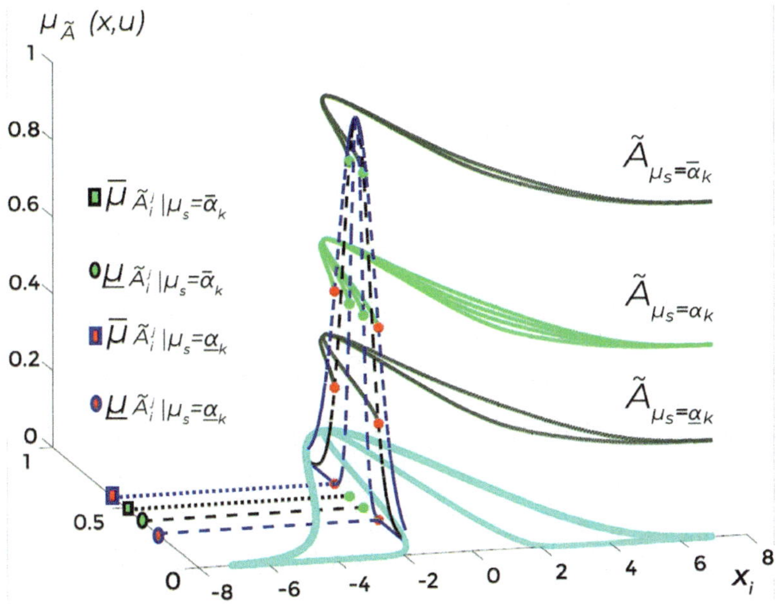

Fig. 3.6 The MF of interval type-3

$$\underline{\mu}_{\tilde{A}_{I_b}^1|\alpha}(I_b) = exp\left(-\frac{\left(I_b - \underline{c}_{\tilde{A}_{I_b}^1}\right)^2}{\underline{\sigma}^2_{\tilde{A}_{I_b}^1|\alpha}}\right) \underline{\mu}_{\tilde{A}_{I_b}^2|\alpha}(I_b) = exp\left(-\frac{\left(I_b - \underline{c}_{\tilde{A}_{I_b}^2}\right)^2}{\underline{\sigma}^2_{\tilde{A}_{I_b}^2|\alpha}}\right) \tag{3.6}$$

$$\underline{\mu}_{\tilde{A}_{I_b}^1|\alpha}(I_b) = exp\left(-\frac{\left(I_b - \underline{c}_{\tilde{A}_{I_b}^1}\right)^2}{\underline{\sigma}^2_{\tilde{A}_{I_b}^1|\alpha}}\right) \underline{\mu}_{\tilde{A}_{I_b}^2|\alpha}(I_b) = exp\left(-\frac{\left(I_b - \underline{c}_{\tilde{A}_{I_b}^2}\right)^2}{\underline{\sigma}^2_{\tilde{A}_{I_b}^2|\alpha}}\right) \tag{3.7}$$

The variables $\tilde{A}_{I_b}^1$ and $\tilde{A}_{I_b}^2$ represent the first and second membership functions for the input I_b. The points $\underline{c}_{\tilde{A}_{I_b}^1}$ and $\underline{c}_{\tilde{A}_{I_b}^2}$ represent the centers of $\tilde{A}_{I_b}^1$ and $\tilde{A}_{I_b}^2$, respectively. The ratios $\underline{\sigma}_{\tilde{A}_{I_b}^1|\alpha}/\underline{\sigma}_{\tilde{A}_{I_b}^1|\alpha}$ and $\underline{\sigma}_{\tilde{A}_{I_b}^2|\alpha}/\underline{\sigma}_{\tilde{A}_{I_b}^2|\alpha}$ represent the standard deviations for the upper and lower bounds of $\tilde{A}_{I_b}^1$ and $\tilde{A}_{I_b}^2$, respectively. Ultimately, the expression for input V_c is as follows:

$$\underline{\mu}_{\tilde{A}_{V_c}^1|\alpha}(V_c) = exp\left(-\frac{\left(V_c - \underline{c}_{\tilde{A}_{V_c}^1}\right)^2}{\underline{\sigma}^2_{\tilde{A}_{V_c}^1|\alpha}}\right) \underline{\mu}_{\tilde{A}_{V_c}^2|\alpha}(V_c) = exp\left(-\frac{\left(V_c - \underline{c}_{\tilde{A}_{V_c}^2}\right)^2}{\underline{\sigma}^2_{\tilde{A}_{V_c}^2|\alpha}}\right) \tag{3.8}$$

$$\underline{\mu}_{\tilde{A}_{V_c}^1|\alpha}(V_c) = exp\left(-\frac{\left(V_c - \underline{c}_{\tilde{A}_{V_c}^1}\right)^2}{\underline{\sigma}^2_{\tilde{A}_{V_c}^1|\alpha}}\right) \underline{\mu}_{\tilde{A}_{V_c}^2|\alpha}(V_c) = exp\left(-\frac{\left(V_c - \underline{c}_{\tilde{A}_{V_c}^2}\right)^2}{\underline{\sigma}^2_{\tilde{A}_{V_c}^2|\alpha}}\right) \tag{3.9}$$

where $\tilde{A}^1_{V_c}$ and $\tilde{A}^2_{V_c}$ represent the first and second membership functions for the inpu V_c. $\underline{c}_{\tilde{A}^1_{V_c}}$ and $\underline{c}_{\tilde{A}^2_{V_c}}$ denote the centers of $\tilde{A}^1_{V_c}$ and $\tilde{A}^2_{V_c}$, respectively. The ratios. $\underline{\sigma}_{\tilde{A}^1_{V_c}|\alpha}/\underline{\sigma}_{\tilde{A}^1_{V_c}|\alpha}$ and $\underline{\sigma}_{\tilde{A}^2_{V_c}|\alpha}/\underline{\sigma}_{\tilde{A}^2_{V_c}|\alpha}$ represent the standard deviations for the upper and lower bounds of $\tilde{A}^1_{V_c}$ and $\tilde{A}^2_{V_c}$, respectively.

(3) The results of $\hat{f}_1$ and $\hat{f}_2$ are:

$$\hat{f}_1 = \theta_1^T \zeta_1 \hat{f}_2 = \theta_2^T \zeta_2 \tag{3.10}$$

where, θ_i and ζ_i are:

$$\theta_i = \left[\underline{w}_{i1}, \ldots, \underline{w}_{iR}, \underline{w}_{i1}, \ldots, \underline{w}_{iR}\right]^T$$

$$\zeta_i = \left[\underline{\zeta}_{i1}, \ldots, \underline{\zeta}_{iR}, \underline{\zeta}_{i1}, \ldots, \underline{\zeta}_{iR}\right]^T \tag{3.11}$$

The parameters of the $l-th$ rule for the $i-th$ IT3-FLS are denoted as $\underline{w}_{il}$ and $\underline{w}_{il}$. R denotes the total number of rules. $\underline{\zeta}_{il}$ and $\underline{\zeta}_{il}$ are calculated as follows:

$$\underline{\zeta}_l = \frac{\sum_{j=1}^{n_\alpha} \underline{\alpha}_j \frac{\underline{z}^l_{\mu_s=\underline{\alpha}_j}}{\sum_{l=1}^R \left(\underline{z}^l_{\mu_s=\underline{\alpha}_j} + \underline{z}^l_{\mu_s=\underline{\alpha}_j}\right)}}{\sum_{j=1}^{n_\alpha}\left(\underline{\alpha}_j + \underline{\alpha}_j\right)} + \frac{\sum_{j=1}^{n_\alpha} \underline{\alpha}_j \frac{\underline{z}^l_{\mu_s=\underline{\alpha}_j}}{\sum_{l=1}^R \left(\underline{z}^l_{\mu_s=\underline{\alpha}_j} + \underline{z}^l_{\mu_s=\underline{\alpha}_j}\right)}}{\sum_{j=1}^{n_\alpha}\left(\underline{\alpha}_j + \underline{\alpha}_j\right)}, l = 1, \ldots, 8 \tag{3.12}$$

$$\underline{\zeta}_l = \frac{\sum_{j=1}^{n_\alpha} \underline{\alpha}_j \frac{\underline{z}^l_{\mu_s=\underline{\alpha}_j}}{\sum_{l=1}^R \left(\underline{z}^l_{\mu_s=\underline{\alpha}_j} + \underline{z}^l_{\mu_s=\underline{\alpha}_j}\right)}}{\sum_{j=1}^{n_\alpha}\left(\underline{\alpha}_j + \underline{\alpha}_j\right)} + \frac{\sum_{j=1}^{n_\alpha} \underline{\alpha}_j \frac{\underline{z}^l_{\mu_s=\underline{\alpha}_j}}{\sum_{l=1}^R \left(\underline{z}^l_{\mu_s=\underline{\alpha}_j} + \underline{z}^l_{\mu_s=\underline{\alpha}_j}\right)}}{\sum_{j=1}^{n_\alpha}\left(\underline{\alpha}_j + \underline{\alpha}_j\right)}, l = 1, \ldots, 8 \tag{3.13}$$

where, n_α is horizontal slices number and:

$$\underline{\zeta}_l = \frac{\sum_{j=1}^{n_\alpha} \underline{\alpha}_j \frac{\underline{z}^l_{\mu_s=\underline{\alpha}_j}}{\sum_{l=1}^R \left(\underline{z}^l_{\mu_s=\underline{\alpha}_j} + \underline{z}^l_{\mu_s=\underline{\alpha}_j}\right)}}{\sum_{j=1}^{n_\alpha}\left(\underline{\alpha}_j + \underline{\alpha}_j\right)} + \frac{\sum_{j=1}^{n_\alpha} \underline{\alpha}_j \frac{\underline{z}^l_{\mu_s=\underline{\alpha}_j}}{\sum_{l=1}^R \left(\underline{z}^l_{\mu_s=\underline{\alpha}_j} + \underline{z}^l_{\mu_s=\underline{\alpha}_j}\right)}}{\sum_{j=1}^{n_\alpha}\left(\underline{\alpha}_j + \underline{\alpha}_j\right)}, l = 1, \ldots, R \tag{3.14}$$

$$\begin{aligned}
\underline{z}^l_{\mu_s=\underline{\alpha}_j} &= \underline{\mu}_{\tilde{A}^{k_{I_p}}_{I_p}|\underline{\alpha}_j}(I_p)\underline{\mu}_{\tilde{A}^{k_{I_b}}_{I_b}|\underline{\alpha}_j}(I_b)\underline{\mu}_{\tilde{A}^{k_{V_c}}_{V_c}|\underline{\alpha}_j}(V_c) \\
\underline{z}^l_{\mu_s=\underline{\alpha}_j} &= \underline{\mu}_{\tilde{A}^{k_{I_p}}_{I_p}|\underline{\alpha}_j}(I_p)\underline{\mu}_{\tilde{A}^{k_{I_b}}_{I_b}|\underline{\alpha}_j}(I_b)\underline{\mu}_{\tilde{A}^{k_{V_c}}_{V_c}|\underline{\alpha}_j}(V_c) \\
\underline{z}^l_{\mu_s=\underline{\alpha}_j} &= \underline{\mu}_{\tilde{A}^{k_{I_p}}_{I_p}|\underline{\alpha}_j}(I_p)\underline{\mu}_{\tilde{A}^{k_{I_b}}_{I_b}|\underline{\alpha}_j}(I_b)\underline{\mu}_{\tilde{A}^{k_{V_c}}_{V_c}|\underline{\alpha}_j}(V_c) \\
\underline{z}^l_{\mu_s=\underline{\alpha}_j} &= \underline{\mu}_{\tilde{A}^{k_{I_p}}_{I_p}|\underline{\alpha}_j}(I_p)\underline{\mu}_{\tilde{A}^{k_{I_b}}_{I_b}|\underline{\alpha}_j}(I_b)\underline{\mu}_{\tilde{A}^{k_{V_c}}_{V_c}|\underline{\alpha}_j}(V_c)
\end{aligned} \tag{3.15}$$

where, $k_{I_p} = 1, 2$, $k_{I_b} = 1, 2$ and $k_{V_c} = 1, 2$. The rules can be written as:

$$\begin{aligned}
&Rule\,\#1: \; If\; I_p\; is\, \tilde{A}^1_{I_p}|\alpha\; and\; I_b\; is\, \tilde{A}^1_{I_b}|\alpha\; and\; V_c\; is\, \tilde{A}^1_{V_c}|\alpha\; Then\; \hat{f}_i \in \\
&[\underline{w}_{i1}, \underline{w}_{i1}]Rule\,\#2: \; If\; I_p\; is\, \tilde{A}^1_{I_p}|\alpha\; and\; I_b\; is\, \tilde{A}^1_{I_b}|\alpha\; and\; V_c\; is\, \tilde{A}^2_{V_c}|\alpha\; Then\; \hat{f}_i \in \\
&[\underline{w}_{i2}, \underline{w}_{i2}]Rule\,\#3: \; If\; I_p\; is\, \tilde{A}^1_{I_p}|\alpha\; and\; I_b\; is\, \tilde{A}^2_{I_b}|\alpha\; and\; V_c\; is\, \tilde{A}^1_{V_c}|\alpha\; Then\; \hat{f}_i \in \\
&[\underline{w}_{i3}, \underline{w}_{i3}]Rule\,\#4: \; If\; I_p\; is\, \tilde{A}^1_{I_p}|\alpha\; and\; I_b\; is\, \tilde{A}^2_{I_b}|\alpha\; and\; V_c\; is\, \tilde{A}^2_{V_c}|\alpha\; Then\; \hat{f}_i \in \\
&[\underline{w}_{i4}, \underline{w}_{i4}]Rule\,\#5: \; If\; I_p\; is\, \tilde{A}^2_{I_p}|\alpha\; and\; I_b\; is\, \tilde{A}^1_{I_b}|\alpha\; and\; V_c\; is\, \tilde{A}^1_{V_c}|\alpha\; Then\; \hat{f}_i \in \\
&[\underline{w}_{i5}, \underline{w}_{i5}]Rule\,\#6: \; If\; I_p\; is\, \tilde{A}^2_{I_p}|\alpha\; and\; I_b\; is\, \tilde{A}^1_{I_b}|\alpha\; and\; V_c\; is\, \tilde{A}^2_{V_c}|\alpha\; Then\; \hat{f}_i \in \\
&[\underline{w}_{i6}, \underline{w}_{i6}]Rule\,\#7: \; If\; I_p\; is\, \tilde{A}^2_{I_p}|\alpha\; and\; I_b\; is\, \tilde{A}^2_{I_b}|\alpha\; and\; V_c\; is\, \tilde{A}^1_{V_c}|\alpha\; Then\; \hat{f}_i \in \\
&[\underline{w}_{i7}, \underline{w}_{i7}]Rule\,\#8: \; If\; I_p\; is\, \tilde{A}^2_{I_p}|\alpha\; and\; I_b\; is\, \tilde{A}^2_{I_b}|\alpha\; and\; V_c\; is\, \tilde{A}^2_{V_c}|\alpha\; Then\; \hat{f}_i \in [\underline{w}_{i8}, \underline{w}_{i8}]
\end{aligned} \tag{3.16}$$

Note 1 *Figure 3.6 demonstrates that, in type-3 membership functions (MFs), the secondary membership is not a precise value but rather a fuzzy set. In the type-2 counterpart, a horizontal slice at level* $\mu_s = \alpha_k$ *is equivalent to two slices at levels* $\mu_s = \underline{\alpha}_k$ *and* $\mu_s = \underline{\alpha}_k$ [6].

3.2.3 Power Management and Stability Analysis

This section presents the formulation of the controllers u_p/u_b. The key findings are summarized below.

Theorem 1 *states that the closed-loop system will exhibit asymptotic stability if the control signals and compensators are built according to Eqs. (3.17–3.18), and if tuning rules for the parameters of IT3-FLSs and the estimated upper bound of the approximation error (AE) are taken into account as described in Eqs. (3.19–3.20).*

$$\begin{aligned}
u_p &= \left(-\hat{f}_1(y, \theta_1) - \lambda_1 e_1 + s_{p_c}\right)/\hat{b}_1 \\
u_b &= \left(-\hat{f}_2(y, \theta_2) - \lambda_2 e_2 + s_{b_c}\right)/\hat{b}_2
\end{aligned} \tag{3.17}$$

$$s_{p_c} = -\underline{\widehat{E}}_1|\tilde{y}_1|\frac{e_1}{|e_1^2| + \varepsilon} s_{b_c} = -\underline{\widehat{E}}_2|\tilde{y}_2|\frac{e_2}{|e_2^2| + \varepsilon} \tag{3.18}$$

$$\dot{\hat{\theta}}_1 = \gamma\tilde{y}_1\xi_1, \dot{\hat{\theta}}_2 = \gamma\tilde{y}_2\xi_2$$

$$\dot{\hat{b}}_1 = \gamma\tilde{y}_1 u_p, \dot{\hat{b}}_2 = \gamma\tilde{y}_2 u_b \tag{3.19}$$

$$\dot{\underline{\hat{E}}}_1 = \gamma|\tilde{y}_1|\dot{\underline{\hat{E}}}_2 = \gamma|\tilde{y}_2| \tag{3.20}$$

In this context, the variables s_{p_c} and s_{b_c} represent compensators, $\lambda_i, i = 1, 2$ are positive constants, and $e_1 = \hat{y}_1 - y_{1d}$ and $e_2 = \hat{y}_2 - y_{2d}$ denote the tracking errors. The signals y_{1d} and y_{2d} are the optimal signals. The adaptation rate is represented by γ, and $\hat{f}_i(y, \theta_i, i = 1, 2)$ refers to IT3-FLSs. The constants $\lambda_i, i = 1, 2$ are positive, and $\underline{E}_i, i = 1, 2$ represent the upper bounds of AE. $\widehat{\underline{E}}_1$ and $\widehat{\underline{E}}_2$ are estimated upper bounds of AE, and ε is a small constant.

Proof The dynamics of y_1 and y_2 in Eq. (3.1) are reformulated as follows:

$$\begin{aligned} \dot{y}_1 &= \frac{1}{L_p}\big(-y_2 + V_p(y_1)\big) + \frac{y_2}{L_p}u_p \\ \dot{y}_2 &= \frac{1}{C}\big(y_1 - y_1 u_p - y_2/R\big) + \frac{y_3}{C}u_b \end{aligned} \tag{3.21}$$

To design u_p and u_b, one can has:

$$\begin{aligned} \dot{\hat{y}}_1 &= \hat{f}_1(y, \theta_1) + \hat{b}_1 u_p \\ \dot{\hat{y}}_2 &= \hat{f}_2(y, \theta_2) + \hat{b}_2 u_b, \end{aligned} \tag{3.22}$$

Let $\hat{y}_1$, $\hat{y}_2$, $\hat{b}_1$ and $\hat{b}_2$ represent the estimated values of y_1, y_2, y_2/L_p and y_3/C, respectively. Similarly, $\hat{f}_1$ and $\hat{f}_2$ denote the IT3-FLSs. The previous section provides an illustration of the structure and parameters of IT3-FLSs $\hat{f}_1$ and $\hat{f}_2$. Based on Eqs. (3.21, 3.22), the behavior of $\tilde{y}_1 = y_1 - \hat{y}_1$ and $\tilde{y}_2 = y_2 - \hat{y}_2$ can be described as follows:

$$\begin{aligned} \dot{\tilde{y}}_1 &= \frac{1}{L_p}\big(-y_2 + V_p(y_1)\big) - \hat{f}_1(y, \theta_1) + \left(\frac{y_2}{L_p} - \hat{b}_1\right)u_p \\ \dot{\tilde{y}}_2 &= \frac{1}{C}\big(y_1 - y_1 u_p - y_2/R\big) - \hat{f}_2(y, \theta_2) + \left(\frac{y_3}{C} - \hat{b}_2\right)u_b \end{aligned} \tag{3.23}$$

From (3.23), one has:

$$\begin{aligned} \dot{\tilde{y}}_1 &= \hat{f}_1^*(y, \theta_1^*) - \hat{f}_1(y, \theta_1) + (\hat{b}_1^* - \hat{b}_1)u_p + \frac{1}{L_p}\big(-y_2 + V_p(y_1)\big)_1 \\ &-\hat{f}_1^*(y, \theta_1^*) + \left(\frac{y_2}{L_p} - \hat{b}_1^*\right)u_p \\ \dot{\tilde{y}}_2 &= \hat{f}_2^*(y, \theta_2^*) - \hat{f}_2(y, \theta_2) + (\hat{b}_2^* - \hat{b}_2)u_b + \frac{1}{C}\big(y_1 - y_1 u_p - y_2/R\big) \\ &-\hat{f}_2^*(y, \theta_2^*) + \left(\frac{y_3}{C} - \hat{b}_2^*\right)u_b \end{aligned} \tag{3.24}$$

The variables $\hat{f}_1^*(y, \theta_1^*)$ and $\hat{f}_2^*(y, \theta_2^*)$ represent the ideal IT3-FLS, whereas $\hat{b}_1^*$ and $\hat{b}_2^*$ represent the optimal gains. Given Eq. (3.24), let's examine the following definitions:

$$\begin{aligned} E_1 &= \frac{1}{L_p}(-y_2 + V_p(y_1))_1 - \hat{f}_1^*(y, \theta_1^*) + \left(\frac{y_2}{L_p} - \hat{b}_1^*\right)u_p \\ E_2 &= \frac{1}{C}(y_1 - y_1 u_p - y_2/R) - \hat{f}_2^*(y, \theta_2^*) + \left(\frac{y_3}{C} - \hat{b}_2^*\right)u_b \end{aligned} \tag{3.25}$$

Equations (3.24, 3.25) lead to the following conclusion:

$$\begin{aligned} \dot{\tilde{y}}_1 &= E_1 + \tilde{\theta}_1^T \xi_1 + \tilde{b}_1 u_p \\ \dot{\tilde{y}}_2 &= E_2 + \tilde{\theta}_2^T \xi_2 + \tilde{b}_2 u_b \end{aligned} \tag{3.26}$$

The variables ξ_1 and ξ_2 specified in Eqs. (3.78, 3.79), and the variables $\tilde{\theta}$ and $\tilde{b}$ are described as follows:

$$\tilde{\theta}_1 = \hat{\theta}_1^* - \hat{\theta}_1, \tilde{\theta}_2 = \hat{\theta}_2^* - \hat{\theta}_2$$

$$\tilde{b}_1 = \hat{b}_1^* - \hat{b}_1, \tilde{b}_2 = \hat{b}_2^* - \hat{b}_2 \tag{3.27}$$

Based on the estimated dynamics of $\hat{y}_1$ and $\hat{y}_2$, the expressions for u_p and u_b can be formulated as:

$$\begin{aligned} u_p &= \left(-\hat{f}_1(y, \theta_1) - \lambda_1 e_1 + s_{p_c}\right)/\hat{b}_1 \\ u_b &= \left(-\hat{f}_2(y, \theta_2) - \lambda_2 e_2 + s_{b_c}\right)/\hat{b}_2 \end{aligned} \tag{3.28}$$

The dynamics of e_1 and e_2 can be expressed as a function of Eqs. (3.28, 3.22):

$$\begin{aligned} \dot{e}_1 &= -\lambda_1 e_1 + s_{p_c} \\ \dot{e}_2 &= -\lambda_2 e_2 + s_{b_c} \end{aligned} \tag{3.29}$$

For the sake of stability study, let us consider a Lyapunov function in the following manner:

$$\begin{aligned} V &= \frac{1}{2}\tilde{y}_1^2 + \frac{1}{2}\tilde{y}_2^2 + \frac{1}{2}e_1^2 + \frac{1}{2}e_2^2 + \frac{1}{2\gamma}\tilde{b}_1^2 + \frac{1}{2\gamma}\tilde{b}_2^2 \\ &+ \frac{1}{2\gamma}\tilde{\theta}_1^T\tilde{\theta}_1 + \frac{1}{2\gamma}\tilde{\theta}_2^T\tilde{\theta}_2 + \frac{1}{2\gamma}\underline{\widetilde{E}}_1^2 + \frac{1}{2\gamma}\underline{\widetilde{E}}_2^2 \end{aligned} \tag{3.30}$$

where, $\underline{\widetilde{E}}_i$, $i = 1, 2$ are defined as:

$$\underline{\widetilde{E}}_i = \underline{E}_i - \underline{\widehat{E}}_i \tag{3.31}$$

Here, $\underline{E}_i$ refers to the maximum value of AE E_i, while $\underline{\widehat{E}}_i$ denotes the estimated value of $\underline{E}_i$. Using Eq. (3.30), the expression for $\dot{V}$ becomes:

$$\dot{V} = \dot{\tilde{y}}_1\tilde{y}_1 + \dot{\tilde{y}}_2\tilde{y}_2 + \dot{e}_1e_1 + \dot{e}_2e_2 - \frac{1}{\gamma}\tilde{b}_1\dot{\hat{b}}_1 - \frac{1}{\gamma}\tilde{b}_2\dot{\hat{b}}_2 - \frac{1}{\gamma}\tilde{\theta}_1^T\dot{\hat{\theta}}_1 - \frac{1}{\gamma}\tilde{\theta}_2^T\dot{\hat{\theta}}_2 - \frac{1}{\gamma}\underline{\tilde{E}}_1\dot{\hat{\underline{E}}}_1 + -\frac{1}{\gamma}\underline{\tilde{E}}_2\dot{\hat{\underline{E}}}_2 \tag{3.32}$$

By utilizing Eqs. (3.29, 3.26), the expression $\dot{V}$ is reformulated as:

$$\begin{aligned}\dot{V} &= \tilde{y}_1\left(\tilde{\theta}_1^T\xi_1 + \tilde{b}_1u_p + E_1\right) + \tilde{y}_2\left(\tilde{\theta}_2^T\xi_2 + \tilde{b}_2u_b + E_2\right) \\ &+ e_1\left(-\lambda_1e_1 + s_{p_c}\right) + e_2\left(-\lambda_2e_2 + s_{b_c}\right) - \frac{1}{\gamma}\tilde{\theta}_1^T\dot{\hat{\theta}}_1 \\ &- \frac{1}{\gamma}\tilde{\theta}_2^T\dot{\hat{\theta}}_2 - \frac{1}{\gamma}\tilde{b}_1\dot{\hat{b}}_1 - \frac{1}{\gamma}\tilde{b}_2\dot{\hat{b}}_2 - \frac{1}{\gamma}\underline{\tilde{E}}_1\dot{\hat{\underline{E}}}_1 + -\frac{1}{\gamma}\underline{\tilde{E}}_2\dot{\hat{\underline{E}}}_2\end{aligned} \tag{3.33}$$

The Eq. (3.33) can be restated as:

$$\begin{aligned}\dot{V} &= s_{p_c}e_1 + s_{b_c}e_2 - \lambda_1e_1^2 - \lambda_2e_2^2 + \tilde{\theta}_1^T\left(\tilde{y}_1\xi_1 - \frac{1}{\eta}\dot{\hat{\theta}}_1\right) + \tilde{\theta}_2^T\left(\tilde{y}_2\xi_2 - \frac{1}{\eta}\dot{\hat{\theta}}_2\right) \\ &+\tilde{b}_1\left(\tilde{y}_1u_p - \frac{1}{\eta}\dot{\hat{b}}_1\right) + \tilde{b}_2\left(\tilde{y}_2u_b - \frac{1}{\eta}\dot{\hat{b}}_2\right) + E_1\tilde{y}_1 + E_2\tilde{y}_2 - \frac{1}{\gamma}\underline{\tilde{E}}_1\dot{\hat{\underline{E}}}_1 + -\frac{1}{\gamma}\underline{\tilde{E}}_2\dot{\hat{\underline{E}}}_2\end{aligned} \tag{3.34}$$

From (3.19), $\dot{V}$ is reformulated as:

$$\dot{V} = -\lambda_1e_1^2 - \lambda_2e_2^2 + s_{p_c}e_1 + s_{b_c}e_2 + E_1\tilde{y}_1 + E_2\tilde{y}_2 - \frac{1}{\gamma}\underline{\tilde{E}}_1\dot{\hat{\underline{E}}}_1 + -\frac{1}{\gamma}\underline{\tilde{E}}_2\dot{\hat{\underline{E}}}_2 \tag{3.35}$$

From Eqs. (3.31, 3.35), $\dot{V}$ can be restated as:

$$\begin{aligned}\dot{V} &= -\lambda_1e_1^2 - \lambda_2e_2^2 + s_{p_c}e_1 + s_{b_c}e_2 + E_1\tilde{y}_1 + E_2\tilde{y}_2 - \frac{1}{\gamma}\left(\underline{E}_1 - \underline{\hat{E}}_1\right)\dot{\hat{\underline{E}}}_1 \\ &+ -\frac{1}{\gamma}\left(\underline{E}_2 - \underline{\hat{E}}_2\right)\dot{\hat{\underline{E}}}_2\end{aligned} \tag{3.36}$$

From Eq. (3.36), it is deduced that:

$$\begin{aligned}\dot{V} &\leq -\lambda_1e_1^2 - \lambda_2e_2^2 + s_{p_c}e_1 + s_{b_c}e_2 + \underline{E}_1|\tilde{y}_1| + \underline{E}_2|\tilde{y}_2| - \frac{1}{\gamma}\left(\underline{E}_1 - \underline{\hat{E}}_1\right)\dot{\hat{\underline{E}}}_1 \\ &+ -\frac{1}{\gamma}\left(\underline{E}_2 - \underline{\hat{E}}_2\right)\dot{\hat{\underline{E}}}_2\end{aligned} \tag{3.37}$$

The inequality (3.37) can be reduced or made simpler as:

$$\dot{V} \leq -\lambda_1e_1^2 - \lambda_2e_2^2 + s_{p_c}e_1 + s_{b_c}e_2 + \underline{E}_1\left(|\tilde{y}_1| - \frac{1}{\gamma}\dot{\hat{\underline{E}}}_1\right) + \underline{E}_2\left(|\tilde{y}_2| - \frac{1}{\gamma}\dot{\hat{\underline{E}}}_2\right)$$

$$+\frac{1}{\gamma}\widehat{\underline{E}}_1\dot{\widehat{\underline{E}}}_1+\frac{1}{\gamma}\widehat{\underline{E}}_2\dot{\widehat{\underline{E}}}_2 \tag{3.38}$$

By inserting Eq. (3.20) into Eq. (3.38), we have:

$$\dot{V}\leq-\lambda_1e_1^2-\lambda_2e_2^2+s_{p_c}e_1+s_{b_c}e_2+\frac{1}{\gamma}\widehat{\underline{E}}_1\dot{\widehat{\underline{E}}}_1+\frac{1}{\gamma}\widehat{\underline{E}}_2\dot{\widehat{\underline{E}}}_2 \tag{3.39}$$

Substituting s_{p_c} and s_{b_c} from Eq. (3.18), yields:

$$\dot{V}\leq-\lambda_1e_1^2-\lambda_2e_2^2+-\widehat{\underline{E}}_1|\tilde{y}_1|\frac{e_1^2}{|e_1^2|+\varepsilon}-\widehat{\underline{E}}_2|\tilde{y}_2|\frac{e_2^2}{|e_2^2|+\varepsilon}+\frac{1}{\gamma}\widehat{\underline{E}}_1\dot{\widehat{\underline{E}}}_1+\frac{1}{\gamma}\widehat{\underline{E}}_2\dot{\widehat{\underline{E}}}_2 \tag{3.40}$$

From Eqs. (3.20, 3.43), we get:

$$\dot{V}\leq-\lambda_1e_1^2-\lambda_2e_2^2+-\widehat{\underline{E}}_1|\tilde{y}_1|\frac{e_1^2}{|e_1^2|+\varepsilon}-\widehat{\underline{E}}_2|\tilde{y}_2|\frac{e_2^2}{|e_2^2|+\varepsilon}+\widehat{\underline{E}}_1|\tilde{y}_1|+\widehat{\underline{E}}_2|\tilde{y}_2| \tag{3.41}$$

After simplification, the inequality (3.41) changes to:

$$\dot{V}\leq-\lambda_1e_1^2-\lambda_2e_2^2-\widehat{\underline{E}}_1|\tilde{y}_1|\left(\frac{e_1^2}{|e_1^2|+\varepsilon}-1\right)\widehat{\underline{E}}_2|\tilde{y}_2|\left(\frac{e_2^2}{|e_2^2|+\varepsilon}-1\right) \tag{3.42}$$

Considering the fact that $e_1^2\approx\left(|e_1^2|+\varepsilon\right)$ and $e_2^2\approx\left(|e_2^2|+\varepsilon\right)$, one has:

$$\dot{V}\leq-\lambda_1e_1^2-\lambda_2e_2^2\leq0 \tag{3.43}$$

From Eq. (3.43), it is clear that $\dot{V}\leq0$, when:

$$e_i=0,\tilde{y}_i\neq0,\tilde{b}_i\neq0,\tilde{\theta}_i\neq0,\widetilde{\underline{E}}_i\neq0 \tag{3.44}$$

Given that the control system aims to achieve $e_i=0$ and the closed-loop system's invariant set does not contain $\left(\tilde{y}_i,\tilde{b}_i,\tilde{\theta}_i,\widetilde{\underline{E}}_i\right)$, we can conclude asymptotic stability based on La Salle's invariant set theorem. The proof has thus been finalized.

Note 2 *The rate at which the suggested controller converges is determined by the values of* λ_1,λ_2 *in Eq. (17), as well as the precision of the estimation of IT3-FLSs.*

Note 3 *Despite the uncertain dynamics of the units, good measurements of PV/Battery currents and output voltage are necessary to apply the suggested controller. For future research, it is worth exploring the feasibility of eliminating the measurement of all states.*

3.2.4 Simulation Studies

The efficacy of the proposed power management system is evaluated through simulations. The parameters for the models under consideration are shown in Table 3.3. The other control parameters are as follows:: $\gamma = 0.5$, $\varepsilon = 0.1$, $\lambda_i = 50$, $i = 1, 2$, $\underline{c}_{\tilde{A}_{I_b}} = 4$, $\underline{c}_{\tilde{A}_{I_P}} = 4$, $\underline{c}_{\tilde{A}_{I_b}} = 0$, $\underline{c}_{\tilde{A}_{I_P}} = 0$, $\underline{c}_{\tilde{A}_{V_c}} = 24$, $\underline{c}_{\tilde{A}_{V_c}} = 10$, $\underline{\sigma}_{\tilde{A}_{I_b}} = 4$, $\underline{\sigma}_{\tilde{A}_{I_P}} = 4$, $\underline{\sigma}_{\tilde{A}_{I_b}} = 2$, $\underline{\sigma}_{\tilde{A}_{I_P}} = 2$, $\underline{\sigma}_{\tilde{A}_{V_c}} = 10$ and $\underline{\sigma}_{\tilde{A}_{V_c}} = 5$.

Note 4 *The centers and standard divisions of membership functions for inputs of IT3-FLSs are computed based on the input range, while the consequent parameters are updated online. The other parameters* λ_i *must be greater than zero,* ε *is a tiny constant, and* $0 < \gamma \leq 1$.

3.2.4.1 Scenario 1

In the first scenario, the radiation is assumed to be constant at a level of 450 w/m^2. The trajectories followed by the signals I_p, P, and V_c can be observed in Figs. 3.7, 3.8 and 3.9, while the controllers u_p/u_b are provided in Figs. 3.10 and 3.11 It can be observed that the system achieves a precise and accurate reaction, and the control signals have a smooth and stable structure without any oscillations.

3.2.4.2 Scenario 2

The second scenario involves testing the performance of the proposed plan in challenging conditions. The load, temperature, and radiation are all subject to time variation. Specifically, the temperature changes from 11 to 50 °C degrees at $t = 30s$, the load varies from 75 to 35 Ω at $t = 100s$, and the irradiation changes from 450 w/m^2 to 251 w/m^2 at $t = 70s$. The signals trajectories of I_p, P, and V_c are illustrated

Table 3.3 Imulation parameters

Parameter	Value	Parameter	Value
L_p	4 (mH)	L_b	10 (mH)
P_b	55 (w)	i_{sc}	3.45 (A)
C	600 (μf)	r_p	30 $(m\Omega)$
T_r	(°C)	k_i	1.2 (A/k)
A	1.2	V_{boc}	9 (v)
i_r	5.99e-8 (A)	E_g	1.12 (ev)
β_1	0.9	β_2	1.1
r_b	80 $(m\Omega)$	k_b	1.381e-23
P_b	21 (w)	W_{Loss}	20 (w)
q	1.60e-19	n	36

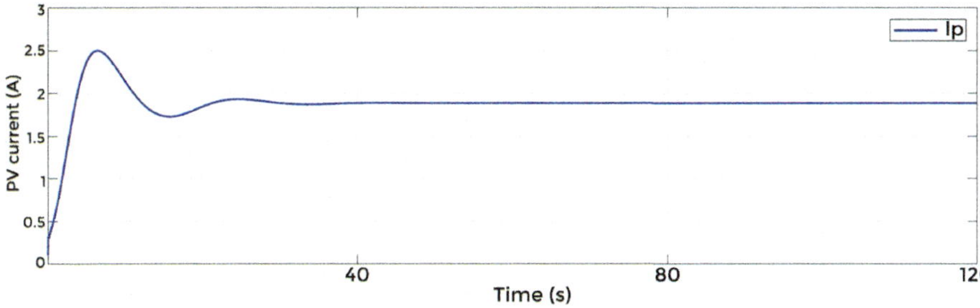

Fig. 3.7 PV current in scenario 1

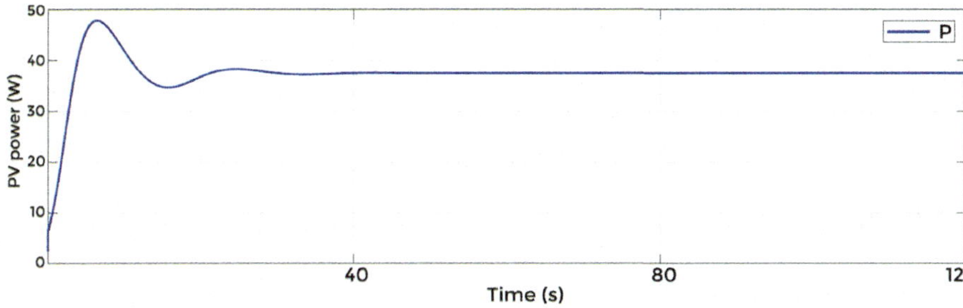

Fig. 3.8 PV power in scenario 1

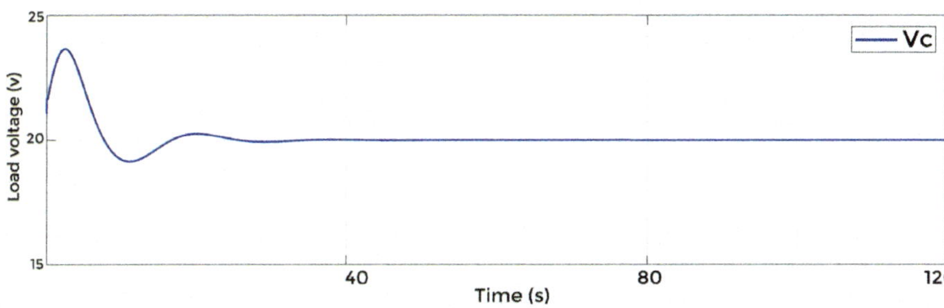

Fig. 3.9 PV load voltage in scenario 1

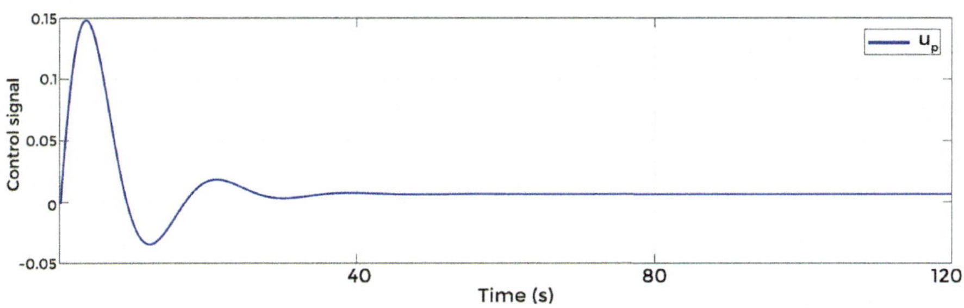

Fig. 3.10 u_p for scenario 1

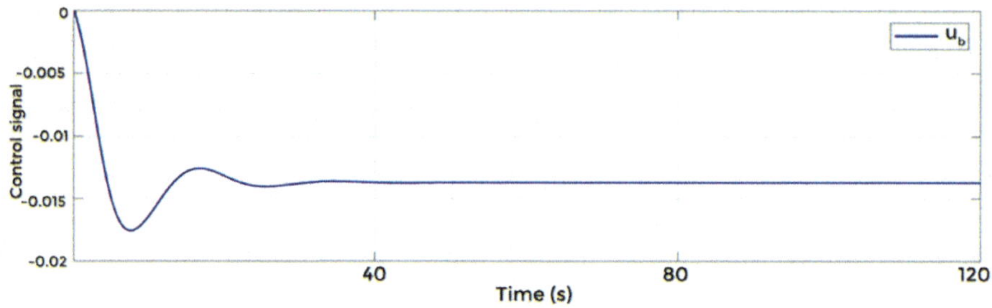

Fig. 3.11 u_b for scenario 1

in Figs. 3.12 and 3.13, while Figs. 3.15 and 3.16 display the signals u_p and u_b. The Figs. 3.12 and 3.13 demonstrate a strong performance despite variations in temperature, load, and radiation over time. Furthermore, the system achieves a commendable level of output voltage regulation, and the power of the photovoltaic (PV) system effectively adapts to variations in conditions Fig. 3.14.

Note 5 *The simulations are conducted using Matlab 2018a on a computer with an i7-4720HQ CPU running at 2.60 GHz. The duration of the simulation is around 129 s.*

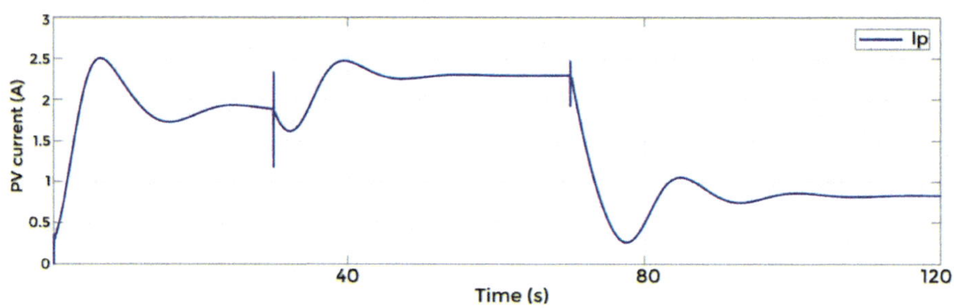

Fig. 3.12 PV current in scenario 2

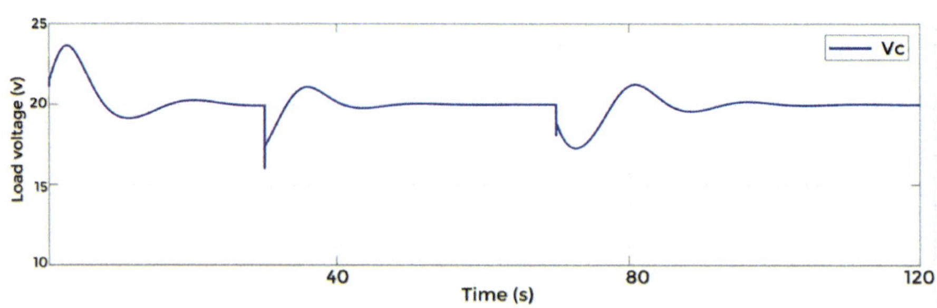

Fig. 3.13 PV load voltage in scenario 2

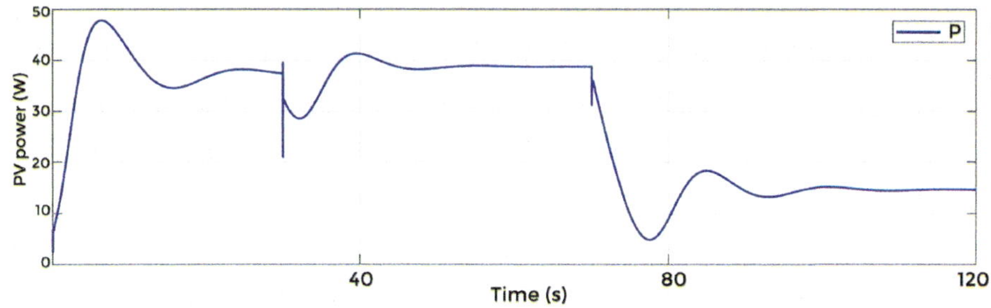

Fig. 3.14 PV power in scenario 2

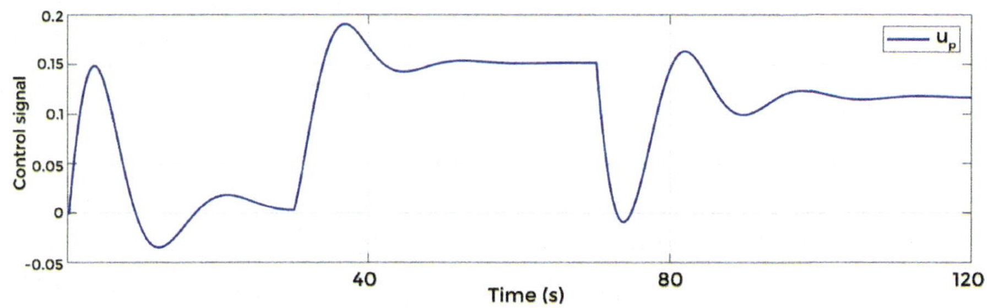

Fig. 3.15 Control signal u_p for scenario 2

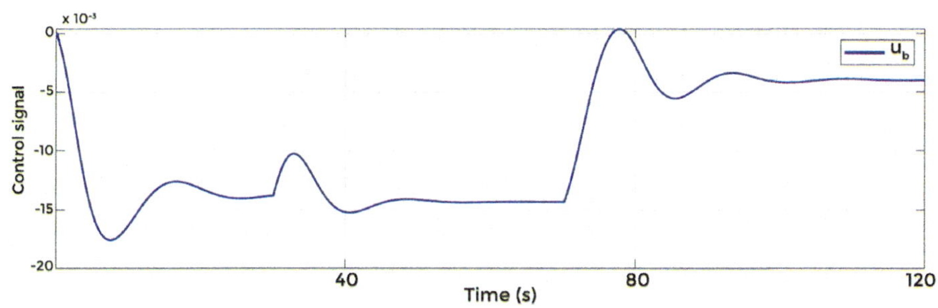

Fig. 3.16 Control signal u_b for scenario 2

3.2.4.3 Comparison

In order to show the effectiveness of the suggested mechanism, a comparison is made with various established regulators like the PID controller [8], LQR [9], passivity based regulator (PBR) [10], and SMC scheme [11]. The comparative findings are presented in Table 3.4, indicating the mean square error (MSE). It is clear that the implemented controller leads to improved regulation performance. In order to further evaluate the efficacy of the proposed Fuzzy Logic System (FLS) based controller, we will compare it with another type of FLS used in the designed controller. After

Table 3.4 MSE comparison regarding different regulators

Signal	Method				
Signal	LQR [9]	PID [8	PBR [10]	SMC [11]	Proposed controller
V_c	148.9197	15.2294	11.2499	9.3327	0.8724
I_p	0.8259	1.1029	0.9579	0.5161	0.3198

Table 3.5 MSE Comparison regarding different fuzzy neural networks

Signal	FLSs					
	Type-1		Type-2		Type-3	
Signal	Min of MSE	Max of MSE	Min of MSE	Max of MSE	Min of MSE	Max of MSE
V_c	1.8731	1.8732	1.1214	1.1216	0.8724	0.8725
I_p	0.9321	0.9321	0.6261	0.6262	0.3198	0.3199

conducting the simulation 20 times, the minimum (Min) and maximum (Max) values of the Mean Squared Error (MSE) are presented in Table 3.5.

Note 6 *The simulation results demonstrate that the implemented control method exhibits desirable performance, maintaining a constant output voltage despite various disturbances, such as fluctuations in irradiation and load* [12].

3.3 Fractional-Order Type-3 Fuzzy Control for Photovoltaic and Battery Systems under Uncertain Conditions

3.3.1 Problem Formulation

3.3.1.1 General View

Figure 3.1 displays the overall block diagram, while Fig. 3.2 provides the specific control block diagram. The currents of PV/battery (I_p/I_b) and the output voltage (V_c) are measured using current and voltage sensors. The online dynamic estimation in Fig. 3.1 utilizes T3-FLSs ($\hat{F}_1, \hat{F}_2$), while the design of the controllers (μ_p, μ_b) is based on the online T3-FLS model. The T3-FLSs parameters (θ_1, θ_1) and control gains ($\hat{g}_1$, $\hat{g}_1$) are adjusted online. The compensators (μ_{cp}, μ_{cb}) are determined through a robustness analysis based on fractional-order calculus. This analysis takes into account tracking errors (e_1, e_2) and upper bounds of approximation errors ($\underline{E}_1$ and $\underline{E}_1$) in order to achieve a high level of robustness against dynamic perturbations, as well as common irradiation and temperature disturbances. The comprehensive explanation is provided in Theorem 1.

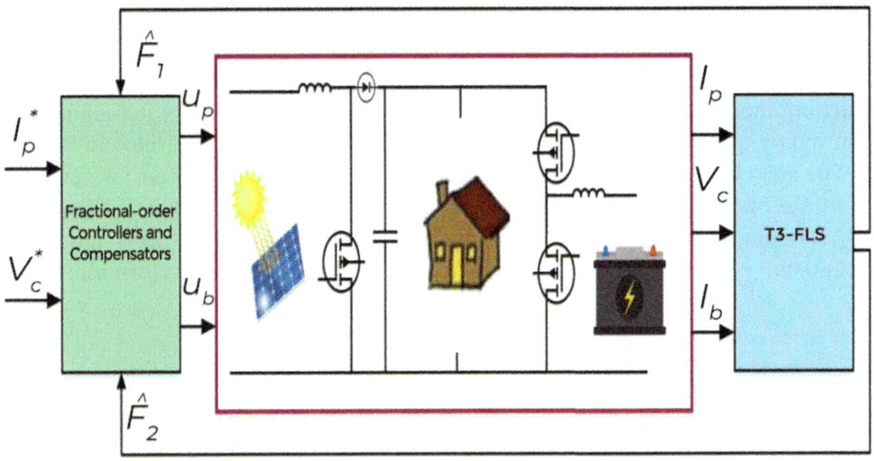

Fig. 3.1 The suggested scheme general block diagram

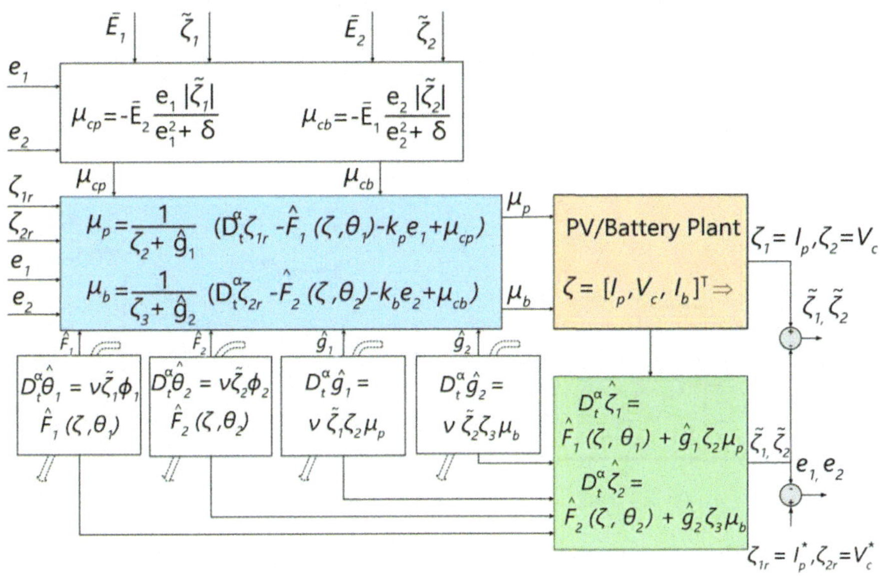

Fig. 3.2 Control diagram in detail

At every sample time, the output load energy is provided by either the photovoltaic (PV) system or the battery. If the PV panel is unable to provide the necessary energy, the battery systems will activate. Additionally, any surplus energy generated by the PV panel is saved in a battery. Converters are utilized to establish a switching mechanism between photovoltaic (PV) systems and batteries. The diagram of the converter is shown in Fig. 3.3. By evaluating the positions of switches SW_1, SW_2 and SW_3, four modes, as illustrated in Fig. 3.4, can be deduced [13].

If SW_1 and SW_3 are in the open position and SW_2 is in the closed position, the switching mode can be achieved as depicted in Fig. 3.4. (a). The switching mode is achieved by closing SW_1 and SW_2, and keeping SW_3 open, as illustrated in Fig. 3.4. (b) The switching mode is achieved as depicted in Fig. 3.4. (c) when SW_1 and SW_2 are in the open position and SW_3 is closed. If both SW_1 and SW_3 are closed and SW_2 is open, the switching mode depicted in Fig. 3.4d is achieved. A state space formulation can be written for each switching mode. A state space formulation can

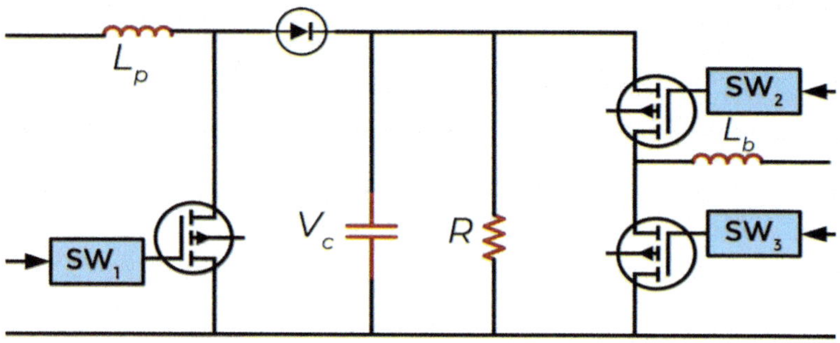

Fig. 3.3 The convertor schematic

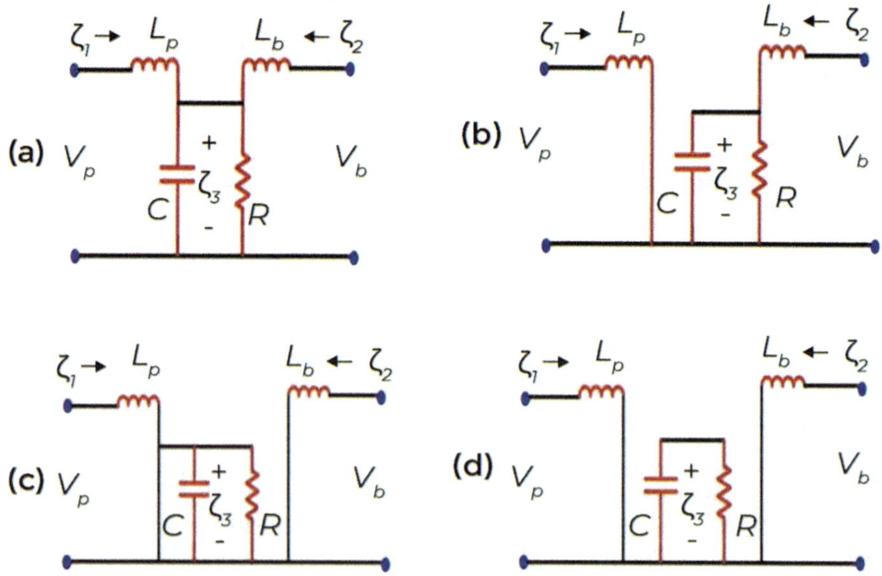

Fig. 3.4 Switching modes of Boost convertors

be expressed for each switching mode. The dynamics of the plant are determined by averaging the state space representations of four different states:

$$\begin{aligned}
\dot{\varsigma}_1 &= (-\varsigma_2 + \nu_p + \varsigma_2\mu_p)/\iota_p \\
\dot{\varsigma}_2 &= \frac{1}{C}(\varsigma_1 - \varsigma_2/R + \varsigma_3\mu_b - \varsigma_1\mu_p) \\
\dot{\varsigma}_3 &= (-\varsigma_2\mu_b + \nu_b)/\iota_b
\end{aligned} \tag{3.1}$$

where ς_1 represents the current of the photovoltaic (PV) system, ς_2 represents the current of the battery, and ς_3 represents the load voltage. l_p/l_b, R, and C represent the values of inductance, resistance, and capacitance in the converter. The voltages of the PV and battery are denoted as ν_p and ν_v, respectively. The control signals are represented by μ_p and μ_b.

3.3.1.2 PV Modeling

The dynamics of PV panels are described using the single-diode modeling technique [14]:

$$l_{ph} = (l_{sc} + (T - T_r)\kappa_i)s \tag{3.2}$$

$$I_p = I_{phg}G - exp((V_p + I_pR_{sg})q/nT\kappa_b - 1)l_o - \frac{1}{R_{shg}}(I_pR_{sg} + V_p) \tag{3.3}$$

$$l_0 = exp\left[qE_g\left(\frac{1}{T_r + 273} - \frac{1}{T + 273}\right)/\kappa_bA\right](T + 273/T_r + 273)^3 l_r \tag{3.4}$$

The definition of parameters are provided in Table 3.1. As illustrated in Fig. 3.5, the power output of a photovoltaic (PV) panel is not constant, but rather varies in response to changes in its current. It is evident that at an ideal current, the most amount of energy may be attained.

3.3.1.3 Battery Modeling

The battery dynamics are obtained using the following equations [14]:

$$E(t) = -\int \eta\nu_{boc}I_b + E_{Loss}dt \tag{3.5}$$

$$\eta = \{\eta_1\, I_b\ \geq\ 0\, \eta_2\, I_b\ <\ 0 \tag{3.6}$$

Table 3.1 PV parameters and descriptions, see Eq. (2)

Parameter	Description
n	Cell number
$G(w/m^2)$	Solar irradiation
κ_b (J/K)	Boltzmann's constant
$E_g(ev)$	Band-gap energy
$R_{sh}/R_s(\Omega)$	Equivalent resistances
$T(°c)$	Temperature of PV
q	Electron charge
A	Constant of the diode ideality
$l_r(A)$	Saturation current
$T_r(°c)$	Reference temperature
$l_{ph}(A)$	Photo generated currents

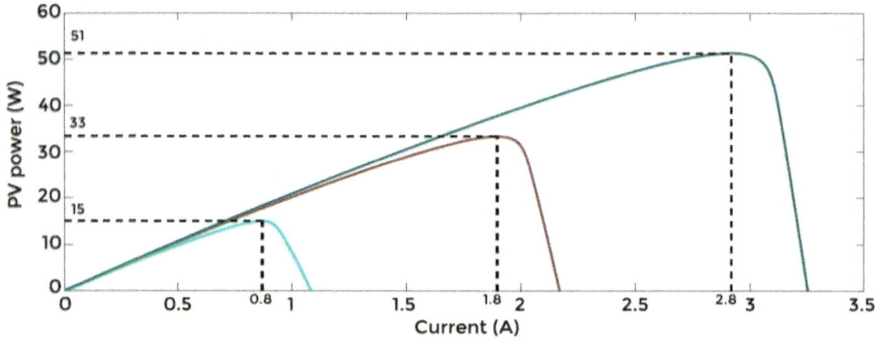

Fig. 3.5 PV panel power trajectory

$$v_b = -R_b I_b + v_{boc} \tag{3.7}$$

$$SoC(t) = E(t)/E_{Max} \tag{3.8}$$

where the definitions of all parameters in Eqs. (3.5–3.8) are gathered in Table 3.2.

3.3.2 Uncertainty Estimation by Type-3 FLS Proposed Type-3 FLS

T3-FLSs are an extension or a broader form of T2-FLSs. The primary concept of T3-FLS is introduced in [15], and it is further expanded in this study to facilitate

Table 3.2 The battery system parameters and descriptions, see Eq. (3.5)

Parameter	Unit	Description
R_b	(Ω)	Battery internal resistance
E_{Loss}	(w)	Power losses
v_{boc}	(v)	Voltage of battery open circuit
η_1 and η_2	–	Rates of charge and discharge
E_{Max}	(J)	Maximum storable energy

the dynamic identification of real-world systems. The fundamental advantage of T3-FLSs over T2-FLSs lies in the fact that type-3 membership functions (MFs) treat both the upper and lower bounds of uncertainties as fuzzy sets, whereas in type-2 MFs, the upper and lower bounds of uncertainty are treated as crisp values. This section provides an explanation of the proposed T3-FLS. The diagram illustrating the construction can be found in Fig. 3.6. T3-FLSs are employed to address the dynamic perturbation caused by photovoltaic (PV) systems, batteries, and other components. The details are provided below.

(1) The inputs of T3-FLSs are considered to be $\varsigma_1 = I_p$, $\varsigma_2 = v$, $\varsigma_3 = I_b$.
(2) Two membership functions (MFs) are examined for each input $\varsigma_j, j = 1, 2, 3$. These MFs have centers $\underline{c}_{\tilde{\psi}^1_{\varsigma_j}}$ and $\underline{c}_{\tilde{\psi}^2_{\varsigma_j}}$ (see [15]). The upper standard division is denoted as $\underline{\sigma}_{\tilde{\psi}^k_{\varsigma_j}}$ and the lower standard division is denoted as $\underline{\sigma}_{\tilde{\psi}^k_{\varsigma_j}}$. The type-3 MFs possess an additional degree of freedom compared to their type-2 counterparts. Secondary membership is observed to be of type-2 MF. For

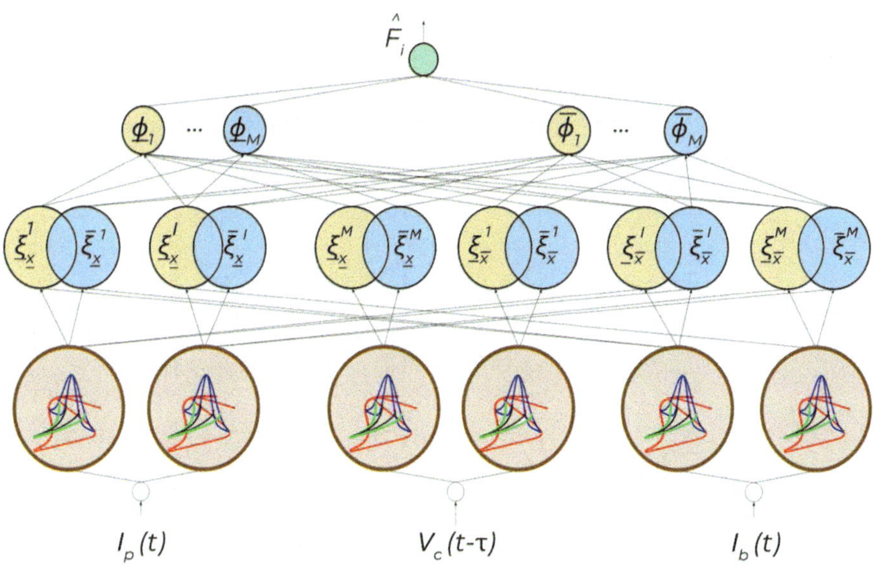

Fig. 3.6 T3-FLS general block diagram

instance, let's examine a type-2 membership function (MF) for the input variable x, where its primary domain lies between $\underline{u}$ and $\underline{u}$. In a type-3 membership function (MF) for input x, the upper and lower limits of the primary domain are not fixed values, but rather they are represented as fuzzy sets. Regarding the input variable I_p, the following applies:

$$\begin{aligned} \underline{\pi}_{\tilde{\psi}_{I_p}^1|\underline{\chi}}(\varsigma_1 = I_p) &= exp\left(-\left(\varsigma_1 - \underline{c}_{\tilde{\psi}_{I_p}^1}\right)^2 / \underline{\sigma}^2_{\tilde{\psi}_{I_p}^1|\underline{\chi}}\right) \\ \underline{\pi}_{\tilde{\psi}_{I_p}^2|\underline{\chi}}(\varsigma_1 = I_p) &= exp\left(-\left(\varsigma_1 - \underline{c}_{\tilde{\psi}_{I_p}^2}\right)^2 / \underline{\sigma}^2_{\tilde{\psi}_{I_p}^2|\underline{\chi}}\right) \end{aligned} \tag{3.9}$$

$$\begin{aligned} \underline{\pi}_{\tilde{\psi}_{I_p}^1|\underline{\chi}}(\varsigma_1 = I_p) &= exp\left(-\left(\varsigma_1 - \underline{c}_{\tilde{\psi}_{I_p}^1}\right)^2 / \underline{\sigma}^2_{\tilde{\psi}_{I_p}^1|\underline{\chi}}\right) \\ \underline{\pi}_{\tilde{\psi}_{I_p}^2|\underline{\chi}}(\varsigma_1 = I_p) &= exp\left(-\left(\varsigma_1 - \underline{c}_{\tilde{\psi}_{I_p}^2}\right)^2 / \underline{\sigma}^2_{\tilde{\psi}_{I_p}^2|\underline{\chi}}\right) \end{aligned} \tag{3.10}$$

Here, χ represents the value of secondary membership. The upper and lower memberships for $\tilde{\psi}_{\varsigma_j}^k$ are represented by $\underline{\pi}_{\tilde{\psi}_{\varsigma_j}^k|\chi}$ and $\underline{\pi}_{\tilde{\psi}_{\varsigma_j}^k|\chi}$ respectively. Similarly to Eqs. (3.9, 3.10), for different values of I_b and V_c, the following relationships hold:

$$\begin{aligned} \underline{\pi}_{\tilde{\psi}_{I_b}^1|\underline{\chi}}(\varsigma_2 = I_b) &= exp\left(-\left(\varsigma_2 - \underline{c}_{\tilde{\psi}_{I_p}^1}\right)^2 / \underline{\sigma}^2_{\tilde{\psi}_{I_b}^1|\underline{\chi}}\right) \\ \underline{\pi}_{\tilde{\psi}_{I_b}^2|\underline{\chi}}(\varsigma_2 = I_b) &= exp\left(-\left(\varsigma_2 - \underline{c}_{\tilde{\psi}_{I_b}^2}\right)^2 / \underline{\sigma}^2_{\tilde{\psi}_{I_b}^2|\underline{\chi}}\right) \end{aligned} \tag{3.11}$$

$$\begin{aligned} \underline{\pi}_{\tilde{\psi}_{I_b}^1|\underline{\chi}}(\varsigma_2 = I_b) &= exp\left(-\left(\varsigma_2 - \underline{c}_{\tilde{\psi}_{I_b}^1}\right)^2 / \underline{\sigma}^2_{\tilde{\psi}_{I_b}^1|\underline{\chi}}\right) \\ \underline{\underline{\pi}}_{\tilde{\psi}_{I_b}^2|\underline{\chi}}(\varsigma_2 = I_b) &= exp\left(-\left(\varsigma_2 - \underline{c}_{\tilde{\psi}_{I_b}^2}\right)^2 / \underline{\sigma}^2_{\tilde{\psi}_{I_b}^2|\underline{\chi}}\right) \end{aligned} \tag{3.12}$$

$$\begin{aligned} \underline{\pi}_{\tilde{\psi}_{V_c}^1|\underline{\chi}}(\varsigma_3 = V_c) &= exp\left(-\left(\varsigma_3 - \underline{c}_{\tilde{\psi}_{V_c}^1}\right)^2 / \underline{\sigma}^2_{\tilde{\psi}_{V_c}^1|\underline{\chi}}\right) \\ \underline{\pi}_{\tilde{\psi}_{V_c}^2|\underline{\chi}}(\varsigma_3 = V_c) &= exp\left(-\left(\varsigma_3 - \underline{c}_{\tilde{\psi}_{V_c}^2}\right)^2 / \underline{\sigma}^2_{\tilde{\psi}_{V_c}^2|\underline{\chi}}\right) \end{aligned} \tag{3.13}$$

$$\begin{aligned} \underline{\pi}_{\tilde{\psi}_{V_c}^1|\underline{\chi}}(\varsigma_3 = V_c) &= exp\left(-\left(\varsigma_3 - \underline{c}_{\tilde{\psi}_{V_c}^1}\right)^2 / \underline{\sigma}^2_{\tilde{\psi}_{V_c}^1|\underline{\chi}}\right) \\ \underline{\underline{\pi}}_{\tilde{\psi}_{V_c}^2|\underline{\chi}}(\varsigma_3 = V_c) &= exp\left(-\left(\varsigma_3 - \underline{c}_{\tilde{\psi}_{V_c}^2}\right)^2 / \underline{\sigma}^2_{\tilde{\psi}_{V_c}^2|\underline{\chi}}\right) \end{aligned} \tag{3.14}$$

(3) The lower rule firing is determined by Eqs. (3.15, 3.16):

$$\underline{\xi}^2_{\underline{\chi}} = \underline{\pi}_{\tilde{\psi}_{I_p}^1|\underline{\chi}} \underline{\pi}_{\tilde{\psi}_{I_b}^1|\underline{\chi}} \underline{\pi}_{\tilde{\psi}_{V_c}^2|\underline{\chi}} \underline{\xi}^3_{\underline{\chi}} = \underline{\pi}_{\tilde{\psi}_{I_p}^1|\underline{\chi}} \underline{\pi}_{\tilde{\psi}_{I_b}^2|\underline{\chi}} \underline{\pi}_{\tilde{\psi}_{V_c}^1|\underline{\chi}} \vdots \underline{\xi}^8_{\underline{\chi}} = \underline{\pi}_{\tilde{\psi}_{I_p}^2|\underline{\chi}} \underline{\pi}_{\tilde{\psi}_{I_b}^2|\underline{\chi}} \underline{\pi}_{\tilde{\psi}_{V_c}^2|\underline{\chi}} \tag{3.15}$$

$$\underline{\xi}^2_{\chi} = \underline{\pi}_{\tilde{\psi}^1_{I_p}|\chi}\underline{\pi}_{\tilde{\psi}^1_{I_b}|\chi}\underline{\pi}_{\tilde{\psi}^2_{V_c}|\chi}\underline{\xi}^3_{\chi} = \underline{\pi}_{\tilde{\psi}^1_{I_p}|\chi}\underline{\pi}_{\tilde{\psi}^2_{I_b}|\chi}\underline{\pi}_{\tilde{\psi}^1_{V_c}|\chi}\vdots\underline{\xi}^8_{\chi} = \underline{\pi}_{\tilde{\psi}^2_{I_p}|\chi}\underline{\pi}_{\tilde{\psi}^2_{I_b}|\chi}\underline{\pi}_{\tilde{\psi}^2_{V_c}|\chi} \quad (3.16)$$

Similarly, for the upper rule firing, we have:

$$\begin{aligned}\underline{\xi}^1_{\chi} &= \underline{\pi}_{\tilde{\psi}^1_{I_p}|\chi}\underline{\pi}_{\tilde{\psi}^1_{I_b}|\chi}\underline{\pi}_{\tilde{\psi}^1_{V_c}|\chi}\underline{\xi}^2_{\chi}\\ &= \underline{\pi}_{\tilde{\psi}^1_{I_p}|\chi}\underline{\pi}_{\tilde{\psi}^1_{I_b}|\chi}\underline{\pi}_{\tilde{\psi}^2_{V_c}|\chi}\underline{\xi}^3_{\chi}\\ &= \underline{\pi}_{\tilde{\psi}^1_{I_p}|\chi}\underline{\pi}_{\tilde{\psi}^2_{I_b}|\chi}\underline{\pi}_{\tilde{\psi}^1_{V_c}|\chi}\vdots\underline{\xi}^8_{\chi}\\ &= \underline{\pi}_{\tilde{\psi}^2_{I_p}|\chi}\underline{\pi}_{\tilde{\psi}^2_{I_b}|\chi}\underline{\pi}_{\tilde{\psi}^2_{V_c}|\chi}\end{aligned} \quad (3.17)$$

$$\begin{aligned}\underline{\xi}^1_{\chi} &= \underline{\pi}_{\tilde{\psi}^1_{I_p}|\chi}\underline{\pi}_{\tilde{\psi}^1_{I_b}|\chi}\underline{\pi}_{\tilde{\psi}^1_{V_c}|\chi}\underline{\xi}^2_{\chi}\\ &= \underline{\pi}_{\tilde{\psi}^1_{I_p}|\chi}\underline{\pi}_{\tilde{\psi}^1_{I_b}|\chi}\underline{\pi}_{\tilde{\psi}^2_{V_c}|\chi}\underline{\xi}^3_{\chi}\\ &= \underline{\pi}_{\tilde{\psi}^1_{I_p}|\chi}\underline{\pi}_{\tilde{\psi}^2_{I_b}|\chi}\underline{\pi}_{\tilde{\psi}^1_{V_c}|\chi}\vdots\underline{\xi}^8_{\chi}\\ &= \underline{\pi}_{\tilde{\psi}^2_{I_p}|\chi}\underline{\pi}_{\tilde{\psi}^2_{I_b}|\chi}\underline{\pi}_{\tilde{\psi}^2_{V_c}|\chi}\end{aligned} \quad (3.18)$$

(4) The output of $\hat{F}_1$ and $\hat{F}_2$ are calculated to be:

$$\begin{aligned}\hat{F}_1 &= \theta_1^T\phi_1\\ \hat{F}_2 &= \theta_2^T\phi_2\end{aligned} \quad (3.19)$$

where, θ_i and ϕ_i are:

$$\theta_i = \left[\underline{\theta}_{i1}, \ldots, \underline{\theta}_{iM}, \underline{\theta}_{i1}, \ldots, \underline{\theta}_{iM}\right]^T$$

$$\phi_i = \left[\underline{\phi}_{i1}, \ldots, \underline{\phi}_{iM}, \underline{\phi}_{i1}, \ldots, \underline{\phi}_{iM}\right]^T \quad (3.20)$$

where, M shows the number of rules. $\underline{\phi}_r$ and $\underline{\phi}_R$ are:

$$\underline{\phi}_r = \frac{\sum_{j=1}^{n_\chi}\underline{\chi}_j\frac{\underline{\xi}^r_{\chi_j}}{\sum_{r=1}^{R}\left(\underline{\xi}^r_{\chi_j}+\underline{\xi}^r_{\chi_j}\right)}}{\sum_{j=1}^{n_\chi}\left(\underline{\chi}_j+\underline{\chi}_j\right)} + \frac{\sum_{j=1}^{n_\chi}\underline{\chi}_j\frac{\underline{\xi}^r_{\chi_j}}{\sum_{r=1}^{R}\left(\underline{\xi}^r_{\chi_j}+\underline{\xi}^r_{\chi_j}\right)}}{\sum_{j=1}^{n_\chi}\left(\underline{\chi}_j+\underline{\chi}_j\right)}, r = 1, \ldots, 8 \quad (3.21)$$

$$\underline{\phi}_r = \frac{\sum_{j=1}^{n_\chi}\underline{\chi}_j\frac{\underline{\xi}^r_{\chi_j}}{\sum_{r=1}^{R}\left(\underline{\xi}^r_{\chi_j}+\underline{\xi}^r_{\chi_j}\right)}}{\sum_{j=1}^{n_\chi}\left(\underline{\chi}_j+\underline{\chi}_j\right)} + \frac{\sum_{j=1}^{n_\chi}\underline{\chi}_j\frac{\underline{\xi}^r_{\chi_j}}{\sum_{r=1}^{R}\left(\underline{\xi}^r_{\chi_j}+\underline{\xi}^r_{\chi_j}\right)}}{\sum_{j=1}^{n_\chi}\left(\underline{\chi}_j+\underline{\chi}_j\right)}, r = 1, \ldots, 8 \quad (3.22)$$

where, n_χ is number of slices.

3.3.3 Main Results

The control signals and tuning rules are succinctly outlined in the subsequent theorem: The given controllers, tuning rules, and compensators, represented as Eqs. (3.23–3.30), ensure asymptotic stability.

$$\mu_p = \frac{1}{\varsigma_2 \hat{g}_1}\left(D_t^\alpha \varsigma_{1r} - \hat{F}_1(\varsigma, \theta_1) - k_p e_1 + \mu_{cp}\right) \tag{3.23}$$

$$\mu_b = \frac{1}{\varsigma_3 \hat{g}_2}\left(D_t^\alpha \varsigma_{2r} - \hat{F}_2(\varsigma, \theta_2) - k_b e_2 + \mu_{cb}\right) \tag{3.24}$$

$$D_t^\alpha \hat{\theta}_1 = \upsilon \tilde{\varsigma}_1 \phi_1 \tag{3.25}$$

$$D_t^\alpha \hat{\theta}_2 = \upsilon \tilde{\varsigma}_2 \phi_2 \tag{3.26}$$

$$D_t^\alpha \hat{g}_1 = \upsilon \tilde{\varsigma}_1 \varsigma_2 \mu_p \tag{3.27}$$

$$D_t^\alpha \hat{g}_2 = \upsilon \tilde{\varsigma}_2 \varsigma_3 \mu_b \tag{3.28}$$

$$\mu_{cp} = -\underline{E}_2 \frac{e_1 |\tilde{\varsigma}_1|}{e_1^2 + \delta} \tag{3.29}$$

$$\mu_{cb} = -\underline{E}_1 \frac{e_2 |\tilde{\varsigma}_2|}{e_2^2 + \delta} \tag{3.30}$$

The variables υ, δ, k_b, k_p, $\underline{E}_i i = 1, 2$ are constant. μ_{cp} and μ_{cb} are symbols that denote compensators. ς_{1r} and ς_{2r} serve as reference signals for ς_1 and ς_2, correspondingly. e_1 and e_2 are errors that can be described as $e_i = \varsigma_i - \varsigma_{ir}$, and $\tilde{\varsigma}_i$ is defined as $\tilde{\varsigma}_i = \varsigma_i - \hat{\varsigma}_i$.

Proof The output dynamics (1) are:

$$\begin{aligned} \dot{\varsigma}_1 &= \left(-\varsigma_2 + V_p(\varsigma_1)\right)/L_p + \frac{\varsigma_2}{L_p}\mu_p \\ \dot{\varsigma}_2 &= \left(\varsigma_1/C - \varsigma_1 \mu_p/C - \varsigma_2/CR\right) + \frac{\varsigma_3}{C}\mu_b \end{aligned} \tag{3.31}$$

The recommended fractional-order T3-FLS model is used to estimate the dynamics of ς_1 and ς_2 in order to design μ_p and μ_b.

$$D_t^{\alpha}\hat{\varsigma}_1 = \hat{F}_1(\varsigma, \theta_1) + \hat{g}_1\varsigma_2\mu_p$$
$$D_t^{\alpha}\hat{\varsigma}_2 = \hat{F}_2(\varsigma, \theta_2) + \hat{g}_2\varsigma_3\mu_b \tag{3.32}$$

The symbols $D_t^{\alpha}\varsigma_i, i = 1, 2$ represent the fractional derivatives, while $\hat{\varsigma}_1$ and $\hat{\varsigma}_2$ represent the estimated values of ς_1 and ς_2, respectively. $\hat{g}_1$ and $\hat{g}_2$ represent the estimated values of $1/L_p$ and $1/C$, respectively. $\hat{F}_1$ and $\hat{F}_2$ are the Type-3 Fuzzy Logic Systems (T3-FLSs). The dynamics of ς_1 and ς_2 in Eq. (3.32) can be expressed based on the universal estimating feature of FLSs.

$$D_t^{\alpha}\varsigma_1 = \hat{F}_1^*(\varsigma, \theta_1) + \hat{g}_1^*\varsigma_2\mu_p + E_1$$
$$D_t^{\alpha}\varsigma_2 = \hat{F}_2^*(\varsigma, \theta_2) + \hat{g}_2^*\varsigma_3\mu_b + E_2 \tag{3.33}$$

Here, $\hat{F}_1^*$ and $\hat{F}_2^*$ represent the optimal T3-FLS values, while $\hat{g}_1^*$ and $\hat{g}_2^*$ represent the optimal values of $\hat{g}_1$ and $\hat{g}_2$, respectively. $E_i, i = 1, 2$ represent the errors in approximation. By using Eqs. (3.32, 3.33), we can get the dynamics of the estimating error ($\tilde{\varsigma}_i = \varsigma_i - \hat{\varsigma}_i$):

$$D_t^{\alpha}\tilde{\varsigma}_1 = \hat{F}_1^*(\varsigma, \theta_1) - \hat{F}_1(\varsigma, \theta_1) + \left(\hat{g}_1^* - \hat{g}_1\right)\varsigma_2\mu_p + E_1$$
$$D_t^{\alpha}\tilde{\varsigma}_2 = \hat{F}_2^*(\varsigma, \theta_2) - \hat{F}_2(\varsigma, \theta_2) + \left(\hat{g}_2^* - \hat{g}_2\right)\varsigma_3\mu_b + E_2 \tag{3.34}$$

Considering definitions:

$$\tilde{\theta}_i = \theta_i^* - \theta_i \tilde{g}_i = g_i^* - g_i \tag{3.35}$$

From Eq. (3.34), we have:

$$D_t^{\alpha}\tilde{\varsigma}_1 = E_1 + \tilde{\theta}_1^T\phi_1 + \tilde{g}_1\varsigma_2\mu_p$$
$$D_t^{\alpha}\tilde{\varsigma}_2 = E_2 + \tilde{\theta}_2^T\phi_2 + \tilde{g}_2\varsigma_3\mu_b \tag{3.36}$$

By applying the control signals of Eqs. (3.23, 3.24, 3.32) to the calculated model Eq. (3.32), the following results are obtained:

$$D_t^{\alpha}e_1 = -k_p e_1 + \mu_{cp} \tag{3.37}$$

$$D_t^{\alpha}e_2 = -k_b e_2 + \mu_{cb} \tag{3.38}$$

In order to assess the vigor and stability, the Lyapunov function is taken into account:

$$V = \frac{1}{2}\tilde{\varsigma}_1^2 + \frac{1}{2}\tilde{\varsigma}_2^2 + \frac{1}{2}e_1^2 + \frac{1}{2}e_2^2\frac{1}{2\upsilon}\tilde{g}_1^2 + \frac{1}{2\upsilon}\tilde{g}_2^2 + \frac{1}{2\upsilon}\tilde{\theta}_1^T\tilde{\theta}_1 + \frac{1}{2\upsilon}\tilde{\theta}_2^T\tilde{\theta}_2 \tag{3.39}$$

By taking fractional time-derivative, $D_t^{\alpha}V$ is formulated as:

$$
\begin{aligned}
D_t^\alpha V &\le D_t^\alpha \tilde{\varsigma}_1 \tilde{\varsigma}_1 + D_t^\alpha \tilde{\varsigma}_2 \tilde{\varsigma}_2 + D_t^\alpha e_1 e_1 + D_t^\alpha e_2 e_2 - \frac{1}{\upsilon}\tilde{g}_1 D_t^\alpha \hat{g}_1 \\
&- \frac{1}{\upsilon}\tilde{g}_2 D_t^\alpha \hat{g}_2 - \frac{1}{\upsilon}\tilde{\theta}_1^T D_t^\alpha \hat{\theta}_1 - \frac{1}{\upsilon}\tilde{\theta}_2^T D_t^\alpha \hat{\theta}_2
\end{aligned}
\tag{3.40}
$$

From Eq. (3.40), $D_t^\alpha V$ changes to:

$$
\begin{aligned}
D_t^\alpha V &\le \tilde{\varsigma}_1\big(\tilde{\theta}_1^T \phi_1 + \varsigma_2 \tilde{g}_1 \mu_p + E_2\big) + \tilde{\varsigma}_2\big(\tilde{\theta}_2^T \phi_2 + \varsigma_3 \tilde{g}_2 \mu_b + E_1\big) \\
&+ e_1\big(-k_p e_1 + \mu_{cp}\big) + e_2(-k_b e_2 + \mu_{cb}) \\
&- \frac{1}{\upsilon}\tilde{g}_1 D_t^\alpha \hat{g}_1 - \frac{1}{\upsilon}\tilde{g}_2 D_t^\alpha \hat{g}_2 - \frac{1}{\upsilon}\tilde{\theta}_1^T D_t^\alpha \hat{\theta}_1 - \frac{1}{\upsilon}\tilde{\theta}_2^T D_t^\alpha \hat{\theta}_2
\end{aligned}
\tag{3.41}
$$

From Eq. (3.41), we have:

$$
\begin{aligned}
D_t^\alpha V &\le -k_p e_1^2 + \mu_{cp} e_1 - k_b e_2^2 + \mu_{cb} e_2 + \tilde{\theta}_1^T\left(\tilde{\varsigma}_1 \phi_1 - \frac{1}{\upsilon} D_t^\alpha \hat{\theta}_1\right) \\
&+ \tilde{\theta}_2^T\left(\tilde{\varsigma}_2 \phi_2 - \frac{1}{\upsilon} D_t^\alpha \hat{\theta}_2\right) + \tilde{\delta}_p\left(\tilde{\varsigma}_1 \varsigma_2 \mu_p - \frac{1}{\upsilon} D_t^\alpha \hat{g}_1\right) \\
&+ \tilde{\delta}_b\left(\tilde{\varsigma}_2 \varsigma_3 \mu_b - \frac{1}{\upsilon} D_t^\alpha \hat{g}_2\right) + E_2 \tilde{\varsigma}_1 + E_1 \tilde{\varsigma}_2
\end{aligned}
\tag{3.42}
$$

Considering tuning rules $D_t^\alpha \hat{\theta}_i = \upsilon \tilde{\varsigma}_i \phi_i, i = 1, 2$, $D_t^\alpha \hat{g}_1 = \upsilon \tilde{\varsigma}_1 \varsigma_2 \mu_p$ and $D_t^\alpha \hat{g}_2 = \upsilon \tilde{\varsigma}_2 \varsigma_3 \mu_b$, $D_t^\alpha V$ becomes:

$$
D_t^\alpha V \le -k_b e_2^2 - k_p e_1^2 + \mu_{cp} e_1 + \mu_{cb} e_2 + E_2 \tilde{\varsigma}_1 + E_1 \tilde{\varsigma}_2 \tag{3.43}
$$

Then, we get:

$$
D_t^\alpha V \le -k_b e_2^2 - k_p e_1^2 + |E_2||\tilde{\varsigma}_1| + \mu_{cp} e_1 + \mu_b e_2 + |E_1||\tilde{\varsigma}_2| \tag{3.44}
$$

Now, implementing the compensators yields the following outcomes:

$$
D_t^\alpha V \le -k_b e_2^2 - k_p e_1^2 + \left[|E_2||\tilde{\varsigma}_1| - \underline{E}_2 |\tilde{\varsigma}_1| \frac{e_1^2}{e_1^2 + \delta}\right] + \left[|E_1||\tilde{\varsigma}_2| - \underline{E}_1 |\tilde{\varsigma}_2| \frac{e_2^2}{e_2^2 + \delta}\right] \tag{3.45}
$$

It has been demonstrated that $D_t^\alpha V \le 0$. The proof is finished by applying Barbalat's Lemma and considering the boundedness of $\ddot{V}$.

3.3.4 Simulation Studies

This section demonstrates the efficacy of the suggested technique in terms of its efficiency, output voltage, and power regulation performance. Tables 3.3 and 3.4 provide the simulation and control settings.

Example 1 *In this example, the baseline situation is considered, assuming that the irradiation remains constant at an intensity of 300* w/m^2. *The trajectories of* $\varsigma_i, i = 1, 2$, *as well as the power of PV* P *and the controllers* μ_p *and* μ_b, *are displayed in Figs. 3.7, 3.8, 3.9, 3.10 and 3.11. From Figs. 3.7–3.11, it is evident that the proposed scenario achieves the necessary regulatory performance and the control signals exhibit a smooth shape. Observations indicate that the paths followed*

Table 3.3 Simulation parameters

Parameter	Value	Parameter	Value
L_p	10 (mH)	L_b	12 (mH)
q	1.60e-19	n	45
P_b	56 (w)	i_{sc}	3.56 (A)
C	521 (μf)	r_p	36 $(m\Omega)$
r_b	70 $(m\Omega)$	k_b	1.481e-23
T_r	$(°C)$	k_i	1.2 (A/k)
A	1.1	V_{boc}	9 (v)
i_r	5.97e-8 (A)	E_g	1.42 (ev)
β_1	0.95	β_2	1.4
P_b	25 (w)	W_{Loss}	25 (w)

Table 3.4 Control parameters

Parameters	Value	Equations
$\underline{E}_i, i = 1, 2$	10	29–30
α	0.9	23–24
k_p, k_b	100	23–24
δ	0.01	29–30
υ	0.1	25–28
$\underline{c}_{\tilde{\psi}^1_{V_c}}, \underline{c}_{\tilde{\psi}^1_{I_p}}$ and $\underline{c}_{\tilde{\psi}^1_{I_b}}$	–10	12–14
$\underline{c}_{\tilde{\psi}^2_{V_c}}, \underline{c}_{\tilde{\psi}^2_{I_p}}$ and $\underline{c}_{\tilde{\psi}^2_{I_b}}$	10	11–13
$\underline{\sigma}_{\tilde{\psi}^j_{V_c}}, \underline{\sigma}_{\tilde{\psi}^j_{I_p}}$ and $\underline{\sigma}_{\tilde{\psi}^j_{I_b}}, j = 1, 2$	5	12–14
$\underline{\sigma}_{\tilde{\psi}^j_{V_c}}, \underline{\sigma}_{\tilde{\psi}^j_{I_p}}$ and $\underline{\sigma}_{\tilde{\psi}^j_{I_b}}, j = 1, 2$	10	11–13
$\underline{\chi}$	[0.6 0.8 1]	12–13
$\underline{\chi}$	[0 0.2 0.4]	12–13

by the current of PV I_p and the output voltage V_c converge to the required level in less than 10 s.

Example 2 *In this example, the irradiation is increased from 200 to 600 w/m^2 at time $t = 55s$, and the temperature is adjusted according to the equation $T = 50-10sin(t)$. The trajectories of ς_i, $i = 1, 2$, as well as the power of PV P and the controllers μ_p and μ_b, can be observed in Figs. 3.12, 3.13, 3.14, 3.15 and 3.16. Despite fluctuations in temperature and irradiation, effective power regulation is consistently achieved.*

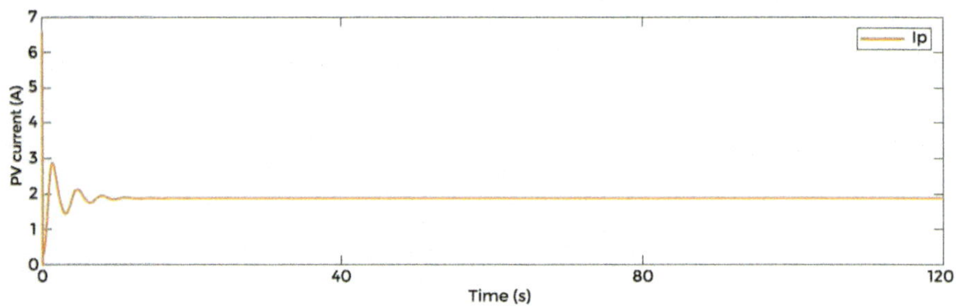

Fig. 3.7 Trajectory of ς_1 in Ex. 1

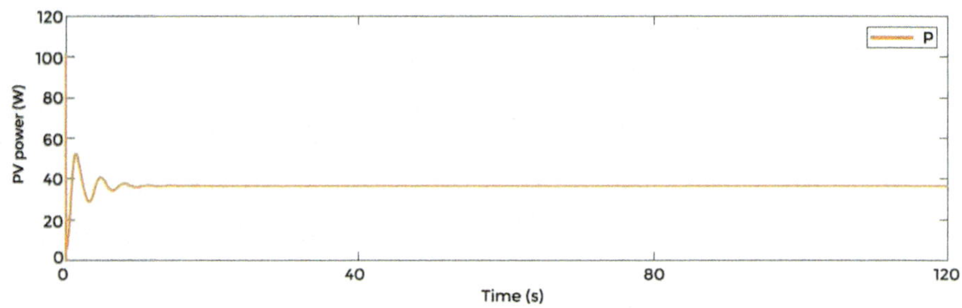

Fig. 3.8 Trajectory of P in Ex. 1

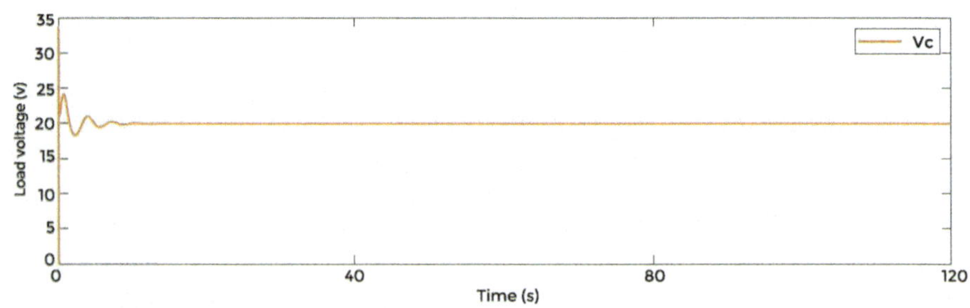

Fig. 3.9 Trajectory of ς_2 in Ex. 1

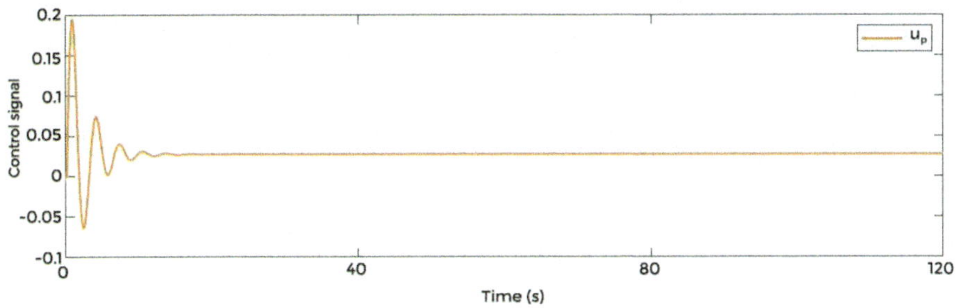

Fig. 3.10 Trajectory of μ_p in Ex. 1

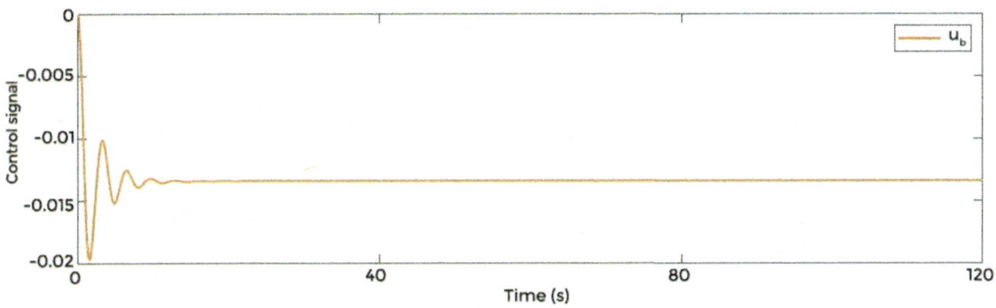

Fig. 3.11 Trajectory of μ_b in Ex. 1

The impact of changes in irradiation is effectively managed, resulting in a well-regulated output voltage at the appropriate level. Additionally, the photovoltaic (PV) track quickly reaches its optimum level within a time frame of less than 10 s.

Example 3 *In this example, a Gaussian noise with a variance of 0.05 is introduced to the temperature variable. The temperature is then randomly altered within the range of 30 to 50. The same procedure as is in Example 2 applied to the irradiation variable. At time* $t = 55s$*, the load abruptly changes from* 70 *to* 40 Ω*. The trajectories*

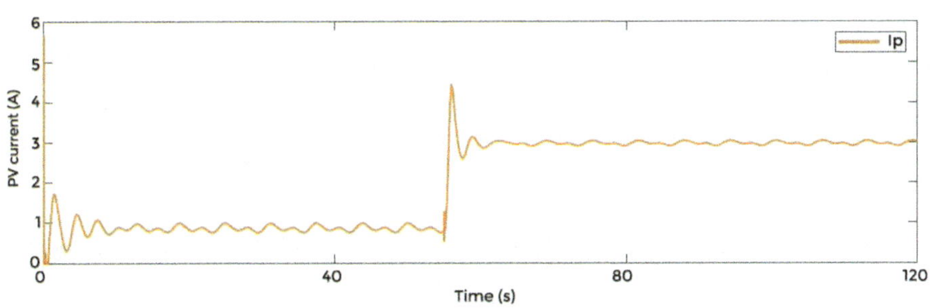

Fig. 3.12 Trajectory of ς_1 in Ex. 2

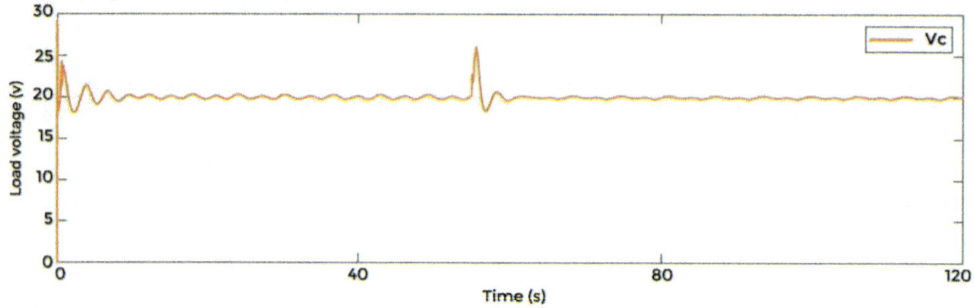

Fig. 3.13 Trajectory of ς_2 in Ex. 2

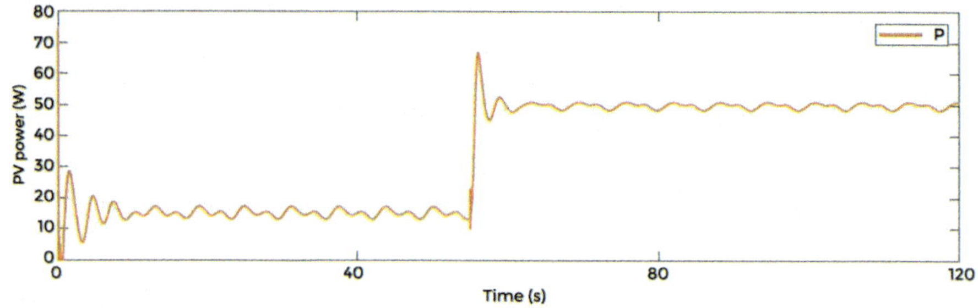

Fig. 3.14 Trajectory of P in Ex. 2

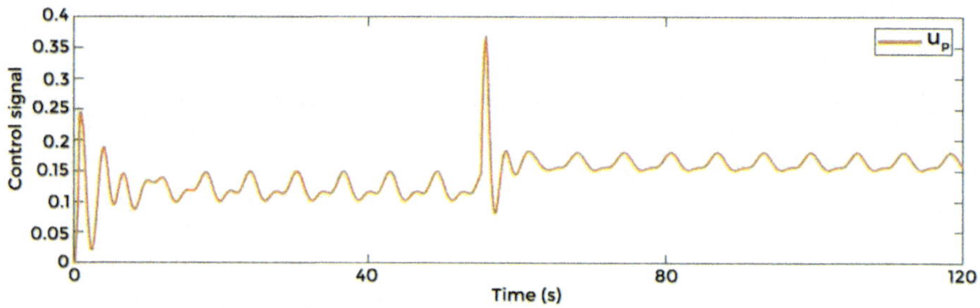

Fig. 3.15 Trajectory of μ_p in Ex. 2

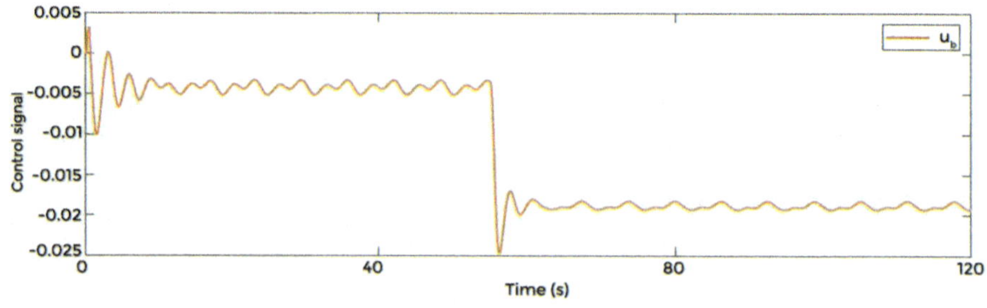

Fig. 3.16 Trajectory of μ_b in Ex. 2

of $\varsigma_i, i = 1, 2$, as well as the power of PV P and the controllers μ_p and μ_b, can be observed in Figs. 3.17, 3.18, 3.19, 3.20 and *3.21. It is evident that the system exhibits excellent tracking performance even in the presence of sudden load changes, fluctuating temperatures, and different levels of irradiation. Additionally, it should be noted that the dynamics of all units are unknown for the controller unit. It is seen that the impact of noise is effectively removed and the output voltage accurately follows the reference signal. It is important to mention that the T3-FLSs exhibit superior performance in noisy situations compared to the type-2 counterparts.*

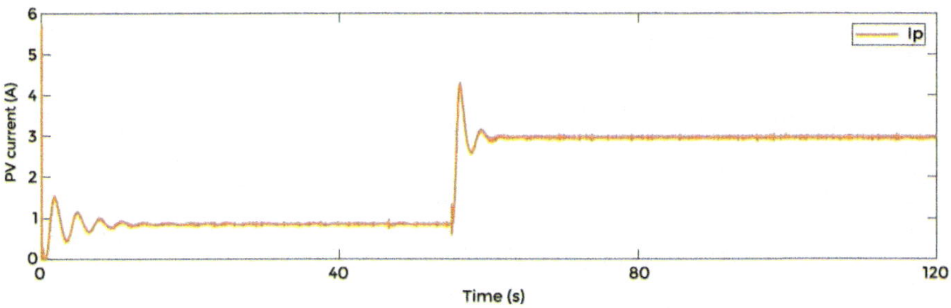

Fig. 3.17 Trajectory of ς_1 in Ex. 3

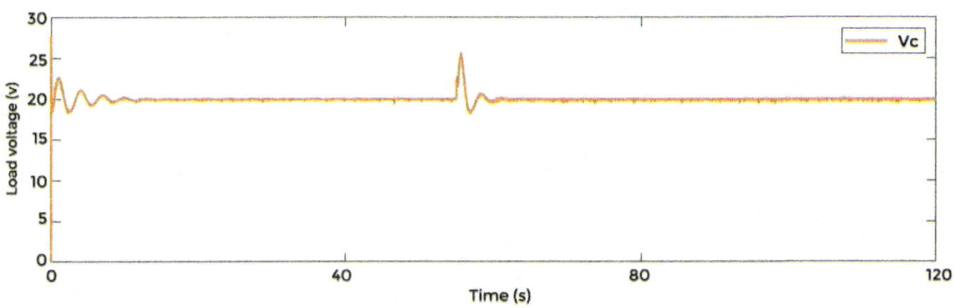

Fig. 3.18 Trajectory of ς_2 in Ex. 3

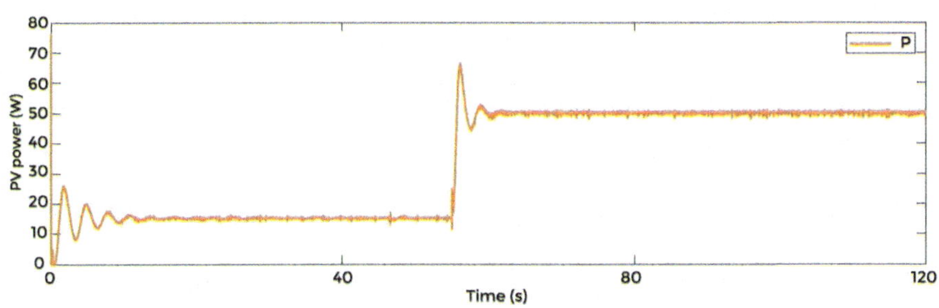

Fig. 3.19 Trajectory of *P* in Ex. 3

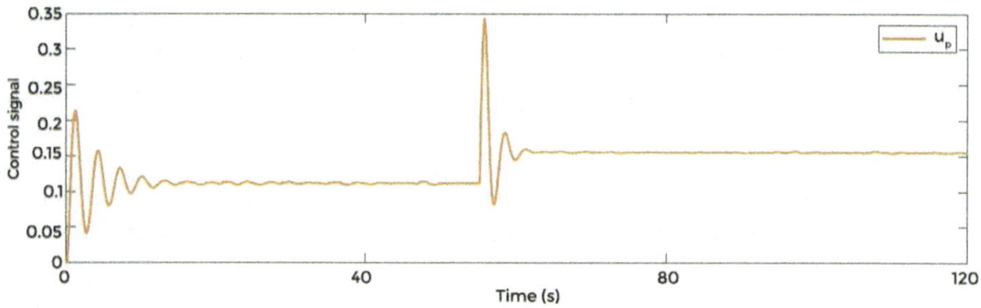

Fig. 3.20 Trajectory of μ_p in Ex. 3

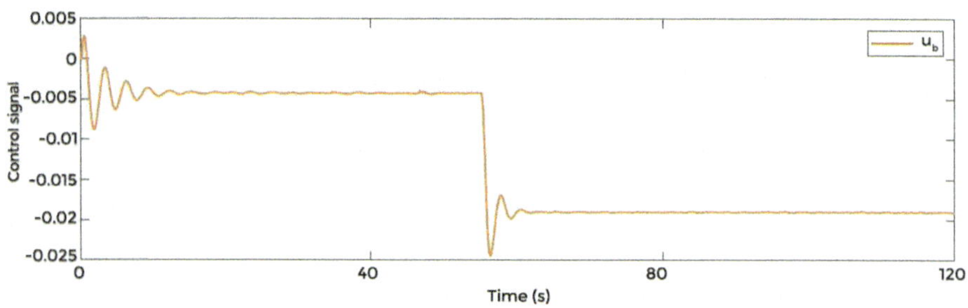

Fig. 3.21 Trajectory of μ_b in Ex. 3

Example 4 *For the last example, we present a numerical and graphical comparison between standard PID* [8], *PBC* [10], *SMC* [11], *LQR* [16], *and our technique based on integer-order calculus (IOC). The simulation circumstances are identical to Example 3, except that the temperature is altered at* $t = 40s$ *and the irradiation is modified at* $t = 80s$. *The mean square errors (MSEs) values in Table 3.5 and* I_p *and* V_c *trajectories in Fig. 3.22 unequivocally demonstrate the superiority of the suggested control scenario. It has been noticed that the suggested strategy, based on fractional-order calculus (FOC), leads to a higher level of accuracy in power/ voltage regulation proficiency. Observations indicate that the LQR and PID classic controllers are unable to accurately follow reference signals when dynamic perturbations occur at 40s and 80s. In addition, the settling time for the recommended controller is shorter than that of the other controllers being examined.*

Note 1 *It is important to acknowledge that the perterbations, such as output load, variations in irradiation, and temperature, are neither anticipated or forecasted. However, these disturbances are regarded as components of dynamic uncertainty. The online optimisation of T3-FLSs and compensators effectively manages the dynamic uncertainty* [17].

Note 2 *In all examples, a slight initial overshoot can be observed from the trajectories of* I_p *and* V_c. *This issue pertains to the assumption of unknown dynamics and its associated costs. Put simply, the recommended T3-FLSs are used to estimate the*

Table 3.5 MSE Comparison in Ex. 4

Method	Signal	
	ς_1	ς_2
LQR [16]	0.8259	148.9198
PBC [10]	0.9579	11.2450
PID [8]	1.1029	15.2295
SMC [11]	0.5161	9.3327
Our method by IOC	0.3101	0.9114
Our method by FOC	0.2021	0.8102

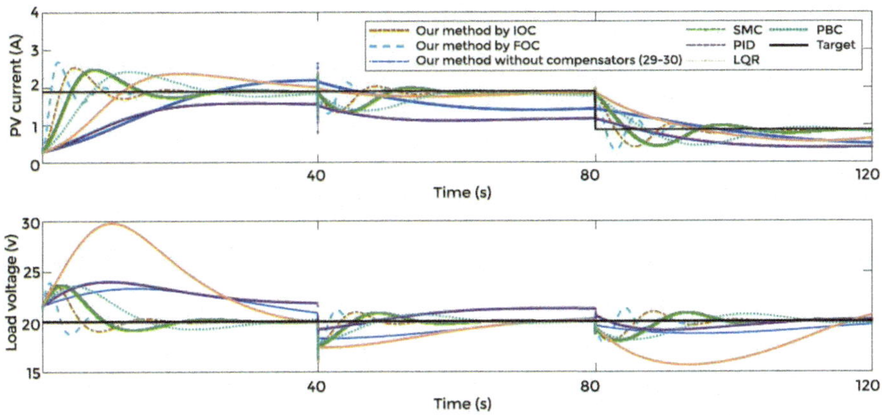

Fig. 3.22 V_c and I_p trajectory by various controllers in Ex. 4

dynamics of all units, which are assumed to be unknown. Additionally, the impacts of dynamic perturbations, such as variations in irradiation, temperature, and output load, are regarded as components of the approximation error. The online optimisation of T3-FLSs and compensators eliminates the impact of approximation errors. Due to stability concerns, the adaption rate of T3-FLSs is selected to be minimal. Conversely, compensators are widely recognised to have a quite large upper limit for approximation errors. It typically takes a few attempts for the regulation error to reach zero, and a slight overshoot is observed at the beginning. Nevertheless, reducing the upper limits of approximation errors leads to a drop in the first peaks. However, it also results in an increase in settling time and a longer duration for the regulation error to subside [17].

Note 3 *The precision and efficiency of the recommended controller are contingent upon the precision and efficiency of the model. Put simply, when the estimating error approaches zero quickly, the regulation error decreases proportionately. Conversely, research has demonstrated that the application of fractional-order calculus allows for more precise modelling of nonlinear dynamics. Due to this motive,* *the proposed controller is formulated using fractional-order calculus and its efficacy is demonstrated in Example 4* [17].

3.4 Advanced Deep Learning Management for Modern Microgrids

3.4.1 Problem Formulation

Figure 3.1 displays the overall configuration of the inverter and the architecture of the MG, as well as the case-study plant. The plant consists of renewable energies, storage systems, and DC/AC generators [18]. Based on the dq-components of currents i_1 and i_2 and voltage v_{cf} (as depicted in Fig. 3.2), the dynamics can be expressed as [19, 20].

$$\dot{\chi} = A\chi + b_u u + b_d dy = C\chi \tag{3.1}$$

where,
$\chi = \left[i_{1d} i_{1q} i_{2d} i_{2q} v_{cfd} v_{cfq}\right]^T$, $y = \left[i_{2d} i_{2q} v_{cfd} v_{cfq}\right]^T$, and

$$A = \begin{bmatrix} -R_{T_1}/L_{f_1} wR_f/L_{f_1} - 1 - R_{T_1}/L_{f_1} 0R_{T_2}/L_{f_2} 01/C_f 00R_f/L_{f_1} 0R_f/L_{f_1} \\ -R_{T_2}/L_{f_2} - w - 1/C_f 00 - 1/L_{f_1} 0R_f/L_{f_1} 0 \\ -1/L_{f_1} w - R_{T_2}/L_{f_2} 0 - 1/C_f 1/L_{f_2} 00 - w1/L_{f_2} w0 \end{bmatrix} \tag{3.2}$$

$$b_u = \frac{v_{dc}}{2}\left[\, 1/L_{f_1}\ 001/L_{f_1}\ 0000\ 0000 \,\right] \tag{3.3}$$

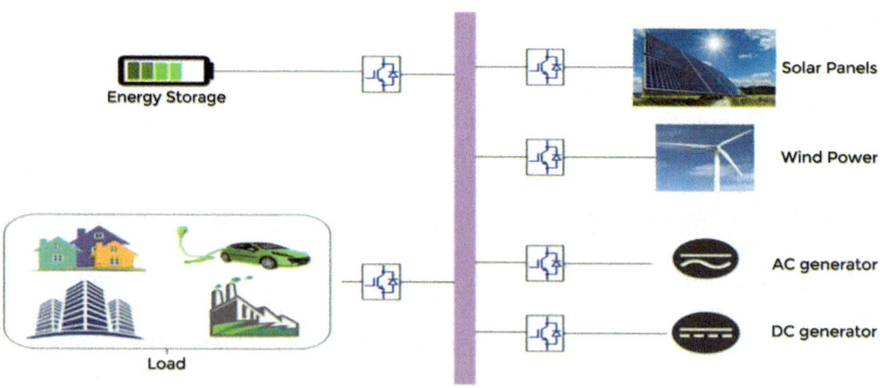

Fig. 3.1 Schematic view of MG topology and inverter

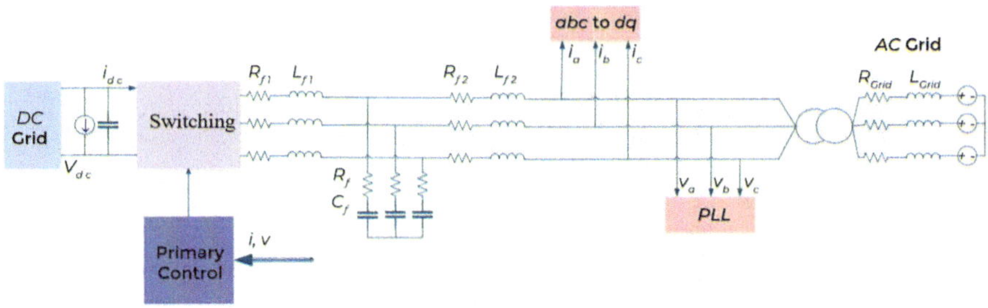

Fig. 3.2 The general scheme for suggested control system; The control signals adjust the switching mechanism

$$b_d = \left[-1/L_{f_2} 00 \ -1/L_{f_2} \ 0000 \ 0000 \right] \tag{3.4}$$

$$C = \left[0000 \ 1001 \ 00 \ 00 \right] \tag{3.5}$$

where, the outputs are denoted as i_{2d} and i_{2q}, while the disturbances are represented by the point of common coupling (PCC) voltages v_{PCCd} and v_{PCCq}. The control signals are indicated by the switching mechanism modulation indices, m_d and m_q, as shown in Fig. 3.2. Time derivative of y_i gives:

$$\ddot{y}_i = f_i(\chi) + g_i u_i, i = 1, ..., 4 \tag{3.6}$$

where,

$$\begin{aligned} f_1 = &\frac{R_{T_2}}{L_{f_2}} \left(-R_{T_1}/L_{f_1} \chi_1 + w\chi_2 + R_f/L_{f_1} \chi_3 - 1/L_{f_1} \chi_5 \right) - R_{T_2}/L_{f_2} \dot{\chi}_3 \\ &+ w\dot{\chi}_4 + 1/L_{f_2} \dot{\chi}_5, g_1 = \frac{R_{T_2} v_{dc}}{2L_{f_2} L_{f_1}} \end{aligned} \tag{3.7}$$

$$\begin{aligned} f_2 = &\frac{R_f}{L_{f_1}} \left(-\chi_1 - \frac{R_{T_1}}{L_{f_1}} \chi_2 + \frac{R_f}{L_{f_1}} \chi_4 - 1/L_{f_1} \chi_6 \right) - \frac{R_{T_2}}{L_{f_2}} \dot{\chi}_4 - w\dot{\chi}_3 \\ &+ 1/L_{f_2} \dot{\chi}_5, g_2 = \frac{R_f v_{dc}}{2L_{f_2} L_{f_1}} \end{aligned} \tag{3.8}$$

Figure 3.3 displays the overall block diagram. The main controller is a FRT-LQR (Finite-time Robust Linear Quadratic Regulator). Put simply, the LQR is constructed using a TSK-based fuzzy model. In order to address the modelling errors, dynamic perturbations, and faults, a compensator based on H_∞ criterion is introduced. This compensator is designed to provide stability. The function f_i, $i = 1, 2$, is estimated using the proposed Deep learned RBMs. The major parts of this estimation process may be found in Algorithm 1.

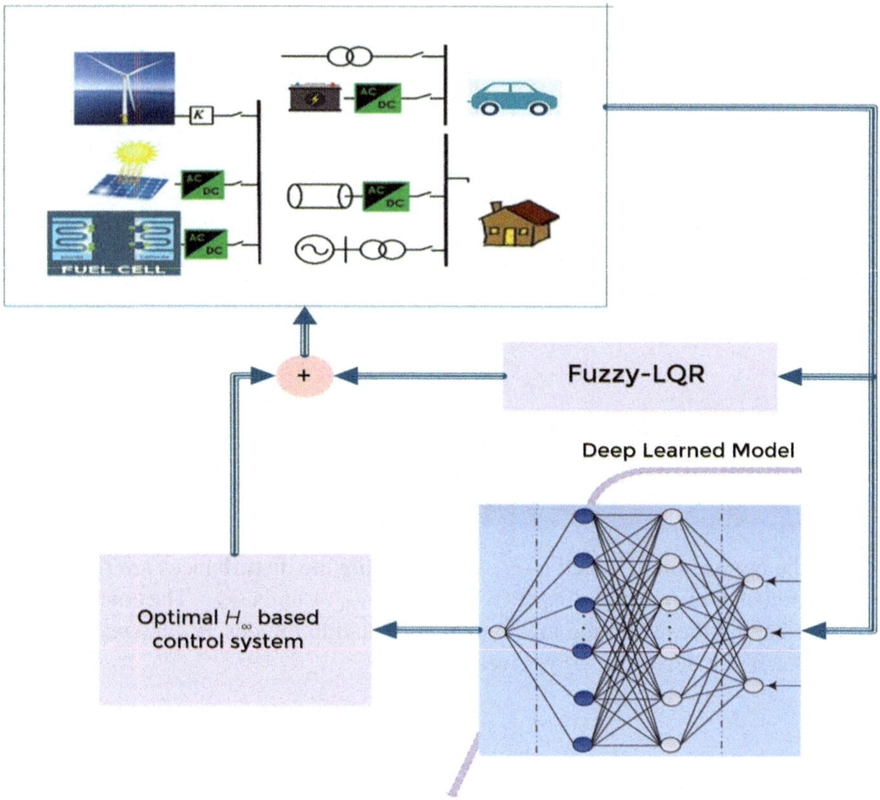

Fig. 3.3 Schematic view for the propsed control system

Algorithm 1: Pseudo code for the proposed controller.

for $t \leq T$.

Get input χ.

Generate the LQR signal Eq. (3.13) based on the fuzzy model Eq. (3.9).

Update the RBM parameters according to Eqs. (3.24, 3.57).

Generate the feedback linearization controller based on Eq. (3.32).

Generate H_∞-based compensator based on Eq. (3.52).

3.4.2 Primary Fuzzy LQR

A fuzzy Linear Quadratic Regulator (LQR) is initially developed to minimize a cost function, as indicated in Eq. (3.43). The impact of perturbations is counterbalanced in the subsequent stage. Figure 3.4 illustrates the block diagram. The LQR controller is developed based on the TSK-fuzzy model of the MG. Given the dynamics shown in Eq. (3.1), the enhanced system is defined as:

$$\dot{\hat{\chi}} = \hat{A}\hat{\chi} + b_u u$$

$$\dot{q} = r - C\hat{\chi} \tag{3.9}$$

The output of FLS is denoted as $\dot{\hat{\chi}}$, and the fuzzy rules and operation points of MG determines $\hat{A}$. Equation (3.9) can be restated as:

$$[\dot{\hat{\chi}}\dot{q}] = [\hat{A}0 - C0][\hat{\chi}q] + [b_u 0]u + [0I]r \tag{3.10}$$

The $l - th$ fuzzy rule is expressed based on work points.

$$l - th\,rule: \; if\; \hat{\chi} \in A_l \; Then\; \dot{\hat{\chi}} = \hat{A}_l\hat{\chi} + b_u u \tag{3.11}$$

Here, A_l denotes the l-th work area. It is important to mention that in order to ensure the strength and reliability of the system, the impact of modelling errors is taken into account, and a supervisor based on the H_∞ method is developed. The LQR control signal (u_{lqr}) is meant to minimize the function of Eq. (3.43).

$$J = \int_0^\infty \hat{\chi}^T Q\hat{\chi} + u_{lqr}^T R u_{lqr} \tag{3.12}$$

where, Q and R are constant matrix. We have:

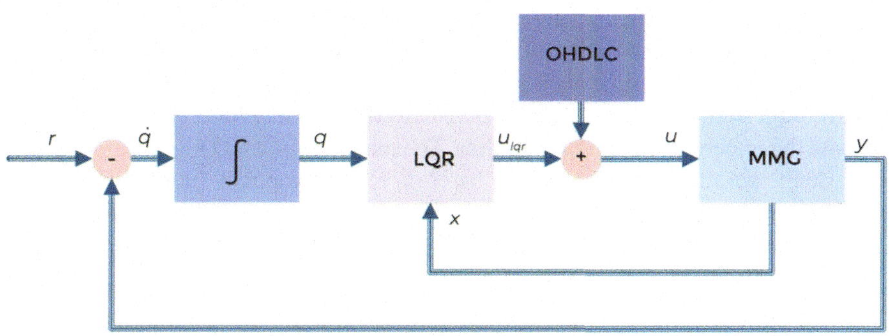

Fig. 3.4 Schematic view of LQR

$$u_{lqr} = -K[\hat{\chi} q] \tag{3.13}$$

where,

$$K = R^{-1}[b_u 0]^T \phi \tag{3.14}$$

and ϕ is the Riccati equation solution shown in the following:

$$[\hat{A}0 - C0]^T \phi + \phi[\hat{A}0 - C0]\phi - \phi[b_u 0]R^{-1}B^T \phi + Q = 0 \tag{3.15}$$

3.4.3 Deep Learned RBM

Figure 3.3 demonstrates the approach of enhancing the accuracy and reliability of the fuzzy-LQR design. This is achieved by estimating the undetermined functions in the system dynamics online, followed by the design of a compensator based on H_∞. In order to improve the accuracy of the estimation, it is recommended to utilize a deep learning method.

Several techniques have been suggested to optimise the parameters of neural networks, and these algorithms can be divided into two categories. In the initial category of algorithms, such as Kalman filter (KF), gradient descent, Levenberg–Marquardt, etc., it is necessary to calculate the derivative of the neural network's output with regard to the adjustable parameters. In these derivation-based approaches, the parameters in the hidden layer are minimally adjusted. Due to the minuscule values of derivatives, the parameters in the hidden layers experience a significant decrease in value. Moreover, the adjustment of these parameters is highly sensitive to the adaption rate. Alternative methods that do not rely on derivatives have been developed in the second approach, including evolutionary-based methods and nonlinear versions of the Kalman Filter. The majority of these approaches are unsuitable for online applications due to their significant computational expense and limited stability. Several deep learning algorithms have been developed to enhance the tuning of neural networks. This work utilizes an unsupervised contrastive divergence (CD) technique for the purpose of deep learning.

The suggested deep learning Restricted Boltzmann Machine (RBM) is used to estimate the uncertain nonlinear function. The unsupervised CD algorithm is used to optimise the parameters of the hidden layer (θ_h). The Lyapunov based tuning rule is used to update the vector of parameters ξ. Figure 3.5 shows the block diagram of the proposed technique. According to Fig. 3.5, the estimation of the system dynamics is as follows:

$$\ddot{y} = \hat{f}(\chi|\xi) + \hat{g}u \tag{3.16}$$

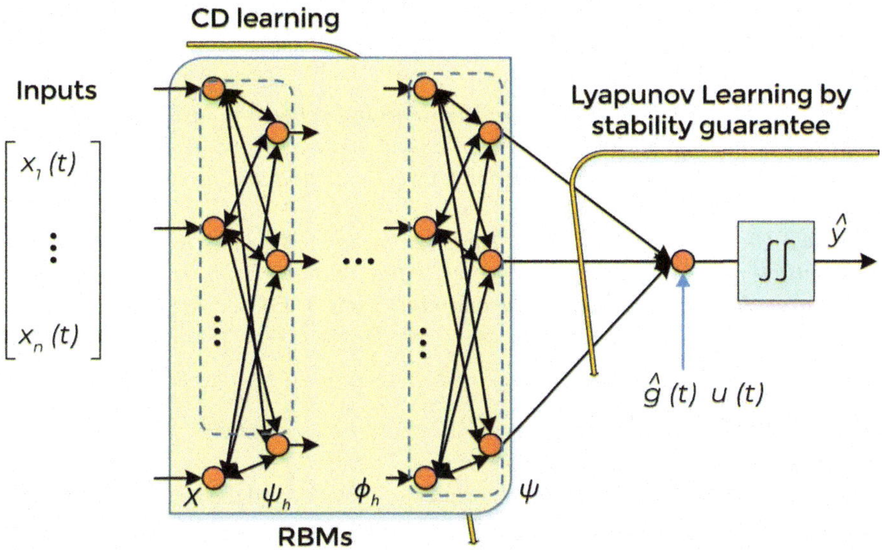

Fig. 3.5 Schematic view for the proposed control system

where,

$$\hat{f}(\chi|\xi) = \xi^T \psi(\chi) \tag{3.17}$$

$$\psi = \mu\left(\theta_h^T \phi_h\right) \tag{3.18}$$

$$\phi_h = \mu(\theta_h \psi_h) \tag{3.19}$$

$$\psi_h = \mu\left(\theta_h^T \chi\right) \tag{3.20}$$

$$\mu(\cdot) = 1 - exp(-\cdot)/1 + exp(-\cdot) \tag{3.21}$$

The hidden vector of parameters θ_h is adjusted according to the minimizing of

$$E(\phi, \psi) = e^{H(\phi,\psi)}/H(\phi, \psi) \tag{3.22}$$

where,

$$H(\phi, \psi) = -\psi^T \theta_h \phi \tag{3.23}$$

The tuning rule of θ_h is stated as:

$$\theta_h(t+1) = \theta_h(t) + \eta\left(\chi\psi_h^T - \phi_h\psi^T\right) \tag{3.24}$$

where, $0 < \eta \leq 1$.

Note 1 *The intricacy of the proposed learning scheme is justifiable, given the proposed model is acquired through an online method. The learning scheme does not involve any iterations. The parameter vector is updated at every sample time using the extracted adaptation law.*

Note 2 *The parameters in the Restricted Boltzmann Machine (RBM) hidden layer are adjusted according to Eq. (3.24), while the parameters in the last layer of the output are adjusted according to the adaptation law of Eq. (3.57).*

3.4.4 Optimal Control

This part focuses on the design of the primery controller and the analysis of stability. The summary of the findings is shown in Theorem 1.

Theorem 1 *The system of Eq. (3.1) achieves asymptotic stability with the implementation of controller shown in Eq. (3.32), tuning rules of Eq. (3.57), and H_∞-based compensator of Eq. (3.52).*

Based on Eq. (3.6), the estimated output dynamics can be determined as follows:

$$\ddot{y}_i = \hat{f}_i(\chi\,|\xi_i) + \hat{g}_i u_i \tag{3.25}$$

Consider the following equation:

$$s_i = \dot{e}_i + \lambda_i e_i \tag{3.26}$$

where,

$$\hat{e}_i = y_{d_i} - \hat{y}_i \tag{3.27}$$

Time derivative of Eq. (3.26) gives:

$$\dot{s}_i = \dot{e}_i + \lambda_i \dot{e}_i \tag{3.28}$$

By substituting $\dot{e}_i$, we get:

$$\dot{s}_i = \ddot{y}_{d_i} - \ddot{y}_i + \lambda_i \dot{e}_i \tag{3.29}$$

By inserting $\ddot{y}_i$ from Eq. (3.25) into Eq. (3.29), we have:

$$\dot{s}_i = \ddot{y}_{d_i} - f_i(\chi) - g_i u_i + \lambda_i \dot{e}_i \tag{3.30}$$

By adding and subtracting $\hat{g}_i u_i$, Eq. (3.30), can be restated as:

$$\dot{s}_i = \ddot{y}_{d_i} - f_i(\chi) + \lambda_i \dot{e}_i - g_i u_i + \hat{g}_i u_i - \hat{g}_i u_i \tag{3.31}$$

Starting from Eq. (3.31), the control signal is defined via feedback linearization as:

$$u_i = \frac{1}{\hat{g}_i}\left[\ddot{y}_{d_i} - \hat{f}_i(\chi|\xi_i) + \lambda_i \dot{e}_i + k_i s_i - \vartheta_i\right] \tag{3.32}$$

The symbol ϑ_i indicates the compensator, while k_i is a positive constant. By implementing the operation shown in Eq. (3.32) on system of Eq. (3.31), we obtain:

$$\dot{s}_i = f_i(\chi) - \hat{f}_i(\chi|\xi_i) + (g_i - \hat{g}_i)u_i - k_i s_i + \vartheta_i \tag{3.33}$$

By adding and subtracting optimal g_i^* and optimal $\hat{f}_i(\chi|\xi_i^*)$ into Eq. (3.33), $\dot{s}$ changes to:

$$\dot{s}_i = \hat{f}_i(\chi|\xi_i^*) - \hat{f}_i(\chi|\xi_i) + (g_i^* - \hat{g}_i)u - k_i s_i + \vartheta_i + f_i(\chi) - \hat{f}_i(\chi|\xi_i^*) + (g_i - g_i^*)u_i \tag{3.34}$$

Considering the following equations:

$$\hat{f}_i(\chi|\xi_i^*) - \hat{f}_i(\chi|\xi_i) = \tilde{\xi}_i^T \psi_i \tilde{\xi}_i = \xi_i^* - \xi_i \tilde{g}_i = g_i^* - \hat{g}_i \tag{3.35}$$

and substituting Eq. (3.35) into Eq. (3.34), $\dot{s}_i$ is reformulated as:

$$\dot{s}_i = \tilde{\xi}_i^T \psi_i + \tilde{g}_i u_i + E_i - k_i s_i + \vartheta_i \tag{3.36}$$

where,

$$E_i = f_i(\chi) - \hat{f}_i(\chi|\xi_i^*) + (g_i - g_i^*)u_i \tag{3.37}$$

The compensator ϑ_i is designed using the following Lemma:

Lemma 1 *Let us consider the schematic representation of the H_∞ control problem, as depicted in Fig. 3.6 ([21–23]). The general H_∞ control method can be expressed as:*

$$\dot{\chi} = A\chi + b_d d + b_u u$$

$$z = C_z \chi + D_u u$$

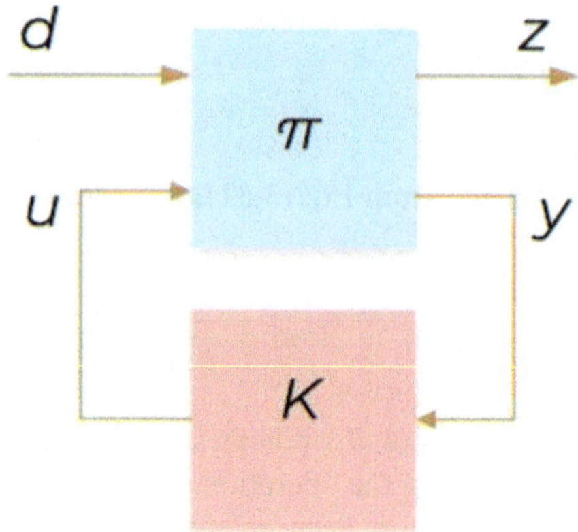

Fig. 3.6 Schematic view of the H_∞ control problem

$$y = C_y\chi + D_d d$$

$$\pi = \left[A\, b_d\, b_u\, C_z\, 0\, D_u\, C_y\, D_d\, 0\right] \tag{3.38}$$

If (A, b_u), (C_1, A), (A, B_2) and are respectively controllable, observable, stabilizable and detectable, and

$$H_\infty \in dom(Ric)\, and\, \chi_\infty := Ric(H_\infty) > 0.J_\infty \in dom(Ric)\, and\, Y_\infty := Ric(J_\infty) > 0.\rho(\chi_\infty Y_\infty) < \eta^2. \tag{3.39}$$

Then $\|T_{zw}\|_\infty < \eta$, if

$$K(s) = [\Lambda_\infty - Z_\infty L_\infty \Phi_\infty 0] \tag{3.40}$$

$$\Lambda_\infty = A + \eta^{-2}B_1B_1^*\chi_\infty + b_u\Phi_\infty + Z_\infty L_\infty C_2$$

$$\Phi_\infty = -b_u^*\chi_\infty, L_\infty = -Y_\infty C_2^*, Z_\infty := \left(I - \eta^{-2}Y_\infty\chi_\infty\right)^{-1} \tag{3.41}$$

$$H_\infty = \left[A\eta^{-2}B_dB_d^* - B_uB_u^* - C_1^*C_1 - A^*\right] \tag{3.42}$$

$$J_\infty = \left[AA^*\eta^{-2}C_1^*C_1 - C_2^*C_2 - B_dB_d^* - A\right] \tag{3.43}$$

Given the upper bound of uncertainty, Eq. (3.36) can be expressed as:

$$\dot{s}_i + k_i s_i = \tilde{E}_i + \vartheta_i \tag{3.44}$$

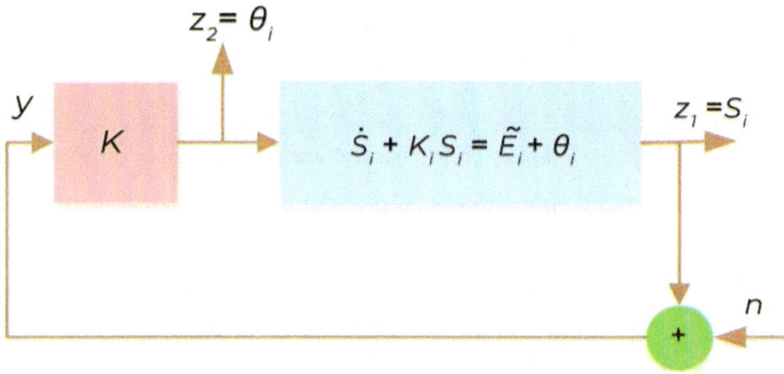

Fig. 3.7 The H_∞ control problem schematic view for designing compensator

where,

$$\tilde{E}_i = E_i + \tilde{\xi}_i^T \psi_i + \tilde{g}_i u_i \tag{3.45}$$

In order to create the compensator, the system is depicted using a block diagram, as illustrated in Fig. 3.7, and described by Eq. (3.46).

$$\dot{s}_i + k_i s_i = \tilde{E}_i + \vartheta_i z_1 = s_i z_2 = \vartheta_i y = s_i + n \tag{3.46}$$

Here, n represents the measurement error. From Lemma 1, we can deduce:

$$\pi = [-k_i|-[10]1|||[10]1 - 0[01] - [01]0] \tag{3.47}$$

$$J_\infty = \left[-k_i \frac{1-\eta^2}{\eta^2} - 1k_i\right], H_\infty = \left[-k_i \frac{1-\eta^2}{\eta^2} - 1k_i\right], \tag{3.48}$$

The eigne values of J_∞ and H_∞ in Eq. (3.48) are considered to be:

$$\lambda_i = \pm \frac{\sqrt{(k_i^2+1)\eta^2 - 1}}{\eta} \tag{3.49}$$

Then if $\eta > \frac{1}{\sqrt{k_i^2+1}}$, it satisfies the conditions $J_\infty \in dom(Ric)$ and $H_\infty \in dom(Ric)$, which yields:

$$\chi_\infty = Y_\infty = \frac{\eta}{\sqrt{(k_i^2+1)\eta^2 - 1} + k_i\eta} > 0 \tag{3.50}$$

$$\rho(\chi_\infty, Y_\infty) = \frac{\eta^2}{\left(\sqrt{(k_i^2+1)\eta^2 - 1} + k_i\eta\right)^2} < \eta^2 \tag{3.51}$$

Then, if $\eta > \sqrt{k_i^2 + 2} - k_i$ Eq. (3.51) is satisfied. By taking account of the optimal η as $\eta_{opt} > \sqrt{k_i^2 + 2} - k_i$, the optimal compensator is restated as:

$$\vartheta_i = -\frac{\eta_{opt}}{\sqrt{(k_i^2+1)\eta_{opt}^2 - 1} + k_i\eta_{opt}} s_i \tag{3.52}$$

In order to examine stability, let us regard Lyapunov as:

$$V = \frac{1}{2}s_i^2 + \frac{1}{2\kappa}\tilde{\xi}_i^T\tilde{\xi}_i + \frac{1}{2\kappa}\tilde{g}_i^2 \tag{3.53}$$

Time derivative of V in Eq. (3.53) gives:

$$\dot{V} = s_i\dot{s}_i - \frac{1}{\kappa}\tilde{\xi}_i^T\xi_i - \frac{1}{\kappa}\tilde{g}_i^2 g_i \tag{3.54}$$

By substituting Eq. (3.36), we get:

$$\dot{V} = s_i\left(\tilde{\xi}_i^T\psi_i + \tilde{g}_i u + E_i - k_i s_i + \vartheta_i\right) - \frac{1}{\kappa}\tilde{\xi}_i^T\xi_i - \frac{1}{\kappa}\tilde{g}_i\dot{\hat{g}}_i \tag{3.55}$$

Equation (3.55) is simplified to:

$$\dot{V} = -k_i s_i^2 + s_i E_i + s_i\vartheta_i + \tilde{\xi}_i^T\left(s_i\psi_i - \frac{1}{\kappa}\xi_i\right) + \tilde{g}_i\left(s_i u_i - \frac{1}{\kappa}\dot{\hat{g}}_i\right) \tag{3.56}$$

By setting the terms in brackets in Eq. (3.56) to zero, the tuning laws can be derived as:

$$\xi_i = \kappa s_i\psi_i \dot{\hat{g}}_i = \kappa s_i u_i \tag{3.57}$$

$\dot{V}$ in Eq. (3.56), becomes:

$$\dot{V} = -k_i s_i^2 + s_i E_i + s_i\vartheta_i \tag{3.58}$$

Considering Eq. (3.52), $\dot{V}$ in Eq. (3.58), is reformulated as:

$$\dot{V} = -k_i s_i^2 + s_i E_i - \frac{\eta_{opt} s_i^2}{\sqrt{(k_i^2+1)\eta_{opt}^2 - 1} + k_i\eta_{opt}} \tag{3.59}$$

Using Eq. (3.59), we get:

$$\{\dot{V} \leq -k_i|s_i| + |s_i||E_i| - \frac{\eta_{opt}s_i^2}{\sqrt{(k_i^2+1)\eta_{opt}^2 - 1 + k_i\eta_{opt}}}, if\,|s_i| < 1\dot{V} \leq$$
$$s_i^2\left(|E_i| - k_i - \frac{\eta_{opt}}{\sqrt{(k_i^2+1)\eta_{opt}^2 - 1 + k_i\eta_{opt}}}\right), if\,|s_i| \geq 1 \tag{3.60}$$

If $k_i > |E_i|$, then the system is guaranteed to be asymptotically stable, which completes the proof.

Note 3 *The significance of the presented approach lies in its implementation within an online framework. Furthermore, the proposed methodology is capable of effectively managing unforeseen fluctuations in dynamics. The suggested deep-learning technique accurately identifies the dynamics, while the designed* H_∞*-based supervisor compensates for the perturbations.*

3.4.5 Simulation

This section evaluates the effectiveness of the developed controller using a case study of MMG. The system parameters are provided in Table 3.1.

3.4.5.1 Balanced Grid in High Short-Circuit Situation

This part evaluates the performance of the intended OHDLC system on a balanced grid subjected to a high short-circuit condition. At $t = 0s$, both the active and reactive powers are zero.

The reference signals are modified at $t = 0.5s$ and $t = 1s$. Figure 3.8 illustrates the trajectories followed by active and reactive powers. It is evident that the active

Table 3.1 Parameter description

Parameter	Value	Unit
L_{f1}	1.10	mH
R_{f1}	0.01	Ω
L_{f2}	1.10	mH
R_{f2}	0.01	Ω
C_f	14.4	μF
R_f	2.07	Ω
V_{dc}	380	v
f_s	8.1	kHz

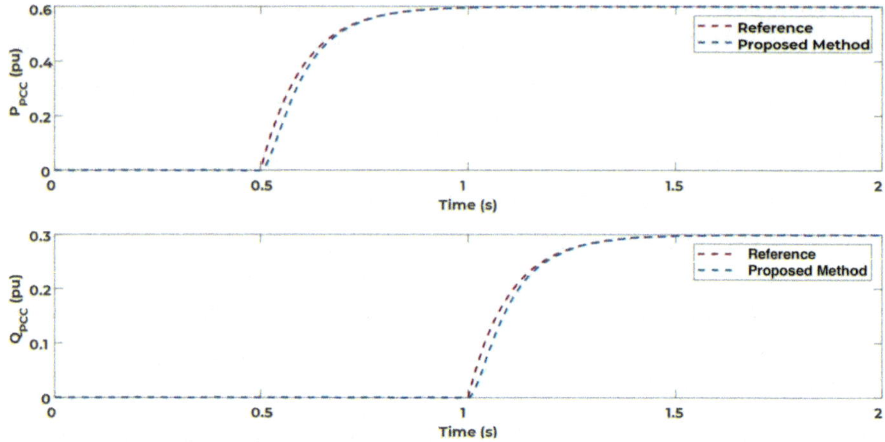

Fig. 3.8 Active and reactive powers trajectories in Sect. 3.4.5.1

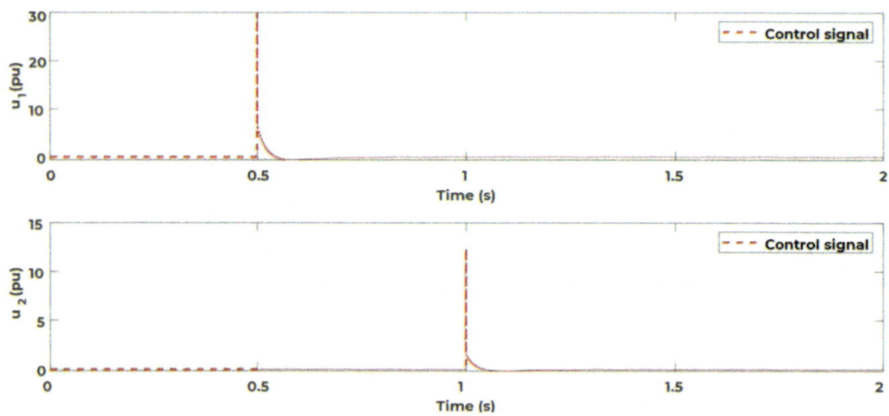

Fig. 3.9 Control signals trajectories in Sect. 3.4.5.1

and reactive power signals closely follow the reference signals. Figure 3.9 displays the control signals that correlate to the given information. The results demonstrate that the control signals obtained using unconventional methods have smoother forms without any fluctuations, in contrast to standard procedures.

3.4.5.2 Unbalanced Grid in High Short-Circuit Situation

This section focuses on analysing the state of imbalance. The voltages of Phase B and C are reduced to 95% and 85% correspondingly. The simulation results presented in Fig. 3.10 confirm the continued validity of the planned control. Similar to the

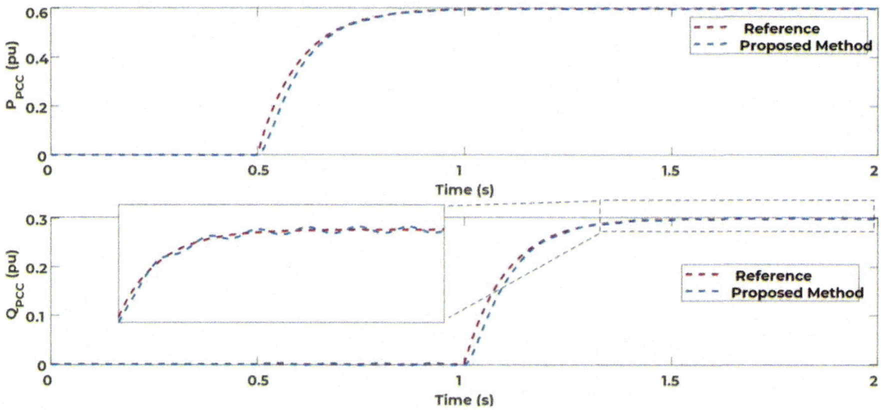

Fig. 3.10 Active and reactive power trajectories in Sect. 3.4.5.2

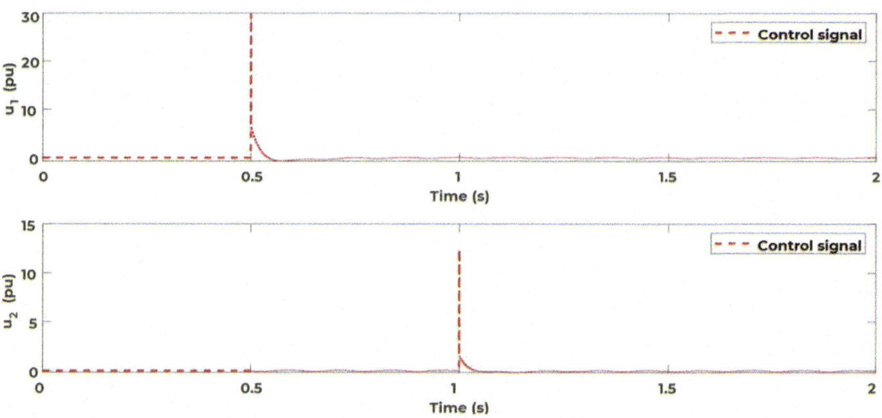

Fig. 3.11 Control signals trajectories in Sect. 3.4.5.2

preceding situation, we observe that the active/reactive power signals closely follow the reference signals. The control trajectories shown in Fig. 3.11 clearly demonstrate the achievable structure of the control signal.

3.4.5.3 Weak Grid Condition

This section examines the impact of weak grid condition. The short-circuit capacity and inertia constant are assumed to have a ratio of 1. The reference values for active and reactive powers are modified at $t = 1.5s$ and $t = 3s$, respectively. Figure 3.12 illustrates the trajectories followed by active and reactive powers. It is noted that the active and reactive power signals closely follow the reference signals. Figure 3.13

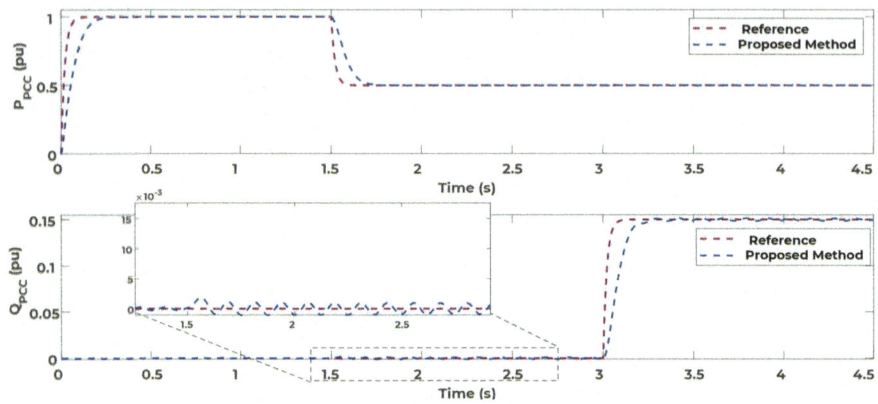

Fig. 3.12 Active and reactive power trajectories in Sect. 3.4.5.3

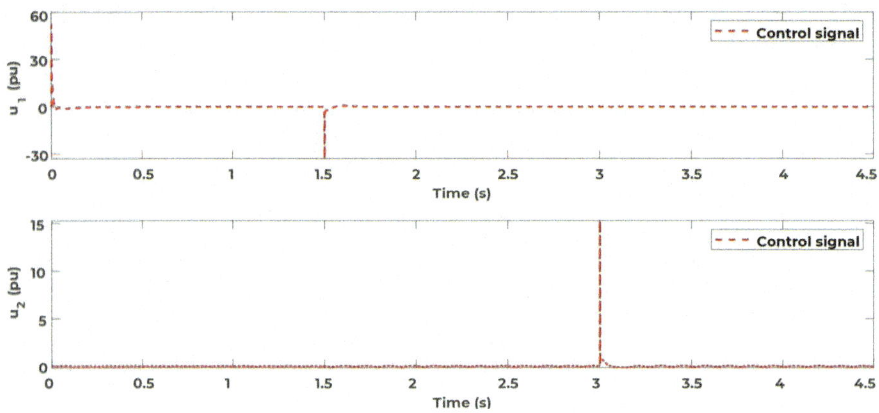

Fig. 3.13 Control signals trajectories in Sect. 3.4.5.3

illustrates the corresponding control signals. It is observed that the control signals of the proposed technique exhibit smooth forms without any chattering issues.

3.4.5.4 High Noisy Condition

In order to evaluate the resilience of the algorithm that has been created, a disturbance is introduced, represented by the equation $d_i = 0.5sin(10t), i = 1, 2$. The active/reactive power and control signals trajectories can be observed in Figs. 3.14 and 3.15. Additionally, in some examinations, a high dynamic perturbation is regarded as white noise with a mean of 0 and a variance of 1.

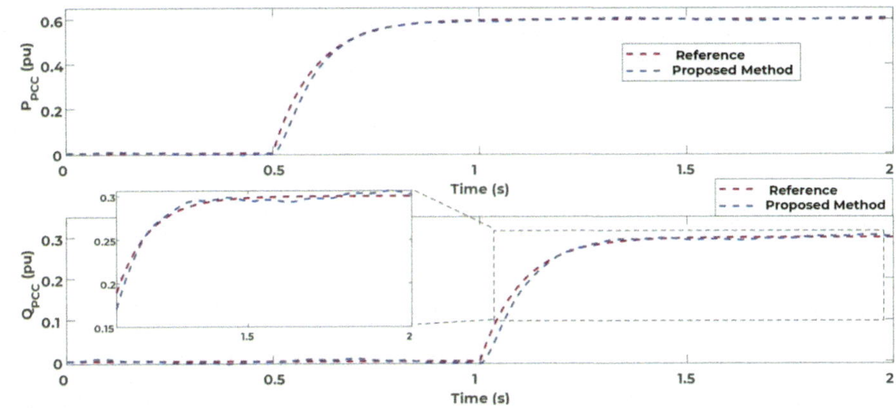

Fig. 3.14 Active and reactive power trajectories, under sinusoidal perturbation in Sect. 3.4.5.4

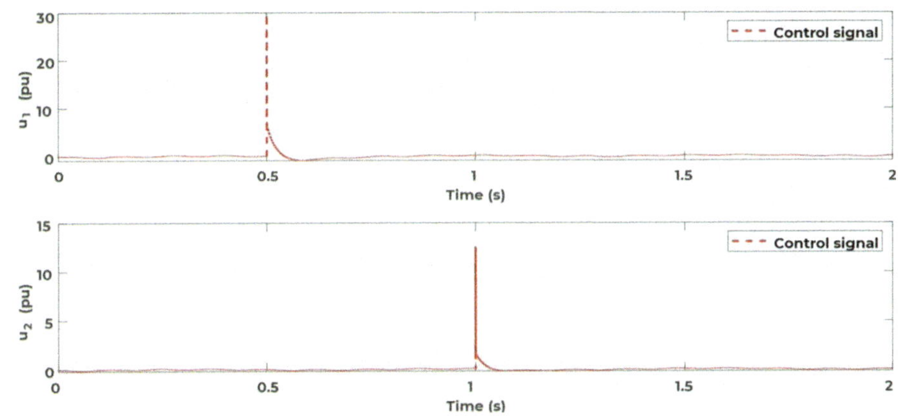

Fig. 3.15 Control signals trajectories, under sinusoidal disturbance in Sect. 3.4.5.4

The active/reactive power and control signals trajectories can be observed in Figs. 3.16 and 3.17. It is evident that the power trajectories exhibit strong resistance against significant disturbances and unpredictable changes in dynamics.

3.4.5.5 Comparison

In order to illustrate the superiority of the suggested OHDLC, a comparison is provided with alternative techniques. The performance is evaluated in comparison to the PI-SMC system [24], the fault-tolerant adaptive controller (FTAC) [19], and the H_∞-based controller [25]. SMC technique has been commonly used to manage the active and reactive power of MGs. The RMSE values are provided in Table 3.2.

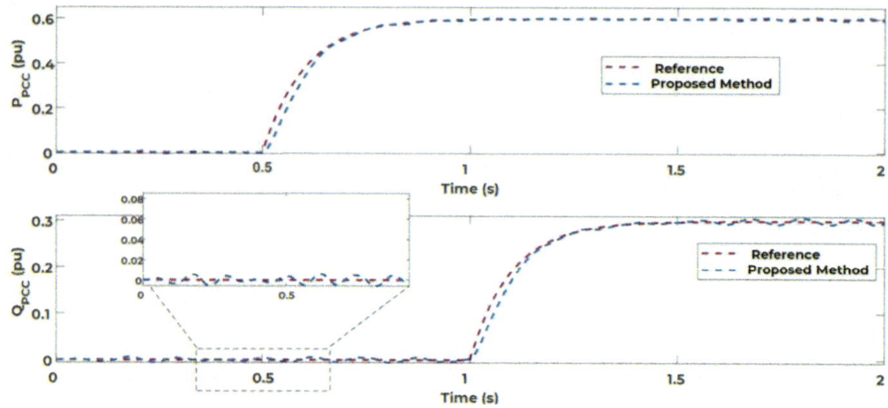

Fig. 3.16 Active and reactive power trajectories, under white noise disturbance in Sect. 3.4.5.4

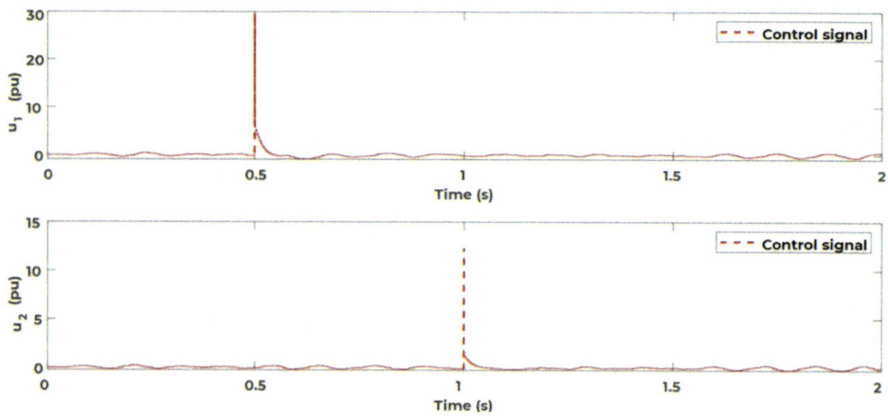

Fig. 3.17 Control signals trajectories, under white noise perturbation in Sect. 3.4.5.4

It is evident that the implemented OHDLC leads to much lower root mean square error (RMSE) and steady state error.

Table 3.2 Performance comparison

	RMSE		Steady-state error	
	e_1	e_2	e_1	e_2
FTAC [19]	0.1740	0.0781	0.0120	0.0101
PI-SMC [24]	0.2354	0.1749	0.0782	0.0514
H_∞[25]	0.3071	0.8210	0.1012	0.2014
Proposed OHDLC	0.0137	0.0072	0.0001	0.0002

3.4.5.6 Experimental Examination

In recent decades, the number of distributed production resources in distribution networks has been increasing. This growth is particularly evident with the rise of new energy sources that have modest power capacity and operate at low voltage levels. Regarding this matter, the utilization of MGs has experienced a substantial growth. The MGs are applicable in commercial malls, industrial parks, and university campuses [26].

As previously stated, the designed technique does not require the dynamics information. In order to demonstrate this characteristic more effectively, a Hardware-in-the-Loop (HiL) simulation is conducted to evaluate the suitability of the developed control technique in a real-time environment. The dynamics are currently uncertain, and the proposed online approach aims to determine the dynamics. Furthermore, the H_∞-based supervisor ensures the stability. In order to achieve this objective, a scheme for the MMG is created on a HiL platform using OPAL-RT, as shown in Fig. 3.18. The HiL configuration consists of multiple sections [26].

(1) The Real Time Simulator (RTS) used to model the MMG system is OPAL-RT.
(2) The programming host is a PC.
(3) A router is used to connect the setup components.

Furthermore, the Ethernet ports serve the purpose of establishing a connection between the OPAL-RT and a DK60 board. The HiL configuration eliminates the delays and faults that frequently occur in real-world systems. Figures 3.19 and 3.20 illustrate the paths followed by active and reactive powers, respectively. We see that a highly accurate reference power is achieved. Figures 3.19 and 3.20 demonstrate that

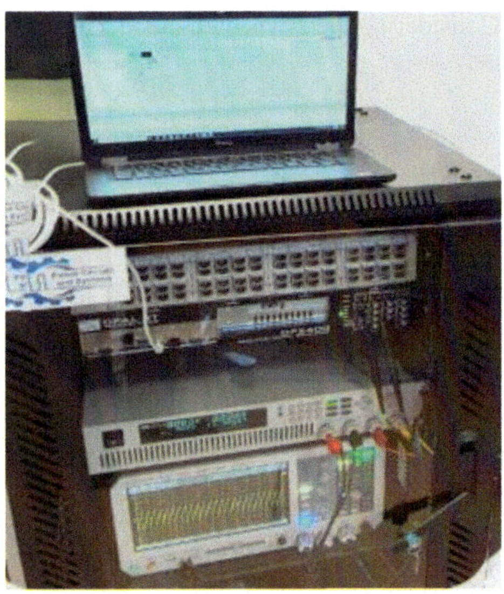

Fig. 3.18 The experimental configuration of the real-time for Sect. 3.4.5.6

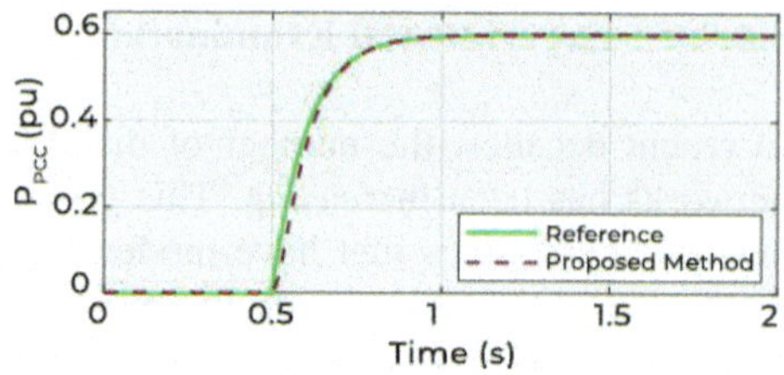

Fig. 3.19 Active power trajectory in the real-time experimental configuration for Sect. 3.4.5.6

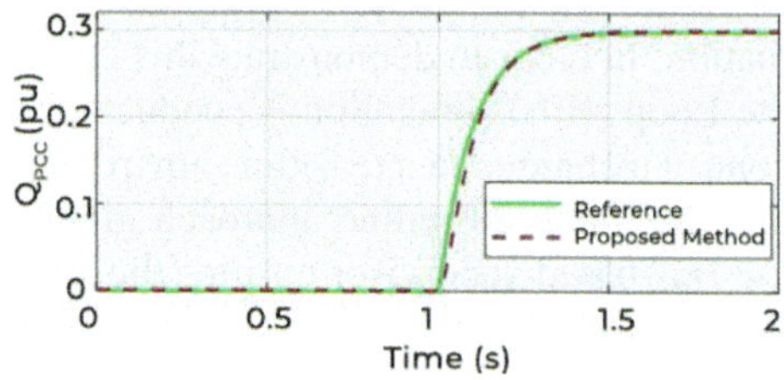

Fig. 3.20 Reactive power trajectory in the real-time experimental configuration for Sect. 3.4.5.6

the proposed strategy effectively withstands uncertainty and achieves satisfactory efficiency even in the absence of information on the dynamics.

3.5 Type-3 Fuzzy Predictive Control for Mobile Robots

3.5.1 Kinematic Model

Considering the 2-wheeled model (see Fig. 3.1), the linear velocity can be calculated as:

$$\vartheta = R(\dot{\phi}_R + \dot{\phi}_L)/2w = R(\dot{\phi}_R - \dot{\phi}_L)/2 \tag{3.1}$$

where, *R*/*L* represents radius by dsitance. Then, it can be stated as:

$$[\dot{x}\dot{y}\dot{P}] = \left[\frac{R}{2}\frac{R}{2}00\frac{R}{2L} - \frac{R}{2L}\right][\dot{\phi}_R\dot{\phi}_L] \tag{3.2}$$

Given the angular velocity and orthogonal rotation, we may conclude:

$$[\dot{x}\dot{y}\dot{P}] = \left[\frac{R}{2}cos(P)\frac{R}{2}cos(P)\frac{R}{2}sin(P)\frac{R}{2}sin(P)\frac{R}{2L} - \frac{R}{2L}\right][\dot{\phi}_R\dot{\phi}_L] \tag{3.3}$$

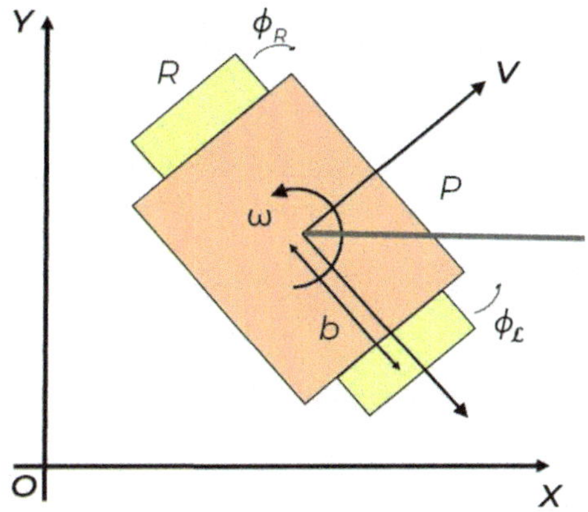

Fig. 3.1 A diagram for the wheeled moving robot

3.5.2 Dynamics

The dynamics are mathematically represented using the Lagrange method in the following manner:

$$\Psi(\eta)\ddot{\eta} + \upsilon(\eta, \dot{\eta})\dot{\eta} + G(\eta) + \tau_\kappa = F(\eta)\tau - A^T\gamma \quad (3.4)$$

where, Ψ is inertia matrix, υ/G is considered to be Coriolis matrix by gravitational, τ_κ/F denotes disturbance by input matrix, and τ is input vector. η shows position. By taking into account of $G = 0$, we have:

$$\Psi(\eta) = [m\,0\,0\,0 - md\sin(\delta)\,0\,0\,md\cos(\delta)\,0\,0 - md\sin(\delta)\,0\,0\,md\cos(\delta)\; I\,0\,0\,0\,0\,I_w\,0\,0\,0\,0\,I_w] \quad (3.5)$$

$$\upsilon(\eta, \dot{\eta}) = [0\,mdcos(\delta)0 - mdsin(\delta)\,000\,000\,0000000\,0000\,0000] \quad (3.6)$$

$$A^T\gamma = [-sin(\delta)\,cos(\delta)\,cos(\delta)\,sin(\delta) - \kappa\,0\,0\,L - R\,0\,cos(\delta)\,sin(\delta) - L\,0 - R] \quad (3.7)$$

Considering ς so that $\dot{\eta} = s\eta$, $\ddot{\eta} = \dot{s}\eta + s\dot{\eta}$, $\eta = [\dot{\phi}_R\,\dot{\phi}_L]^T$, $\varsigma^T A^T = 0$, $A^T\gamma$ is replaced and:

$$[\dot{x}\,\dot{y}\,\dot{\delta}\,\dot{\phi}_R\,\dot{\phi}_L] = \left[\frac{R}{2}cos(\delta)\frac{R}{2}cos(\delta)\frac{R}{2}sin(\delta)\frac{R}{2}sin(\delta)\frac{R}{2L}10 - \frac{R}{2L}01\right][\dot{\phi}_R\,\dot{\phi}_L] \quad (3.8)$$

$$\Psi(\eta)(\dot{\varsigma}\eta + \varsigma\dot{\eta}) + \upsilon(\eta, \dot{\eta})\varsigma = F(\eta)\tau - A^T\gamma \quad (3.9)$$

From Eq. (3.9), we can write:

$$\underline{\Psi}(\eta)\dot{\eta} + \underline{\upsilon}(\eta, \dot{\eta})\eta = \underline{F}(\eta)\tau \tag{3.10}$$

where,

$$\underline{\Psi}(\eta) = \left[I_w + \frac{R^2}{4L^2}\left(mL^2 + I\right)\frac{R^2}{4L^2}\left(mL^2 - I\right)\frac{R^2}{4L^2}\left(mL^2 - I\right)I_w + \frac{R^2}{4L^2}\left(mL^2 + I\right)\right] \tag{3.11}$$

$$\underline{\upsilon}(\eta) = \left[0\frac{R^2}{2L^2}\left(m_c\kappa\dot{\delta}\right) - \frac{R^2}{2L^2}\left(m_c\kappa\dot{\delta}\right)0\right] \tag{3.12}$$

$$\underline{F}(\eta) = [1\,0\,0\,1] \tag{3.13}$$

$$m = m_c + 2m_w I = I_c + m_c\kappa^2 + 2m_w L^2 + 2I_m \tag{3.14}$$

Considering Eq. (3.4) to gether with Eq. (3.10), we obtain:

$$\begin{aligned}&\left[0\frac{R^2}{2L^2}\left(m_c\kappa\dot{\delta}\right) - \frac{R^2}{2L^2}\left(m_c\kappa\dot{\delta}\right)0\right]\left(\dot{\vartheta}\dot{w}\right) + [0\; - \; m_c\, dw\, m_c\, dw\, 0](\vartheta\; w)\\&= \left[\frac{1}{R}\,0\,0\,\frac{1}{R}\right](u_1\; u_2)\end{aligned} \tag{3.15}$$

3.5.3 General View on the Proposed Method

Figure 3.2 provides a comprehensive depiction of the intended strategy. The problem lies in the fact that controllers u_1 and u_2 are able to minimize the tracking errors to nearly zero.

$$\chi_i(t) = \eta_i(t) - \eta_{d_i}(t), i = 1, 2 \tag{3.16}$$

Here, η_i represents the position of the robot. The dynamics Eq. (3.15) are calculated or determined as:

$$\dot{\hat{\eta}}_1 = \Xi_1(\eta|P_1) + \mu_1\dot{\hat{\eta}}_2 = \Xi_2(\eta|P_2) + \mu_2 \tag{3.17}$$

The recommended FLSs are denoted as $\Xi_1(\eta|P_1)$, and $\Xi_2(\eta|P_2)$, where P_1, and P_2 are the learnable parameters. The controller should adapt to unknown dynamics and input nonlinearities.

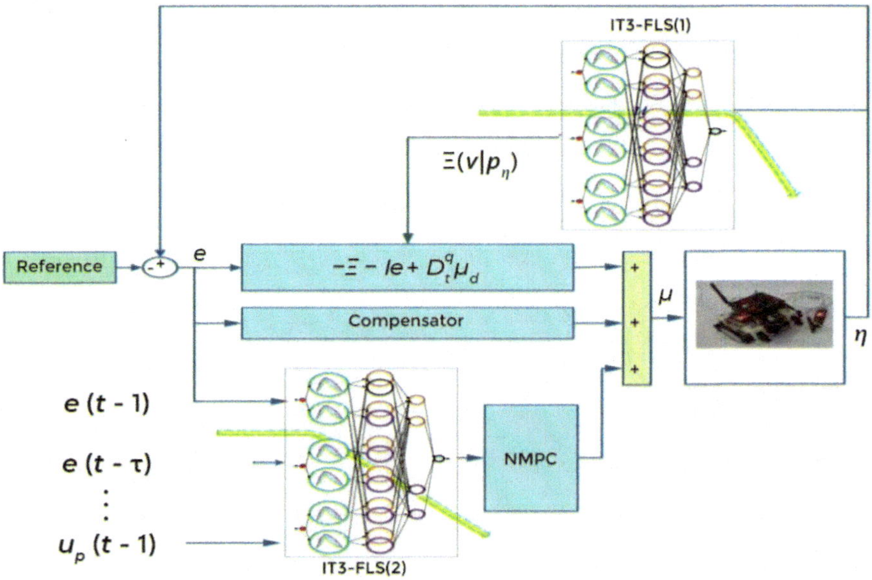

Fig. 3.2 Schematic view of the proposed control

3.5.4 Type-3 FLS

The behavior of the controlled system is disrupted by multiple causes. Next, a robust estimator based on T3-FLSs is proposed to address the perturbations. Figure 3.3 displays the primary framework.

(1) The inputs of FLS can be taken into account as $\zeta_1, \zeta_2, \frac{d}{dt}\zeta_1, \frac{d}{dt}\zeta_2$.
(2) The computation of the memberships of the fuzzy set (FS) should be performed in this phase. The memberships for $\tilde{\Lambda}_i^j$ ($j-th$ FS for input ζ_i) are [27]:

$$\underline{\xi}_{\tilde{\Lambda}^j_{i|\underline{\alpha}_i}} = \{1 - \left(\frac{\left|\zeta_i - c_{\tilde{\Lambda}_i^j}\right|}{\underline{\Upsilon}_{\tilde{\Lambda}_i^j}}\right)^{\underline{\alpha}_i} \; if\; c_{\tilde{\Lambda}_i^j} - \underline{\Upsilon}_{\tilde{\Lambda}_i^j} < \zeta_i \le c_{\tilde{\Lambda}_i^j} 1$$
$$- \left(\frac{\left|\zeta_i - c_{\tilde{\Lambda}_i^j}\right|}{\underline{\Upsilon}_{\tilde{\Lambda}_i^j}}\right)^{\underline{\alpha}_i} \; if\; c_{\tilde{\Lambda}_i^j} < \zeta_i \le c_{\tilde{\Lambda}_i^j} +$$
$$\underline{\Upsilon}_{\tilde{\Lambda}_i^j} 0 \quad if\, \zeta_i > c_{\tilde{\Lambda}_i^j} + \underline{\Upsilon}_{\tilde{\Lambda}_i^j}\, or\, \zeta_i \le c_{\tilde{\Lambda}_i^j} - \underline{\Upsilon}_{\tilde{\Lambda}_i^j} \tag{3.18}$$

$$\underline{\xi}_{\tilde{\Lambda}^j_{i|\underline{\alpha}_i}} = \{1 - \left(\frac{\left|\zeta_i - c_{\tilde{\Lambda}_i^j}\right|}{\underline{\Upsilon}_{\tilde{\Lambda}_i^j}}\right)^{\underline{\alpha}_i} \; if\; c_{\tilde{\Lambda}_i^j} - \underline{\Upsilon}_{\tilde{\Lambda}_i^j} < \zeta_i \le c_{\tilde{\Lambda}_i^j}$$

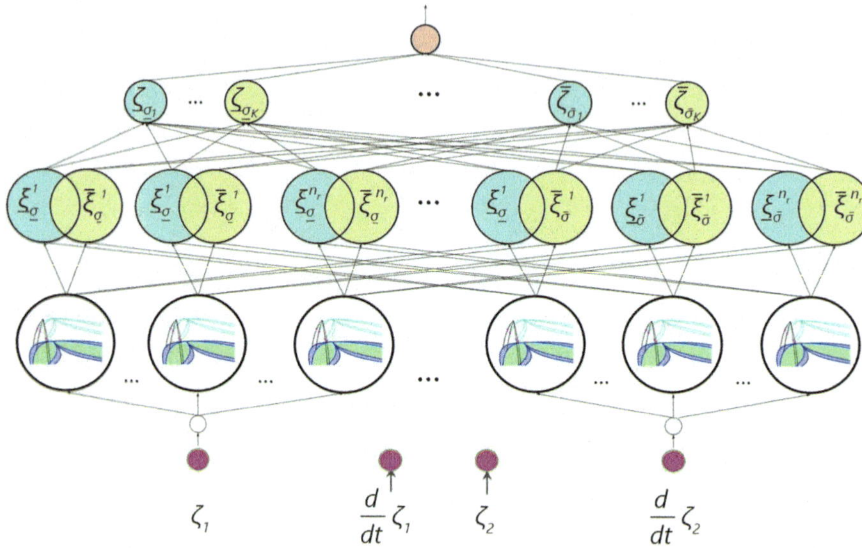

Fig. 3.3 The primary diagram for T3-FLS

$$1-\left(\frac{\left|\zeta_i-c_{\tilde{\Lambda}_i^j}\right|}{\underline{\Upsilon}_{\tilde{\Lambda}_i^j}}\right)^{\underline{\sigma}_i}$$

$$if\ c_{\tilde{\Lambda}_i^j}<\zeta_i\leq c_{\tilde{\Lambda}_i^j}+\underline{\Upsilon}_{\tilde{\Lambda}_i^j}0\quad if\ \zeta_i>c_{\tilde{\Lambda}_i^j}+\underline{\Upsilon}_{\tilde{\Lambda}_i^j}\ or\ \zeta_i\leq c_{\tilde{\Lambda}_i^j}-\underline{\Upsilon}_{\tilde{\Lambda}_i^j} \tag{3.19}$$

$$\begin{aligned}\underline{\xi}_{\tilde{\Lambda}^j_{i|\underline{\sigma}_i}}&=\{1-\left(\frac{\left|\zeta_i-c_{\tilde{\Lambda}_i^j}\right|}{\underline{\Upsilon}_{\tilde{\Lambda}_i^j}}\right)^{\frac{1}{\underline{\sigma}_i}}\ if\ c_{\tilde{\Lambda}_i^j}-\underline{\Upsilon}_{\tilde{\Lambda}_i^j}<\zeta_i\leq c_{\tilde{\Lambda}_i^j}1\\&-\left(\frac{\left|\zeta_i-c_{\tilde{\Lambda}_i^j}\right|}{\underline{\Upsilon}_{\tilde{\Lambda}_i^j}}\right)^{\frac{1}{\underline{\sigma}_i}}\ if\ c_{\tilde{\Lambda}_i^j}<\zeta_i\leq c_{\tilde{\Lambda}_i^j}+\\&\underline{\Upsilon}_{\tilde{\Lambda}_i^j}0\quad if\ \zeta_i>c_{\tilde{\Lambda}_i^j}+\underline{\Upsilon}_{\tilde{\Lambda}_i^j}\ or\ \zeta_i\leq c_{\tilde{\Lambda}_i^j}-\underline{\Upsilon}_{\tilde{\Lambda}_i^j}\end{aligned} \tag{3.20}$$

$$\begin{aligned}&\underline{\xi}_{\tilde{\Lambda}^j_{i|\underline{\sigma}_i}}=\{1-\left(\frac{\left|\zeta_i-c_{\tilde{\Lambda}_i^j}\right|}{\underline{\Upsilon}_{\tilde{\Lambda}_i^j}}\right)^{\frac{1}{\underline{\sigma}_i}}\ if\ c_{\tilde{\Lambda}_i^j}-\underline{\Upsilon}_{\tilde{\Lambda}_i^j}<\zeta_i\leq c_{\tilde{\Lambda}_i^j}1\\&-\left(\frac{\left|\zeta_i-c_{\tilde{\Lambda}_i^j}\right|}{\underline{\Upsilon}_{\tilde{\Lambda}_i^j}}\right)^{\frac{1}{\underline{\sigma}_i}}\ if\ c_{\tilde{\Lambda}_i^j}<\zeta_i\leq c_{\tilde{\Lambda}_i^j}+\\&\underline{\Upsilon}_{\tilde{\Lambda}_i^j}0\quad if\ \zeta_i>c_{\tilde{\Lambda}_i^j}+\underline{\Upsilon}_{\tilde{\Lambda}_i^j}\ or\ \zeta_i\leq c_{\tilde{\Lambda}_i^j}-\underline{\Upsilon}_{\tilde{\Lambda}_i^j}\end{aligned} \tag{3.21}$$

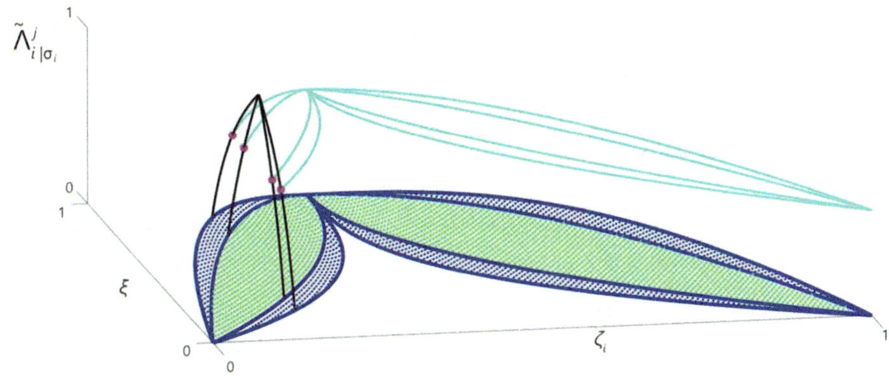

Fig. 3.4 Type-3 FS

where, the upper and lower memberships for $\tilde{\Lambda}^j_i$ are represented by $\underline{\xi}_{\tilde{\Lambda}^j_{i|\underline{\sigma}_i}}/\underline{\xi}_{\tilde{\Lambda}^j_{i|\underline{\sigma}_i}}$ and $\underline{\xi}_{\tilde{\Lambda}^j_{i|\underline{\sigma}_i}}/\underline{\xi}_{\tilde{\Lambda}^j_{i|\underline{\sigma}_i}}$, respectively. The symbole $c_{\tilde{\Lambda}^j_i}$, shows the center of $\tilde{\Lambda}^j_i$. $\underline{\Upsilon}_{\tilde{\Lambda}^j_i}$, and $\underline{\Upsilon}_{\tilde{\Lambda}^j_i}$ represents the distances from $c_{\tilde{\Lambda}^j_i}$ to the left or right points of $\tilde{\Lambda}^j_i$ (see Fig. 3.4).

(3) The rule firing degrees are:

$$\underline{\zeta}^l_{\underline{\sigma}_i} = \underline{\xi}_{\tilde{\Lambda}^{p_1}_{1|\underline{\sigma}_i}} \cdot \underline{\xi}_{\tilde{\Lambda}^{p_2}_{1|\underline{\sigma}_i}} \cdots \underline{\xi}_{\tilde{\Lambda}^{p_n}_{1|\underline{\sigma}_i}} \tag{3.22}$$

$$\underline{\zeta}^l_{\underline{\sigma}_i} = \underline{\xi}_{\tilde{\Lambda}^{p_1}_{1|\underline{\sigma}_i}} \cdot \underline{\xi}_{\tilde{\Lambda}^{p_2}_{1|\underline{\sigma}_i}} \cdots \underline{\xi}_{\tilde{\Lambda}^{p_n}_{1|\underline{\sigma}_i}} \tag{3.23}$$

$$\underline{\zeta}^l_{\underline{\sigma}_i} = \underline{\xi}_{\tilde{\Lambda}^{p_1}_{1|\underline{\sigma}_i}} \cdot \underline{\xi}_{\tilde{\Lambda}^{p_2}_{1|\underline{\sigma}_i}} \cdots \underline{\xi}_{\tilde{\Lambda}^{p_n}_{1|\underline{\sigma}_i}} \tag{3.24}$$

$$\underline{\zeta}^l_{\underline{\sigma}_i} = \underline{\xi}_{\tilde{\Lambda}^{p_1}_{1|\underline{\sigma}_i}} \cdot \underline{\xi}_{\tilde{\Lambda}^{p_2}_{1|\underline{\sigma}_i}} \cdots \underline{\xi}_{\tilde{\Lambda}^{p_n}_{1|\underline{\sigma}_i}} \tag{3.25}$$

(4) The FLS output is considered as:

$$h = \frac{\sum_{i=1}^{n_\sigma}\left(\underline{\sigma}_i \underline{h}_i + \underline{\sigma}_i \underline{h}_i\right)}{\sum_{i=1}^{n_\sigma}\left(\underline{\sigma}_i + \underline{\sigma}_i\right)} \tag{3.26}$$

where,

$$\underline{h}_i = \frac{\sum_{l=1}^{n_r}\left(\underline{\zeta}^l_{\underline{\sigma}_i} \underline{\upsilon}_l + \underline{\zeta}^l_{\underline{\sigma}_i} \underline{\upsilon}_l\right)}{\sum_{l=1}^{n_r}\left(\underline{\zeta}^l_{\underline{\sigma}_i} + \underline{\zeta}^l_{\underline{\sigma}_i}\right)} \tag{3.27}$$

$$\underline{h}_i = \frac{\sum_{l=1}^{n_r} \left(\zeta^l_{\underline{\sigma}_i} \underline{\upsilon}_l + \zeta^l_{\underline{\sigma}_i} \underline{\upsilon}_l \right)}{\sum_{l=1}^{n_r} \left(\zeta^l_{\underline{\sigma}_i} + \zeta^l_{\underline{\sigma}_i} \right)} \tag{3.28}$$

The output is restated as:

$$\hat{O}(\zeta | P) = P^T C \tag{3.29}$$

where,

$$C^T = \left[\underline{C}_1, \ldots, \underline{C}_{n_r}, \underline{C}_1, \ldots, \underline{C}_{n_r} \right] \tag{3.30}$$

$$P^T = \left[\underline{\upsilon}_1, \ldots, \underline{\upsilon}_{n_r}, \underline{\upsilon}_1, \ldots, \underline{\upsilon}_{n_r} \right] \tag{3.31}$$

$$\underline{C}_l = \frac{\sum_{i=1}^{n_\sigma} \underline{\sigma}_i \zeta^l_{\underline{\sigma}_i}}{\sum_{i=1}^{n_\sigma} (\underline{\sigma}_i + \underline{\sigma}_i) \sum_{l=1}^{n_r} \left(\zeta^l_{\underline{\sigma}_i} + \zeta^l_{\underline{\sigma}_i} \right)} + \frac{\sum_{i=1}^{n_\sigma} \underline{\sigma}_i \zeta^l_{\underline{\sigma}_i}}{\sum_{i=1}^{n_\sigma} (\underline{\sigma}_i + \underline{\sigma}_i) \sum_{l=1}^{n_r} \left(\zeta^l_{\underline{\sigma}_i} + \zeta^l_{\underline{\sigma}_i} \right)} \tag{3.32}$$

$$\underline{C}_l = \frac{\sum_{i=1}^{n_\sigma} \underline{\sigma}_i \zeta^l_{\underline{\sigma}_i}}{\sum_{i=1}^{n_\sigma} (\underline{\sigma}_i + \underline{\sigma}_i) \sum_{l=1}^{n_r} \left(\zeta^l_{\underline{\sigma}_i} + \zeta^l_{\underline{\sigma}_i} \right)} + \frac{\sum_{i=1}^{n_\sigma} \underline{\sigma}_i \zeta^l_{\underline{\sigma}_i}}{\sum_{i=1}^{n_\sigma} (\underline{\sigma}_i + \underline{\sigma}_i) \sum_{l=1}^{n_r} \left(\zeta^l_{\underline{\sigma}_i} + \zeta^l_{\underline{\sigma}_i} \right)} \tag{3.33}$$

3.5.5 *Predictive Controller*

The predictive control scheme is defined as:

$$\min_{\mu_{pz}(h), \ldots, \mu_{pz}(h+n_C)} J = \sum_{h=i}^{i+n_P} \iota \cdot \hat{e}_z^2(h) + q \cdot \Delta \mu_{p_z}^2(h) \tag{3.34}$$

where,

$$\hat{e}_z(i) = y_{T3-FLS(2)} \left(\underline{\hat{e}}_z | P_z \right) \tag{3.35}$$

where, ι and q are constants, moreover

$$\underline{\hat{e}}_z = \left[\hat{e}_z(i-1), \ldots, \hat{e}_z(i-\tau), \mu_{p_z}(i-1) \right]^T \tag{3.36}$$

P_z is tuned as:

$$P_z(i) = P_z(i-1) + \kappa\left(\underline{e}_z - \hat{\underline{e}}_z\right)C_z \tag{3.37}$$

Here, $\kappa \in [0, 1]$, and C_z is calculated to be:

$$C_z^T = \frac{0.5}{\sum_{j=1}^{K}\underline{\sigma}_j}\begin{bmatrix} \frac{\underline{\sigma}_1\left(\underline{C}^1_{\underline{\sigma}_1} + \underline{C}^1_{\underline{\sigma}_1}\right)}{\sum_{l=1}^{R}\left(\underline{C}^1_{\underline{\sigma}_j} + \underline{C}^1_{\underline{\sigma}_j}\right)}, \ldots, \frac{\underline{\sigma}_{K_z}\left(\underline{C}^1_{\underline{\sigma}_{K_z}} + \underline{C}^1_{\underline{\sigma}_{K_z}}\right)}{\sum_{l=1}^{R}\left(\underline{C}^1_{\underline{\sigma}_j} + \underline{C}^1_{\underline{\sigma}_j}\right)}, \\ \frac{\underline{\sigma}_1\left(\underline{C}^{R_z}_{\underline{\sigma}_1} + \underline{C}^{R_z}_{\underline{\sigma}_1}\right)}{\sum_{l=1}^{R}\left(\underline{C}^1_{\underline{\sigma}_j} + \underline{C}^1_{\underline{\sigma}_j}\right)}, \ldots, \frac{\underline{\sigma}_{K}\left(\underline{C}^{R_z}_{\underline{\sigma}_{K_z}} + \underline{C}^{R_z}_{\underline{\sigma}_{K_z}}\right)}{\sum_{l=1}^{R}\left(\underline{C}^1_{\underline{\sigma}_j} + \underline{C}^1_{\underline{\sigma}_j}\right)}, \\ \underline{\sigma}_1\frac{\left(\underline{C}^1_{\underline{\sigma}_1} + \underline{C}^1_{\underline{\sigma}_1}\right)}{\sum_{l=1}^{R}\left(\underline{C}^l_{\underline{\sigma}_j} + \underline{C}^l_{\underline{\sigma}_j}\right)}, \ldots, \underline{\sigma}_K\frac{\left(\underline{C}^1_{\underline{\sigma}_{K_z}} + \underline{C}^1_{\underline{\sigma}_{K_z}}\right)}{\sum_{l=1}^{R}\left(\underline{C}^l_{\underline{\sigma}_j} + \underline{C}^l_{\underline{\sigma}_j}\right)}, \\ \underline{\sigma}_1\frac{\left(\underline{C}^{R_z}_{\underline{\sigma}_1} + \underline{C}^{R_z}_{\underline{\sigma}_1}\right)}{\sum_{l=1}^{R}\left(\underline{C}^l_{\underline{\sigma}_j} + \underline{C}^l_{\underline{\sigma}_j}\right)}, \ldots, \underline{\sigma}_K\frac{\left(\underline{C}^{R_z}_{\underline{\sigma}_{K_z}} + \underline{C}^{R_z}_{\underline{\sigma}_{K_z}}\right)}{\sum_{l=1}^{R}\left(\underline{C}^l_{\underline{\sigma}_j} + \underline{C}^l_{\underline{\sigma}_j}\right)} \end{bmatrix} \tag{3.38}$$

The IT3-FLS (2) output is stated as:

$$\hat{e}_z(i+h) = \hat{e}_{z,forced}(i+h) + \hat{e}_{z,free}(i+h) \tag{3.39}$$

where,

$$\hat{e}_{z,free}(i+h) = y_{IT3-FNN(2)}\left(\hat{\underline{e}}_z(i+h)|P_z\right) \tag{3.40}$$

$$\hat{e}_{z,forced}(i+h) = \sum_{\iota=0}^{h-1} l_\iota \Delta\mu_{p_z}(i+h-\iota+1) \tag{3.41}$$

$$s(i-1) = \left[\hat{e}_{z,step}(i+h) - \hat{e}_{z,free}(i+h)\right]/d\mu_{p_z}(i) \tag{3.42}$$

$$\hat{e}_{z,step}(i+h) = y_{IT3-FNN(2)}\left(\hat{\underline{e}}_z(i+h)|P_z\right) \tag{3.43}$$

$$\hat{\underline{e}}_z(i+h) = \left[\hat{e}_z(i+h-1)\ldots, \hat{e}_z(i+h-\tau), \mu_{p_z}(i+h-1)\right]^T \tag{3.44}$$

$$\mu_{p_z}(i) = \{\mu_{p_z}(i-1) + d\mu_{p_z}(i)\ if\ h > i-1, \mu_{p_z}(i-1) \quad else \tag{3.45}$$

The magnitude of the step is represented by the symbol $d\mu_{p_z}(i)$. The equations mentioned earlier can be restated as:

$$Q = S\Delta U_{p_z} + Q_{free} \tag{3.46}$$

where,

$$Q = \left[\hat{e}_z(i+1)\hat{e}_z(i+2)\vdots\hat{e}_z\left(i+n_p\right)\right] \tag{3.47}$$

$$S = \left[s_0 0 s_1 s_0 \vdots s_{n_p} - 1 \vdots s_{n_p} - 1 \cdots 0 \cdots 0 \ddots \cdots \vdots s_0\right] \tag{3.48}$$

$$\Delta U_{p_z} = \left[\delta\mu_{p_z}(i)\delta\mu_{p_z}(i+1)\vdots\delta\mu_{p_z}\left(i+n_p-1\right)\right] \tag{3.49}$$

$$Q_{free} = \left[\hat{e}_{z,free}(i+1)\hat{e}_{z,free}(i+2)\vdots\hat{e}_{z,free}\left(i+n_p\right)\right] \tag{3.50}$$

$$\hat{e}_{z,free}(i+h) = y_{IT3-FNN(2)}\left(\underline{\hat{e}}_z(i+h)|P_z\right) \tag{3.51}$$

$$\underline{\hat{e}}_z(i+h) = \left[\hat{e}_z(i+h-1)\ldots,\hat{e}_z(i+h-\tau),\mu_{p_z}(i+h-1)\right]^T \tag{3.52}$$

$$\mu_{p_z}(i) = \{\mu_{p_z}(i-1)\ if\ h > i-1,\ \mu_{p_z}(i)\quad else \tag{3.53}$$

From Eqs. (3.34) and (3.46), we get:

$$J = \iota Q^T Q + q\Delta U_{p_z}^T \Delta U_{p_z} = \iota\left(S\Delta U_{p_z} + Q_{free}\right)^T\left(S\Delta U_{p_z} + Q_{free}\right) + q\Delta U_{p_z}^T \Delta U_{p_z} \tag{3.54}$$

From Eq. (3.54), $\partial J/\partial\Delta U_{p_z}$ is calculated as:

$$\partial J/\partial\Delta U_{p_z} = 0 \Rightarrow \iota S^T Q_{free} + \Delta U_p\left(\iota S^T S + q\right) = 0 \tag{3.55}$$

Using Eq. (3.55), ΔU_{p_z} is formulated as:

$$\Delta U_{p_z} = \iota\left(\iota S^T S + q\right)^{-1} S^T Q_{free} \tag{3.56}$$

The control signal of $\mu_{p_z}(i)$ is arranged as:

$$\mu_{p_z}(i) = \mu_{p_z}(i-1) + \Delta\mu_{p_z}(i) \tag{3.57}$$

In Eq. (3.56), $\Delta\mu_{p_z}(i)$ represents the first term in vector ΔU_{p_z}.

3.5.6 Stability Analysis

The primary findings about stability are presented in Theorem 1.

Theorem 1 The stability of the system is guaranteed by the implementation of controllers Eqs. (3.58, 3.59), compensators Eqs. (3.60, 3.61), and adaptation rules Eqs. (3.62, 3.63).

$$\mu_1 = -\Xi_1\left(\nu|P_{\eta_1}\right) - \iota_1 e_1 + D_t^\beta \eta_{d_1} + \mu_{c_1} + \mu_{p_1} \tag{3.58}$$

$$\mu_2 = -\Xi_2\left(\nu|P_{\eta_2}\right) - \iota_1 e_2 + D_t^\beta \eta_{d_2} + \mu_{c_2} + \mu_{p_2} \tag{3.59}$$

$$\mu_{c_1} = -tanh(e_1)\left(\underline{w}_1 + \underline{\mu}_{p_1}\right) \tag{3.60}$$

$$\mu_{c_2} = -tanh(e_2)\left(\underline{w}_2 + \underline{\mu}_{p_2}\right) \tag{3.61}$$

$$D_t^\beta P_{\eta_1} = \gamma e_1 C_{\eta_1} \tag{3.62}$$

$$D_t^\beta P_{\eta_2} = \gamma e_2 C_{\eta_2} \tag{3.63}$$

where $0 < \gamma \leq 1$, and $\underline{\mu}_{p_1}$ and $\underline{\mu}_{p_2}$ are, respectively, the upper bound of μ_{p_1} and μ_{p_2}.

Proof Given IT3-FLSs(1) with optimal parameters, the following can be concluded:

$$D_t^\beta \eta_1 = \Xi_1^*\left(\nu|P_{\eta_1}^*\right) + \mu_1 + w_1 \tag{3.64}$$

$$D_t^\beta \eta_2 = \Xi_2^*\left(\nu|P_{\eta_2}^*\right) + \mu_2 + w_2 \tag{3.65}$$

Estimation errors are considered to be w_1 and w_2. From control signals Eqs. (3.58, 3.59), we have:

$$D_t^\beta e_1 = \Xi_1^*\left(\nu|P_{\eta_1}^*\right) - \Xi_1\left(\nu|P_{\eta_1}\right) - \iota_1 e_1 + w_1 + \mu_{c_1} + \mu_{p_1}, \tag{3.66}$$

$$D_t^\beta e_2 = \Xi_2^*\left(\nu|P_{\eta_2}^*\right) - \Xi_2\left(\nu|P_{\eta_2}\right) - \iota_2 e_2 + w_2 + \mu_{c_2} + \mu_{p_2}. \tag{3.67}$$

From Eqs. (3.66, 3.67), one has:

$$\Xi_1^*\left(\nu|P_{\eta_1}^*\right) - \Xi_1\left(\nu|P_{\eta_1}\right) = \tilde{P}_{\eta_1}^T C_{\eta_1} \tag{3.68}$$

$$\Xi_2^*\left(\nu|P_{\eta_2}^*\right) - \Xi_2\left(\nu|P_{\eta_2}\right) = \tilde{P}_{\eta_2}^T C_{\eta_2} \tag{3.69}$$

From Eq. (3.68, 3.69), Eqs. (3.66, 3.67) are reformulated as:

$$D_t^\beta e_1 = \tilde{P}_{\eta_1}^T C_{\eta_1} - \iota_1 e_1 w_1 + \mu_{c_1} + \mu_{p_1} \tag{3.70}$$

$$D_t^{\beta} e_2 = \tilde{P}_{\eta_2}^T C_{\eta_2} - \iota_2 e_2 w_2 + \mu_{c_2} + \mu_{p_2} \tag{3.71}$$

Considering Lyapunov function, we can write:

$$V = \frac{1}{2} e_1^2 + \frac{1}{2} e_2^2 + \frac{1}{2\gamma} \tilde{P}_{\eta_1}^T \tilde{P}_{\eta_1} + \frac{1}{2\gamma} \tilde{P}_{\eta_2}^T \tilde{P}_{\eta_2} \tag{3.72}$$

Then $D_t^q V$ is:

$$D_t^q V \leq e_1 D_t^q e_1 + e_2 D_t^q e_2 - \frac{1}{\gamma} \tilde{P}_{\eta_1}^T D_t^{\beta} P_{\eta_1} - \frac{1}{\gamma} \tilde{P}_{\eta_2}^T D_t^{\beta} P_{\eta_2} \tag{3.73}$$

Substituting Eqs. (3.71, 3.72) into Eq. (3.73), we get:

$$\begin{aligned} D_t^q V &\leq -\iota_1 e_1^2 - \iota_2 e_2^2 + w_1 e_1 + e_1 \mu_{c_1} + e_1 \mu_{p_1} + w_2 e_2 + e_2 \mu_{c_2} + e_1 \mu_{p_2} \\ &+ \tilde{P}_{\eta_1}^T C_{\eta_1} e_1 - \frac{1}{\gamma} \tilde{P}_{\eta_1}^T D_t^{\beta} P_{\eta_1} + \tilde{P}_{\eta_2}^T C_{\eta_2} e_2 - \frac{1}{\gamma} \tilde{P}_{\eta_2}^T D_t^{\beta} P_{\eta_2} \end{aligned} \tag{3.74}$$

From Eqs. (3.72, 3.73), the inequality Eq. (3.74) changes to:

$$D_t^q V \leq -\iota_1 e_1^2 - \iota_2 e_2^2 + w_1 e_1 + e_1 \mu_{c_1} + e_1 \mu_{p_1} + w_2 e_2 + e_2 \mu_{c_2} + e_1 \mu_{p_2} \tag{3.75}$$

By taking account of the upper bounds of w_1, w_2, μ_{p_1} and μ_{p_2} as $\underline{w}_1$, $\underline{w}_2$, $\underline{\mu}_{p_1}$ and $\underline{\mu}_{p_2}$, we have:

$$D_t^q V \leq -\iota_1 e_1^2 - \iota_2 e_2^2 + \underline{w}_1 |e_1| + e_1 \mu_{c_1} + |e_1| \underline{\mu}_{p_1} + \underline{w}_2 |e_2| + e_2 \mu_{c_2} + |e_2| \underline{\mu}_{p_2} \tag{3.76}$$

Inserting the compensators from Eqs. (3.60, 3.61) culminates in:

$$D_t^q V \leq -\iota_1 e_1^2 - \iota_2 e_2^2 \tag{3.77}$$

From Eq. (3.77), it can be deduced that $e_1 \in l^2$ and $e_2 \in l^2$. By applying Barbalat's lemma, it can be guaranteed that the system is stable in the asymptotic sense.

Note 1 This paper proposes the use of a T3-FLS for the purpose of estimating uncertainty. T3-FLSs exhibit superior accuracy in noisy practical situations when compared to type-2 FLSs. Furthermore, the complexity of T3-FLSs is just slightly higher than that of the type-2 counterpart, which does not pose a significant issue.

Note 2 In the proposed approach, the dynamics of the robot are assumed to be unknown and are calculated using the recommended T3-FLS. Subsequently, the developed controller might be utilized for other systems that possess identical dimensionality.

3.5.7 Simulations

The dynamics of the reference plant has been taken into account as [28]:

$$
\begin{aligned}
\{D_t^{0.98}\chi_{11} &= 35(\chi_{12}-\chi_{11}) + 35\chi_{12}\chi_{13}D_t^{0.98}\chi_{12} \\
&= 25\chi_{11} - 5\chi_{11}\chi_{13} + \chi_{12} + \chi_{14}D_t^{0.98}\chi_{13} \\
&= \chi_{11}\chi_{12} - 4\chi_{13}D_t^{0.98}\chi_{14} = -100\chi_{12}
\end{aligned}
\tag{3.78}
$$

Figure 3.5 illustrates the four switching modes. Figure 3.6 illustrates the output signals ζ_1 and ζ_2, as well as the reference signals and demonstrates a high level of tacking efficiency. Figure 3.7 displays the control signals that correlate to the given information. The estimated signals and path portraits in all modes are presented in Figs. 3.8 and 3.9, demonstrating accurate tracking. The intentionally devised chaotic trajectory, characterized by abrupt changes in criteria, yields a very robust route for patrol robots.

For the purpose of comparing different types of FLSs, the dynamics estimation of Eq. 15 is taken into account without the presence of a controller [28]. Table 3.1

$$
\left\{\dot{\hat{\zeta}}_1 = \hat{H}_1(\zeta|\theta_1)\dot{\hat{\zeta}}_2 = \hat{H}_2(\zeta|\theta_2)\right. \tag{3.79}
$$

where, $c_{\tilde{A}^l_{i|\alpha}}$, $l \in [-70, 70]$, $\underline{\alpha} \in \{0.8, 1\}$, $\underline{\alpha} \in \{0.9, 1\}$ and $\underline{\alpha} \in \{0.5, 0.6\}$. In addition, a noise is introduced in the following manner: the variance is set at 0.04 and 0.15, while the mean is set at zero. The calculated root mean square errors (RMSEs) are provided in Table 3.2. The RMSEs presented in Table 3.18 demonstrate the superior efficiency of the proposed approach.

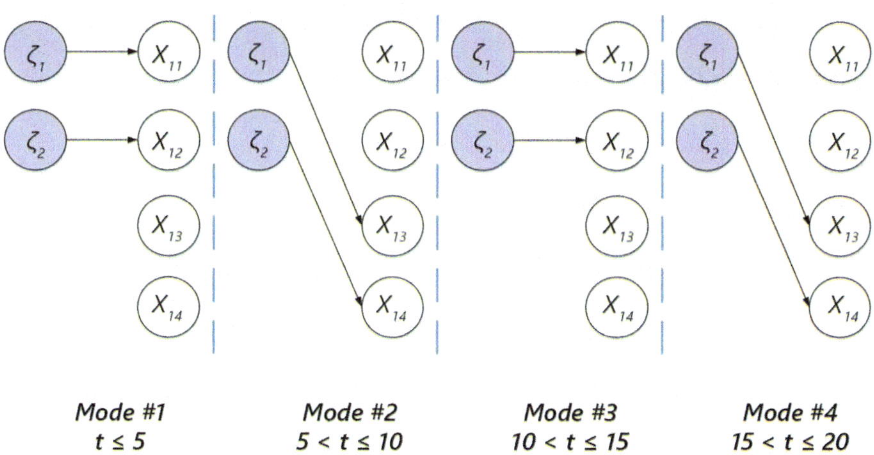

Fig. 3.5 The multi-switching modes

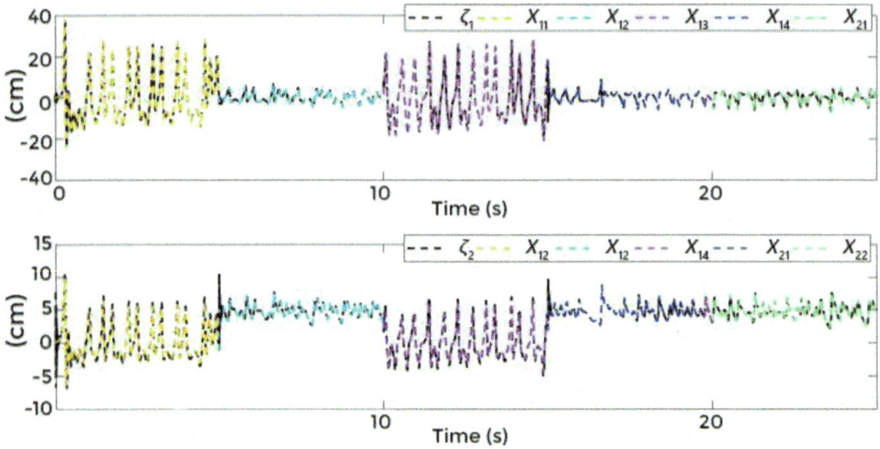

Fig. 3.6 Outputs and references

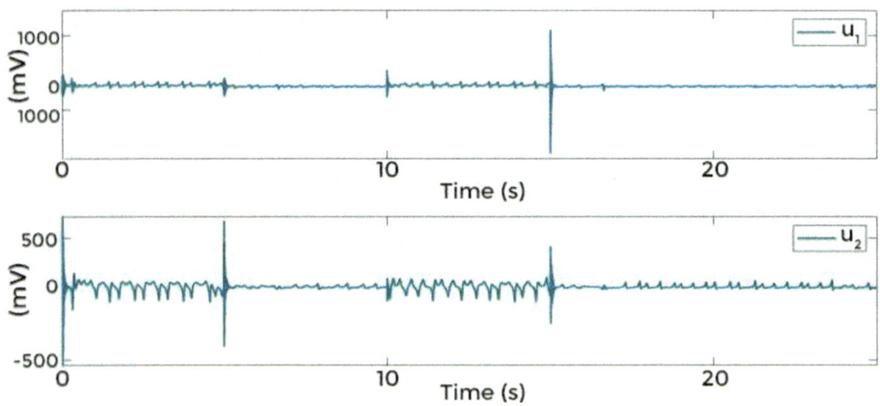

Fig. 3.7 Control signals

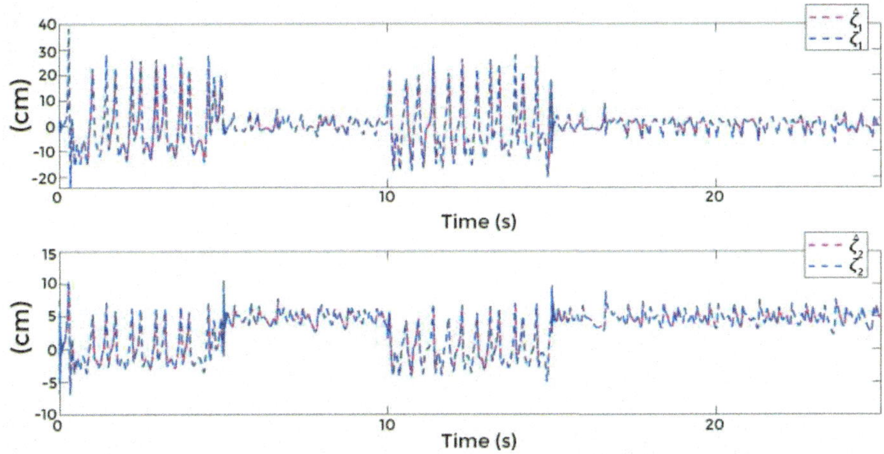

Fig. 3.8 Target and estimated signals

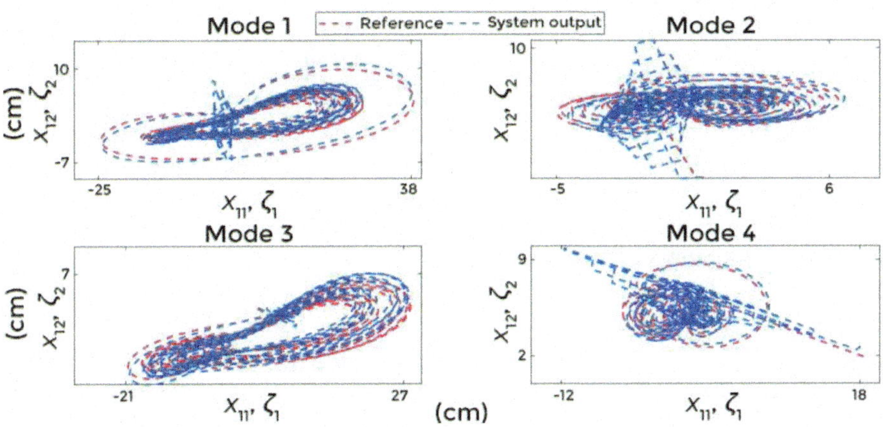

Fig. 3.9 The phase portrait

Table 3.1 Simulation conditions

R	m	0.10	–
L	m	0.50	–
w	rad/s	–1	1
d	m	0.01	0.30
I_c	$Kg.m^2$	6	12
m_c	kg	87	181
m_w	kg	1.50	–
I_m	$Kg.m^2$	0.003	–
u	m/s	0	1
I_w	$Kg.m^2$	0.008	–

Table 3.2 Comparison of RMSE

Variance	FLS	$\hat{\zeta}_1 - \zeta_1$	$\hat{\zeta}_2 - \zeta_2$
*0	IT2-FLS	4.2414	5.1804
	T3-FLS	1.8174	2.1140
*0.05	IT2-FLS	2.2527	3.2128
	T3-FLS	1.7624	1.2441
*0.1	IT2-FLS	5.3274	6.2230
	T3-FLS	1.79984	1.7779

3.6 Type-3 Fuzzy Control with Boltzmann-Based Learning for Robotics

3.6.1 Problem Description

3.6.1.1 A General View

Figure 3.1 provides an overall representation of the proposed controller. The dynamics of magnetic resonance (MR) are modeled using Nonlinear Time-Three Frequency Spectra (NT3FSs), and a principal controller is developed. Subsequently, the Boltzmann machine is utilized to develop a nonlinear model predictive control (MPC) system in order to enhance the precision. Ultimately, with the creation of a compensator and the modification of the regulations governing NT3FSs to prioritize stability, the stability of the closed-loop system is guaranteed. The NMPC is optimized using a Boltzmann machine, while the NT3FSs are trained by a Lyapunov technique.

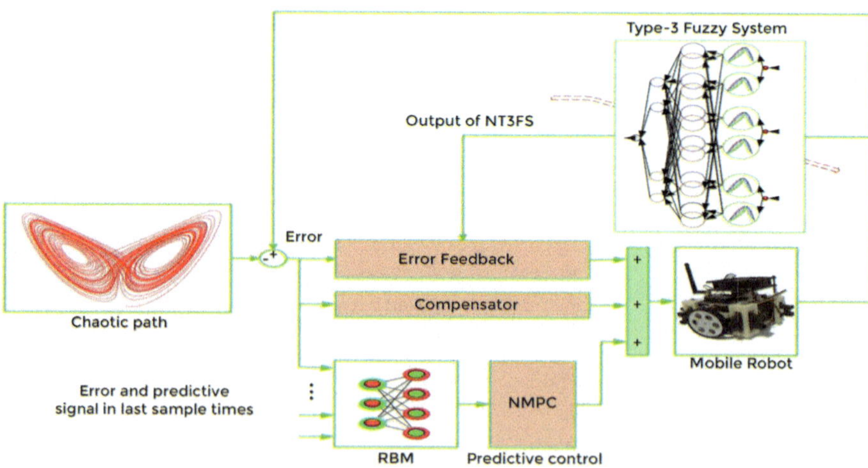

Fig. 3.1 Schematic diagram of control

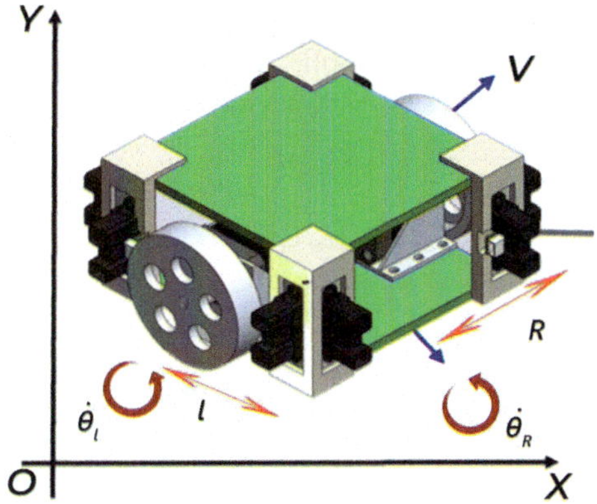

Fig. 3.2 Schematic diagram of wheeled MR

3.6.1.2 Kinematic Model

This paper focuses on analyzing a case study including a mobile robot model with two wheels. Figure 3.2 illustrates the overall schematic of a 2-wheeled model. The equations for velocity are expressed as [29]:

$$v = R(\dot{\theta}_R + \dot{\theta}_\iota)/2$$

$$\omega = R(\dot{\theta}_R - \dot{\theta}_\iota)/2\iota \tag{3.1}$$

In this equation, v and ω represent the linear velocities, R represents the radius, and ι represents the distance between the wheels. The equations governing rotation are as follows [30]:

$$[\dot{x}\,\dot{y}\,\dot{P}] = \left[\frac{R}{2}\,\frac{R}{2}\,0\,0\,\frac{R}{2\iota} - \frac{R}{2\iota}\right][\dot{\theta}_R\,\dot{\theta}_\iota] \tag{3.2}$$

where, P represents the rotation angle. Considering the angular velocity, we have:

$$[\dot{x}\,\dot{y}\,\dot{P}] = \left[\frac{R}{2}cos(P)\frac{R}{2}cos(P)\frac{R}{2}sin(P)\frac{R}{2}sin(P)\frac{R}{2\iota} - \frac{R}{2\iota}\right][\dot{\theta}_R\,\dot{\theta}_\iota] \tag{3.3}$$

3.6.1.3 Dynamics

The dynamics are formulated by applying the Lagrange method, as stated in the following [31]:

$$\theta(q)\ddot{q} + \eta(q,\dot{q})\dot{q} + \Psi(q) + \tau_\vartheta = W(q)\tau - Q^T\gamma \tag{3.4}$$

where, $\theta(q)$, $W(q)$, τ_{ϑ}, Ψ, and η, are, respectively, inertia, input, disturbance, gravitational and Coriolis matrices. τ and q show the input vector and the position. Substituting $\Psi = 0$, we have:

$$\theta(q) = [m000 - mdsin(P)00mdcos(P)00 - mdsin(P)00mdcos(P) \\ I0000I_w0000I_w] \tag{3.5}$$

$$\eta(q, \dot{q}) = [0mdcos(P)0 - mdsin(P)000\,000\,0000000\,0000\,0000] \tag{3.6}$$

$$Q^T\gamma = [-sin(P)cos(P)cos(P)sin(P) - \vartheta00\iota - R0cos(P)sin(P) - \iota0 - R] \tag{3.7}$$

The values of all parameters are shown in Table 3.1. By taking into account the transformation matrix ς as the null space of Q so that $\dot{\chi} = \varsigma\chi$, $\ddot{\chi} = \dot{\varsigma}\chi + \varsigma\dot{\chi}$, $\chi = \left[\dot{\theta}_R\,\dot{\theta}_\iota\right]^T$, $\varsigma^T Q^T = 0$ (ς is the null space of Q), we have:

$$\left[\dot{x}\,\dot{y}\,\dot{\varphi}\,\dot{\theta}_R\,\dot{\theta}_\iota\right] = \left[\frac{R}{2}cos(P)\frac{R}{2}cos(P)\frac{R}{2}sin(P)\frac{R}{2}sin(P)\frac{R}{2\iota}10 - \frac{R}{2\iota}01\right]\left[\dot{\theta}_R\,\dot{\theta}_\iota\right] \tag{3.8}$$

$$\theta(q)(\dot{\varsigma}\chi + \varsigma\dot{\chi}) + \eta(q, \dot{q})\varsigma = W(q)\tau - Q^T\gamma \tag{3.9}$$

Using Eq. (3.9), by multiplying both side in ς^T, considering $\underline{\theta}(q) = \varsigma^T\theta(q)\varsigma$, $\underline{\eta}(q, \dot{q}) = \varsigma^T\theta(q)\dot{\varsigma} + \varsigma^T\eta(q, \dot{q})\varsigma$, and $\underline{W}(q) = \varsigma^T W(q)$, we can write:

$$\underline{\theta}(q)\dot{\chi} + \underline{\eta}(q, \dot{q})\chi = \underline{W}(q)\tau \tag{3.10}$$

where,

$$\underline{\theta}(q) = \left[I_\omega + \frac{R^2}{4\iota^2}\left(m\iota^2 + I\right)\frac{R^2}{4\iota^2}\left(m\iota^2 - I\right)\frac{R^2}{4\iota^2}\left(m\iota^2 - I\right)I_\omega + \frac{R^2}{4\iota^2}\left(m\iota^2 + I\right)\right] \tag{3.11}$$

$$\underline{\eta}(q) = \left[0\frac{R^2}{2\iota^2}\left(m_c\vartheta\dot{P}\right) - \frac{R^2}{2\iota^2}\left(m_c\vartheta\dot{P}\right)0\right] \tag{3.12}$$

Table 3.1 Parameters value

ω	rad/s	2
d	m	0.31
ι	m	0.40
m_ω	kg	1.30
m_c	kg	88
I	$Kg.m^2$	0.004
R	m	0.12
I_ω	$Kg.m^2$	0.0098

$$\underline{W}(q) = [1\,0\,0\,1] \tag{3.13}$$

$$m = m_c + 2m_\omega I = I_c + m_c\vartheta^2 + 2m_\omega\iota^2 + 2I_m \tag{3.14}$$

From Eqs. (3.4) and (3.10), we have:

$$\left[0\frac{R^2}{2\iota^2}\left(m_c\vartheta\,\dot{P}\right)\frac{R^2}{-}2\iota^2\left(m_c\,\vartheta\dot{P}\,\right)0\right](\dot{v}\,\dot{\omega}) + [0\, -\, m_c\,dw\,m_c\,dw\,0](v\,\omega)$$
$$= \left[\frac{1}{R}\,0\,0\,\frac{1}{R}\right](u_1\,u_2) \tag{3.15}$$

The developed theorems are based on the following assumptions:

Assumption 1 *The mechanisms of magnetic resonance (MR) are not defined, and they can be disrupted by many disturbances.*

Assumption 2 *The perturbations are bounded.*

Assumption 3 *The alterations in the reference trajectory are not excessively abrupt, and they can be accurately tracked by a magnetic resonance system.*

3.6.2 Non-Singleton T3-FS

The paper acknowledges that the dynamics of MR and the disturbances are not known. The NT3FSs serve as estimators for uncertainty. Figure 3.3 presents the overall structure. The calculation of NT3FS is described in further below.

(1) The inputs are $U_1 = \chi_1$, and $U_2 = \chi_2$.
(2) Using the non-singleton fuzzifications, the input U is changed to $\underline{U}$ and $\underline{U}$, as follows:

$$\underline{U}_{i,\underline{\varsigma}_j}(t) = \frac{U_i(t)\underline{\beta}^2_{\Lambda_i^h,\underline{\varsigma}_j} + o_{\Lambda_i^h,\underline{\varsigma}_j}\underline{\beta}^2_U}{\underline{\beta}^2_{\Lambda_i^h,\underline{\varsigma}_j} + \underline{\beta}^2_U} \tag{3.16}$$

$$\underline{U}_{i,\underline{\varsigma}_j}(t) = \frac{U_i(t)\underline{\beta}^2_{\Lambda_i^h,\underline{\varsigma}_j} + o_{\Lambda_i^h,\underline{\varsigma}_j}\underline{\beta}^2_U}{\underline{\beta}^2_{\Lambda_i^h,\underline{\varsigma}_j} + \underline{\beta}^2_U} \tag{3.17}$$

$$\underline{U}_{i,\underline{\varsigma}_j}(t) = \frac{U_i(t)\underline{\beta}^2_{\Lambda_i^h,\underline{\varsigma}_j} + o_{\Lambda_i^h,\underline{\varsigma}_j}\underline{\beta}^2_U}{\underline{\beta}^2_{\Lambda_i^h,\underline{\varsigma}_j} + \underline{\beta}^2_U} \tag{3.18}$$

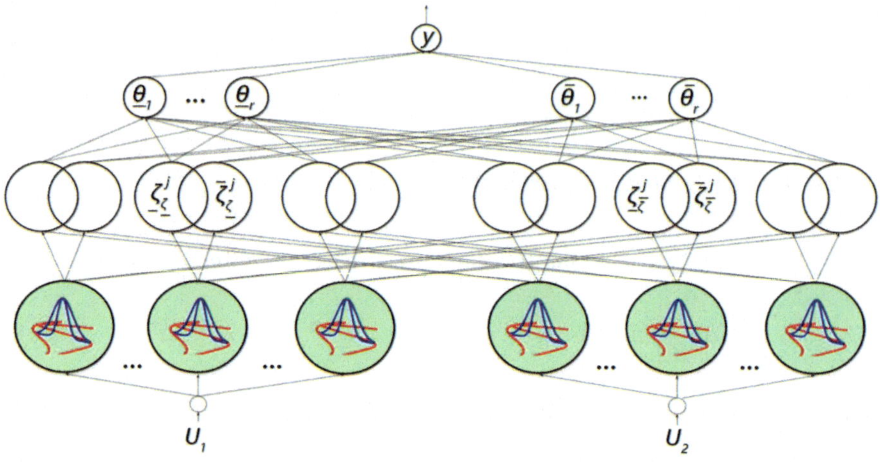

Fig. 3.3 Schematic diagram of NT3FS

$$\underline{U}_{i,\underline{\varsigma}_j}(t) = \frac{U_i(t)\underline{\beta}^2_{\Lambda_i^h,\underline{\varsigma}_j} + o_{\Lambda_i^h,\underline{\varsigma}_j}\underline{\beta}^2_U}{\underline{\beta}^2_{\Lambda_i^h,\underline{\varsigma}_j} + \underline{\beta}^2_U} \tag{3.19}$$

where, $U_i(t)$ represents the $i-th$ input of NT3FS, and Λ_i^h denotes the membership function (MF) (h-th MF for $U_i(t)$). $o_{\Lambda_i^h,\underline{\varsigma}_j}$ shows the center of MFs, and the terms $\underline{\beta}_{\Lambda_i^h,\underline{\varsigma}_j}, \underline{\beta}_{\Lambda_i^h,\underline{\varsigma}_j}, \underline{\beta}_{\Lambda_i^h,\underline{\varsigma}_j}, \underline{\beta}_{\Lambda_i^h,\underline{\varsigma}_j}$ represent the standard divisions of Gaussian MFs. Finally, $\underline{\beta}_U$ shows the level of fuzzification.

(3) The memberships are calculated as follows according to Fig. 3.4.

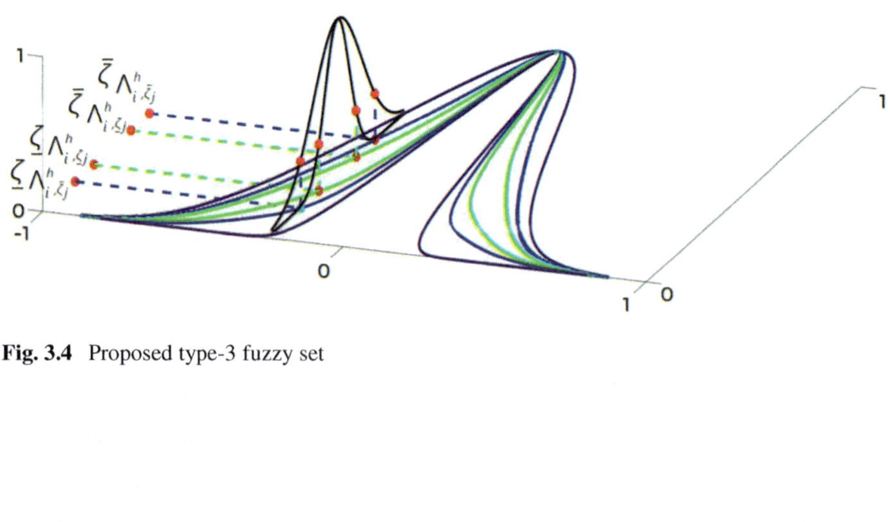

Fig. 3.4 Proposed type-3 fuzzy set

$$\underline{\zeta}_{\Lambda_i^h,\underline{\varsigma}_j}(U_i(t)) = exp\left(-\frac{\left(\underline{U}_{i,\underline{\varsigma}_j}(t) - o_{\Lambda_i^h,\underline{\varsigma}_j}\right)}{\underline{\beta}^2_{\Lambda_i^h,\underline{\varsigma}_j}}\right) \tag{3.20}$$

$$\underline{\zeta}_{\Lambda_i^h,\underline{\varsigma}_j}(U_i(t)) = exp\left(-\frac{\left(\underline{U}_{i,\underline{\varsigma}_j}(t) - o_{\Lambda_i^h,\underline{\varsigma}_j}\right)}{\underline{\beta}^2_{\Lambda_i^h,\underline{\varsigma}_j}}\right) \tag{3.21}$$

$$\underline{\zeta}_{\Lambda_i^h,\underline{\varsigma}_j}(U_i(t)) = exp\left(-\frac{\left(\underline{U}_{i,\underline{\varsigma}_j}(t) - o_{\Lambda_i^h,\underline{\varsigma}_j}\right)}{\underline{\beta}^2_{\Lambda_i^h,\underline{\varsigma}_j}}\right) \tag{3.22}$$

$$\underline{\zeta}_{\Lambda_i^h,\underline{\varsigma}_j}(U_i(t)) = exp\left(-\frac{\left(\underline{U}_{i,\underline{\varsigma}_j}(t) - o_{\Lambda_i^h,\underline{\varsigma}_j}\right)}{\underline{\beta}^2_{\Lambda_i^h,\underline{\varsigma}_j}}\right) \tag{3.23}$$

(4) For the rules firing, we have:

$$\underline{\Omega}^l_{\underline{\varsigma}_j} = \underline{\zeta}_{\Lambda_1^{h_1},\underline{\varsigma}_j} \cdot \underline{\zeta}_{\Lambda_2^{h_2},\underline{\varsigma}_j} \cdots \underline{\zeta}_{\Lambda_n^{h_n},\underline{\varsigma}_j} \tag{3.24}$$

$$\underline{\Omega}^l_{\underline{\varsigma}_j} = \underline{\zeta}_{\Lambda_1^{h_1},\underline{\varsigma}_j} \cdot \underline{\zeta}_{\Lambda_2^{h_2},\underline{\varsigma}_j} \cdots \underline{\zeta}_{\Lambda_n^{h_n},\underline{\varsigma}_j} \tag{3.25}$$

$$\underline{\Omega}^l_{\underline{\varsigma}_j} = \underline{\zeta}_{\Lambda_1^{h_1},\underline{\varsigma}_j} \cdot \underline{\zeta}_{\Lambda_2^{h_2},\underline{\varsigma}_j} \cdots \underline{\zeta}_{\Lambda_n^{h_n},\underline{\varsigma}_j} \tag{3.26}$$

$$\underline{\Omega}^l_{\underline{\varsigma}_j} = \underline{\zeta}_{\Lambda_1^{h_1},\underline{\varsigma}_j} \cdot \underline{\zeta}_{\Lambda_2^{h_2},\underline{\varsigma}_j} \cdots \underline{\zeta}_{\Lambda_n^{h_n},\underline{\varsigma}_j} \tag{3.27}$$

The form of $l - th$ rule is supposed to be given as:

$$\begin{gathered} if\ \chi_1\ is\ \Lambda^{h_l}_{1,\varsigma_j}\ and\ \chi_2 is \Lambda^{h_l}_{2,\varsigma_j}\ and \\ \chi_n\ is\ \Lambda^{h_l}_{n,\varsigma_j}\ Then\, y_l \in \left[\underline{\alpha}_{l,j}, \underline{\alpha}_{l,j}\right] \end{gathered} \tag{3.28}$$

where, $\underline{\alpha}_{l,j}$ and $\underline{\alpha}_{l,j}$ show the rule coefficients, and y_l shows the fuzzy system output at $l - th$ rule.

(5) The output of NT3FS is formulated as [1]:

$$y = \frac{\sum_{j=1}^{\lambda}}{\sum_{j=1}^{\lambda} \underline{\varsigma}_j} \left(\frac{\sum_{l=1}^{r} \underline{\varsigma}_j \left(\underline{\Omega}^l_{\underline{\varsigma}_j} + \underline{\Omega}^l_{\underline{\varsigma}_j}\right) \underline{\alpha}_{l,j}/2}{\sum_{l=1}^{r} \left(\underline{\Omega}^l_{\underline{\varsigma}_j} + \underline{\Omega}^l_{\underline{\varsigma}_j}\right)} + \frac{\underline{\varsigma}_j \sum_{l=1}^{r} \left(\underline{\Omega}^l_{\underline{\varsigma}_j} + \underline{\Omega}^l_{\underline{\varsigma}_j}\right) \underline{\alpha}_{l,j}/2}{\sum_{l=1}^{r} \left(\underline{\Omega}^l_{\underline{\varsigma}_j} + \underline{\Omega}^l_{\underline{\varsigma}_j}\right)} \right) \tag{3.29}$$

where, $\underline{\varsigma}_j$ and $\underline{\varsigma}_j$ denote slice levels, λ show the slice number, and r represents the number of rules. The MR dynamics of are stated as:

$$\dot{\chi} = g(U|\alpha) + u \tag{3.30}$$

where the input vector is denoted by U, the control signal is demonstrated by u, $g(U|\alpha)$ shows a fuzzy system so that, using Eq. (3.29), $g(U|\alpha)$ is stated as:

$$g(U|\alpha) = \alpha^T \Phi \tag{3.31}$$

where,

$$\alpha^T = \left[\underline{\alpha}_{1,1}, \ldots, \underline{\alpha}_{1,\lambda}, \ldots, \underline{\alpha}_{r,1}, \ldots, \underline{\alpha}_{r,\lambda}, \underline{\alpha}_{1,1}, \ldots, \underline{\alpha}_{1,\lambda}, \ldots, \underline{\alpha}_{1,1}, \ldots, \underline{\alpha}_{1,\lambda}\right] \tag{3.32}$$

$$\Phi^T = \frac{0.5}{\sum_{j=1}^{\lambda} \underline{\varsigma}_j} \begin{bmatrix} \dfrac{\underline{\varsigma}_1\left(\underline{\Omega}^1_{\underline{\varsigma}_1} + \underline{\Omega}^1_{\underline{\varsigma}_1}\right)}{\sum_{l=1}^{r}\left(\underline{\Omega}^1_{\underline{\varsigma}_j} + \underline{\Omega}^1_{\underline{\varsigma}_j}\right)}, \ldots, \dfrac{\underline{\varsigma}_\lambda\left(\underline{\Omega}^1_{\underline{\varsigma}_\lambda} + \underline{\Omega}^1_{\underline{\varsigma}_\lambda}\right)}{\sum_{l=1}^{r}\left(\underline{\Omega}^1_{\underline{\varsigma}_j} + \underline{\Omega}^1_{\underline{\varsigma}_j}\right)}, \\ \dfrac{\underline{\varsigma}_1\left(\underline{\Omega}^r_{\underline{\varsigma}_1} + \underline{\Omega}^r_{\underline{\varsigma}_1}\right)}{\sum_{l=1}^{r}\left(\underline{\Omega}^1_{\underline{\varsigma}_j} + \underline{\Omega}^1_{\underline{\varsigma}_j}\right)}, \ldots, \dfrac{\underline{\varsigma}_\lambda\left(\underline{\Omega}^r_{\underline{\varsigma}_\lambda} + \underline{\Omega}^r_{\underline{\varsigma}_\lambda}\right)}{\sum_{l=1}^{r}\left(\underline{\Omega}^1_{\underline{\varsigma}_j} + \underline{\Omega}^1_{\underline{\varsigma}_j}\right)}, \\ \underline{\varsigma}_1\dfrac{\left(\underline{\Omega}^1_{\underline{\varsigma}_1} + \underline{\Omega}^1_{\underline{\varsigma}_1}\right)}{\sum_{l=1}^{r}\left(\underline{\Omega}^l_{\underline{\varsigma}_j} + \underline{\Omega}^l_{\underline{\varsigma}_j}\right)}, \ldots, \underline{\varsigma}_\lambda\dfrac{\left(\underline{\Omega}^1_{\underline{\varsigma}_\lambda} + \underline{\Omega}^1_{\underline{\varsigma}_\lambda}\right)}{\sum_{l=1}^{r}\left(\underline{\Omega}^l_{\underline{\varsigma}_j} + \underline{\Omega}^l_{\underline{\varsigma}_j}\right)}, \\ \underline{\varsigma}_1\dfrac{\left(\underline{\Omega}^r_{\underline{\varsigma}_1} + \underline{\Omega}^r_{\underline{\varsigma}_1}\right)}{\sum_{l=1}^{r}\left(\underline{\Omega}^l_{\underline{\varsigma}_j} + \underline{\Omega}^l_{\underline{\varsigma}_j}\right)}, \ldots, \underline{\varsigma}_\lambda\dfrac{\left(\underline{\Omega}^r_{\underline{\varsigma}_\lambda} + \underline{\Omega}^r_{\underline{\varsigma}_\lambda}\right)}{\sum_{l=1}^{r}\left(\underline{\Omega}^l_{\underline{\varsigma}_j} + \underline{\Omega}^l_{\underline{\varsigma}_j}\right)} \end{bmatrix} \tag{3.33}$$

3.6.3 Nonlinear MPC

A predictive controller is a widely employed technique in several industries. One of the benefits of Model Predictive Control (MPC) is its capacity to take into account various system restrictions. The anticipated behavior of the system is predicted based on the assumed model at each sample moment within the specified horizon. Within the algorithm, the MPC optimizer is tasked with identifying the most advantageous control input within the pre-established time frame. This work is completed using the established control approach and anticipated outputs determined by the prediction block. Optimization has consistently been a crucial stage in the processing and execution of various control systems. Performing calculations is crucial in the context of MPC to attain an optimal and instantaneous response. The MPC issue is defined as the task of reducing the cost function [32, 33]:

$$\min_{v_{p_z}(j),\dots,v_{p_z}(j+n_C)} J = \sum_{j=t}^{t+n_P} \mu \cdot \hat{\delta}_z^2(j) + \pi \cdot \Delta v_{p_z}^2(j) \ \textit{subject to} \ \hat{\delta}_z(t) = y_{BM}\left(\underline{\hat{\delta}}_z | X_z\right) \tag{3.34}$$

where, $z = 1, 2$, π and μ show the constant values, v_{p_z} is the $z - th$ predictive controller, the prediction horizon is denoted by n_P, the differential of v_{p_z} is shown by Δv_{p_z}. $\underline{\hat{\delta}}_z$ is stated as:

$$\underline{\hat{\delta}}_z = \left[\hat{\delta}_z(t-1), \dots, \hat{\delta}_z(t-\tau), v_{p_z}(t-1)\right]^T \tag{3.35}$$

$y_{BM}\left(\underline{\hat{\delta}}_z | X_z\right)$ is the output of Boltzmann machine (BM) (refere to Fig. 3.5), and X_z consists of output and hidden layer weights (X_v and X_{z0}). $y_{BM}\left(\underline{\hat{\delta}}_z | X_z\right)$ is formulated as:

$$y_{BM}\left(\underline{\hat{\delta}}_z | X_z\right) = X_{z0}^T \mu_1 \tag{3.36}$$

where,

$$\mu_1 = \Xi\left(X_v^T \eta_1\right) \tag{3.37}$$

In Eq. (3.39), η_1 is obtained as:

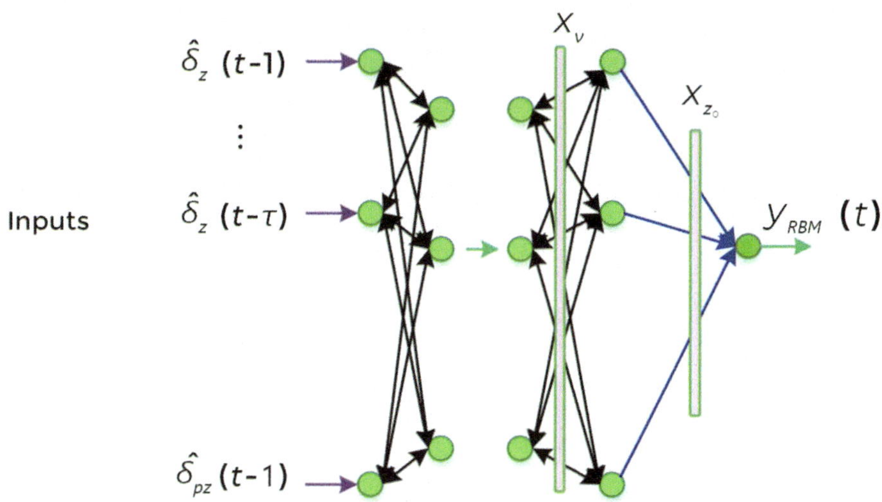

Fig. 3.5 Schematic diagram of BM block

$$\eta_1 = \Xi(X_v \mu_0) \tag{3.38}$$

In Eq. (3.38), η_0 is formulated as:

$$\mu_0 = \Xi\left(X_v^T \eta_0\right) \tag{3.39}$$

In this context, η_0 shows input vector of BM, and

$$\Xi(y) = \left[1 - exp(-y)\right] / \left[1 + exp(-y)\right] \tag{3.40}$$

The CD training approach [34, 35] is used to optimize X_v, ensuring that the cost function Eq. (3.60) is reduced.

$$J'(\eta, \mu) = \delta^{E(\eta, \mu)} / E(\eta, \mu) \tag{3.41}$$

where,

$$E(\eta, \mu) = -\mu^T X_v \eta \tag{3.42}$$

Then X_v is arranged as:

$$X_v(t+1) = X_v(t) + \gamma\left(\eta_0 \mu_0^T - \eta_1 \mu_1^T\right) \tag{3.43}$$

where $0 < \gamma \leq 1$, and X_{z_0} is adjusted as:

$$X_{z_o}(t+1) = X_{z_o}(t) + \gamma\left(\delta_z - \hat{\delta}_z\right)\mu_1 \tag{3.44}$$

BM output at preceding stages (j-steps) is:

$$\hat{\delta}_z(t+j) = \hat{\delta}_{z,forced}(t+j) + \hat{\delta}_{z,free}(t+j) \tag{3.45}$$

where,

$$\hat{\delta}_{z,free}(t+j) = y_{BM}\left(\hat{\underline{\delta}}_z(t+j) | X_z\right) \tag{3.46}$$

$$\hat{\delta}_{z,forced}(t+j) = \sum_{\iota=0}^{j-1} l_\iota \Delta v_{p_z}(t+j-\iota+1) \tag{3.47}$$

The step response coefficients are showed by $l_\iota, \iota = 0, \ldots, j-1$. The estimated step response $s(t-1)$ is expressed as:

$$s(t-1) = \left[\hat{\delta}_{z,step}(t+j) - \hat{\delta}_{z,free}(t+j)\right] / dv_{p_z}(t) \tag{3.48}$$

where, $dv_{p_z}(t)$ shows the size of the step, and we have:

$$\hat{\delta}_{z,step}(t+j) = y_{BM}\left(\hat{\underline{\delta}}_z(t+j)|X_z\right), \tag{3.49}$$

$$\hat{\underline{\delta}}_z = \left[\hat{\delta}_z(t+j-1)..., \hat{\delta}_z(t+j-\tau), v_{p_z}(t+j-1)\right]^T, \tag{3.50}$$

$$v_{p_z}(t) = \{v_{p_z}(t-1) + dv_{p_z}(t)\, if\, j > t-1 v_{p_z}(t-1) else \tag{3.51}$$

The problem described before may be designated in matrix form as [34, 35]:

$$\Psi = S\Delta v_{p_z} + \Psi_{free} \tag{3.52}$$

where,

$$\Psi = \left[\hat{\delta}_z(t+1)\hat{\delta}_z(t+2)\vdots\hat{\delta}_z(t+n_p)\right] \tag{3.53}$$

$$S = \left[s_0 0 s_1 s_0 \vdots s_{n_p} - 1 \vdots s_{n_p} - 1 \cdots 0 \cdots 0 \ddots \cdots \vdots s_0\right] \tag{3.54}$$

$$\Delta v_{p_z} = \left[\varphi v_{p_z}(t)\varphi v_{p_z}(t+1)\vdots\varphi v_{p_z}(t+n_p-1)\right] \tag{3.55}$$

$$\Psi_{free} = \left[\hat{\delta}_{z,free}(t+1)\hat{\delta}_{z,free}(t+2)\vdots\hat{\delta}_{z,free}(t+n_p)\right] \tag{3.56}$$

$$\hat{\delta}_{z,free}(t+j) = y_{BM}\left(\hat{\underline{\delta}}_z(t+j)|X_z\right) \tag{3.57}$$

$$\hat{\underline{\delta}}_z(t+j) = \left[\hat{\delta}_z(t+j-1)..., \hat{\delta}_z(t+j-\tau), v_{p_z}(t+j-1)\right]^T \tag{3.58}$$

$$v_{p_z}(t) = \{v_{p_z}(t-1)\, if\, j > t-1 v_{p_z}(t)\, else \tag{3.59}$$

Using Eqs. (3.34, 3.52), we have:

$$J = \mu\Psi^T\Psi + \pi\Delta v_{p_z}^T\Delta v_{p_z} = \mu\left(S\Delta v_{p_z} + \Psi_{free}\right)^T\left(S\Delta v_{p_z} + \Psi_{free}\right) + \pi\Delta v_{p_z}^T\Delta v_{p_z} \tag{3.60}$$

Using Eq. (3.60), $\partial J/\partial\Delta v_{p_z}$ is stated as:

$$\partial J/\partial\Delta v_{p_z} = 0 \Rightarrow \mu S^T\Psi_{free} + \Delta v_p\left(\mu S^T S + \pi\right) = 0 \tag{3.61}$$

From Eq. (3.61), Δv_{p_z} is formulated as:

$$\Delta v_{p_z} = \mu\left(\mu S^T S + \pi\right)^{-1} S^T\Psi_{free} \tag{3.62}$$

The controller $v_{p_z}(t)$ is:

$$v_{p_z}(t) = v_{p_z}(t-1) + \Delta v_{p_z}(t) \tag{3.63}$$

where, the first element in Δv_{p_z} in Eq. (3.62) is $\Delta v_{p_z}(t)$.

3.6.4 Stability Analysis

Considering the optimal NT3FSs, the robot dynamics are calculated to be:

$$\dot{\chi}_1 = g_1^*(U|\alpha_1^*) + u_1 + \omega_1 \tag{3.64}$$

$$\dot{\chi}_2 = g_2^*(U|\alpha_2^*) + u_2 + \omega_2 \tag{3.65}$$

The symbols $\omega_i, i = 1, 2$ represent estimation errors. The controllers u_1 and u_2 consist of three components: the principal error feedback, predictive signals v_p (refer to Eq. (3.63)), and compensators v_c (refer to Eqs. (3.81, 3.82)). As depicted in Fig. 3.1, the controllers are:

$$u_1 = -g_1(\theta|\alpha_1) - K_1\delta_1 + \omega_1 + v_{c_1} + v_{p_1} \tag{3.66}$$

$$u_2 = -g_2(\theta|\alpha_2) - K_2\delta_2 + \omega_2 + v_{c_2} + v_{p_2} \tag{3.67}$$

where, K_1 and K_2 are considered to be constant, and $\dot{\delta}_1$ and $\dot{\delta}_2$ denote errors. Then $\dot{\delta}_1$ and $\dot{\delta}_2$ can be stated as:

$$\dot{\delta}_1 = g_1^*(\theta|\alpha_1^*) - g_1(\theta|\alpha_1) - K_1\delta_1 + \omega_1 + v_{c_1} + v_{p_1} \tag{3.68}$$

$$\dot{\delta}_2 = g_2^*(\theta|\alpha_2^*) - g_2(\theta|\alpha_2) - K_2\delta_2 + \omega_2 + v_{c_2} + v_{p_2} \tag{3.69}$$

From Eq. (3.31), we can write:

$$g_1^*(\theta|\alpha_1^*) - g_1(\theta|\alpha_1) = \tilde{\alpha}_1^T \Phi_{\xi_1} \tag{3.70}$$

$$g_2^*(\theta|\alpha_2^*) - g_2(\theta|\alpha_2) = \tilde{\alpha}_2^T \Phi_{\xi_2} \tag{3.71}$$

From Eqs. (3.70, 3.71), the Eqs. (3.68, 3.69) are stated as:

$$\dot{\delta}_1 = \tilde{\alpha}_1^T \Phi_{\xi_1} - K_1\delta_1 + \omega_1 + v_{c_1} + v_{p_1} \tag{3.72}$$

$$\dot{\delta}_2 = \tilde{\alpha}_2^T \Phi_{\xi_2} - K_2\delta_2 + \omega_2 + v_{c_2} + v_{p2} \tag{3.73}$$

We consider the Lyapunov function as:

$$V = \frac{1}{2}\delta_1^2 + \frac{1}{2}\delta_2^2 + \frac{1}{2\gamma}\tilde{\alpha}_1^T\tilde{\alpha}_1 + \frac{1}{2\gamma}\tilde{\alpha}_2^T\tilde{\alpha}_2 \tag{3.74}$$

Derivative of Eq. (3.74), culminates in:

$$\dot{V} = \delta_1\dot{\delta}_1 + \delta_2\dot{\delta}_2 - \frac{1}{\gamma}\tilde{\alpha}_1^T\dot{\alpha}_1 - \frac{1}{\gamma}\tilde{\alpha}_2^T\dot{\alpha}_2 \tag{3.75}$$

From Eqs. (3.72, 3.73, 3.75), $\dot{V}$ is stated as:

$$\begin{aligned}\dot{V} &= -K_1\delta_1^2 - K_2\delta_2^2 + \omega_1\delta_1 + \delta_1 v_{c_1} + \delta_1 v_{p_1} + \omega_2\delta_2 + \delta_2 v_{c_2} + \delta_1 v_{p_2} \\ &+\tilde{\alpha}_1^T\Phi_{\xi_1}\delta_1 - \frac{1}{\gamma}\tilde{\alpha}_1^T\dot{\alpha}_1 + \tilde{\alpha}_2^T\Phi_{\xi_2}\delta_2 - \frac{1}{\gamma}\tilde{\alpha}_2^T\dot{\alpha}_2\end{aligned} \tag{3.76}$$

Then the tuning rules are stated as:

$$\dot{\alpha}_1 = \gamma\Phi_{\xi_1}\delta_1 \tag{3.77}$$

$$\dot{\alpha}_2 = \gamma\Phi_{\xi_2}\delta_2 \tag{3.78}$$

Using Eqs. (3.77, 3.78), Eq. (3.76), becomes:

$$\dot{V} = -K_1\delta_1^2 - K_2\delta_2^2 + \omega_1\delta_1 + \delta_1 v_{c_1} + \delta_1 v_{p_1} + \omega_2\delta_2 + \delta_2 v_{c_2} + \delta_1 v_{p_2} \tag{3.79}$$

When discussing the features of MR, we can examine the maximum limits for ω_i and $v_{p_i} i = 1, 2$, and express it as follows:

$$\dot{V} \leq -K_1\delta_1^2 - K_2\delta_2^2 + \underline{\omega}_1|\delta_1| + \delta_1 v_{c_1} + |\delta_1|\underline{v}_{p_1} + \underline{\omega}_2|\delta_2| + \delta_2 v_{c_2} + |\delta_2|\underline{v}_{p_2} \tag{3.80}$$

Using Eq. (3.80), in order to delete the effect of terms $\underline{\omega}_1|\delta_1| + |\delta_1|\underline{v}_{p_1}$ and $\underline{\omega}_2|\delta_2| + |\delta_2|\underline{v}_{p_2}$, the compensators are:

$$v_{c_1} = -tanh(\delta_1)\left(\underline{\omega}_1 + \underline{v}_{p_1}\right) \tag{3.81}$$

$$v_{c_2} = -tanh(\delta_2)\left(\underline{\omega}_2 + \underline{v}_{p_2}\right) \tag{3.82}$$

Considering Eqs. (3.81, 3.82), we have:

$$\dot{V} \leq -K_1\delta_1^2 - K_2\delta_2^2 \tag{3.83}$$

Then,

$$V \leq \int_{0}^{t} -K_1 \|\delta_1(\tau)\|^2 - K_2 \|\delta_2(\tau)\|^2 d\tau \prec \infty \tag{3.84}$$

Inequality (3.84), proves that $\delta_1 \in l^2$ and $\delta_2 \in l^2$, and thus the proof of asymptotic stability has been established.

3.6.5 Implementation

This section scrutinizes the proposed methodology by implementing it on an actual robot. Figure 3.6 illustrates the robot that has been designed. The robot is wirelessly connected to a laptop computer for monitoring purposes. Figure 3.7 displays the arrangement. The robot is equipped with two stepper motors that have great precision and are connected to two standard wheels with a diameter of 7 cm. Additionally, there are two idle pins to maintain the robot's equilibrium. The robot is equipped with a total of thirty-two Sharp sensors, with eight sensors placed on each of its four sides. Additionally, it is equipped with an MPU6050 angle acceleration sensor positioned at its center of mass. The robot's bottom plate, which houses the motors, is constructed from 1.5 mm aluminum. The angular accuracy of two-phase stepper motors is 1.8 degrees per step. The PCB is mounted onto the chassis surface using four 7 cm spacers. The NRF24L01 radio transmitter module facilitates communication between the laptop and the robot. To assess the viability, we will analyze a complex chaotic path-tracking method. Figure 3.7 displays the arrangement, and Fig. 3.8 displays the results. The developed technique demonstrates precise performance, even in highly noisy situations. The robot effectively navigates a complex and unpredictable trajectory. The proposed predictive methodology reduces errors in tracking and addresses the inherent disturbances.

In order to observe the unpredictable nature of the system, the phase portrait is displayed in Fig. 3.9. It is evident that the robot is traversing a chaotic trajectory. The intricate routes can be utilized in diverse security applications. For instance, the recommended robot can be utilized in a patrolling application. In this problem, it is necessary for the robot's path to be both safe and unpredictable.

3.6.6 Simulations

The reference path is determined by tracking a chaotic system [36]:

$$\begin{aligned} \{D_t^{0.98} y_1 &= 35\left[y_2 y_3 + (y_2 - y_1)\right] D_t^{0.98} y_2 = -5 y_1 y_3 25 y_1 + y_2 + y_4 D_t^{0.98} y_3 \\ &= -4 y_3 + y_1 y_2 D_t^{0.98} y_4 = -100 y_2 \end{aligned} \tag{3.85}$$

Fig. 3.6 The designed robot

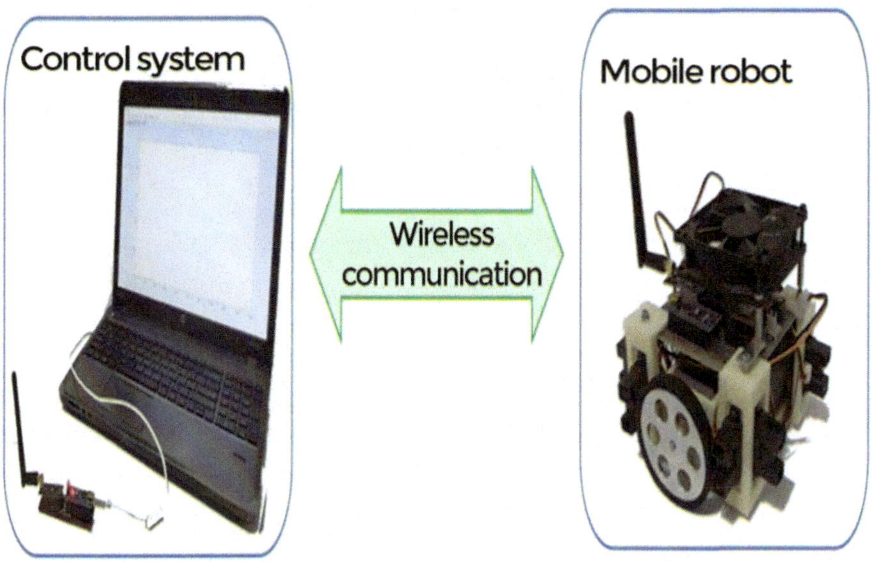

Fig. 3.7 The designed setup

Figure 3.10 displays the several switching modes. The output signals referenced in Fig. 3.10 have been modified four times. The simulation circumstances are illustrated in Tables 3.1 and 3.2. It is important to mention that the criteria are regarded as identical for both IT2-FLS and NT3FS, including the number of membership functions, regulations, and values of the centers of membership functions. The results are illustrated in Fig. 3.11, which demonstrates an exemplary tracking response. It is evident that the output signals closely follow the target signals. Despite the presence

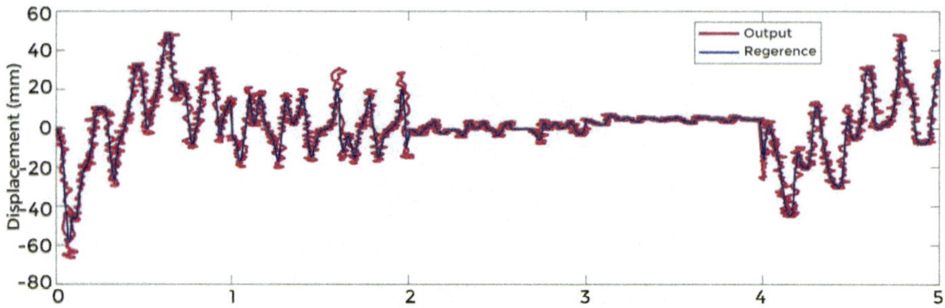

Fig. 3.8 Implementation performance

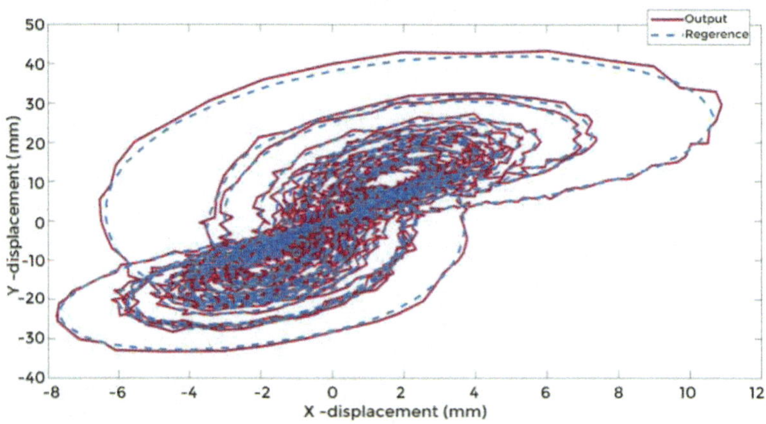

Fig. 3.9 Phase portrait of the implementation performance

of time-varying references and frequent switching between distinct signals, the robot effectively follows the planned course. Figure 3.12 illustrates the controllers. The control signals obtained do not exhibit any high-frequency variations. The calculated signals are presented in Fig. 3.13. The phase portrait is depicted in Fig. 3.14. We observe a satisfactory level of precision in tracking. The chaotic trajectory offers a reliable route for robots, particularly, when taking into account the abrupt alteration of the recommended route from one path to another.

In order to provide an unbiased comparison, the influence of the controller is eliminated, and the evaluation of Eq. (3.15) is taken into account in the absence of a controller. Thus, the following results are obtained [36]:

$$\{\dot{\hat{\chi}}_1 = g_1(U|\alpha_1)\dot{\hat{\chi}}_2 = g_2(U|\alpha_2) \tag{3.86}$$

In addition, a noise variance of 0.04 and a noise mean of 0 are included. The root mean square errors (RMSEs) can be found in Table 3.3. Furthermore, Fig. 3.15 illustrates the phase portrait for the comparison. Based on the Root Mean Square Error

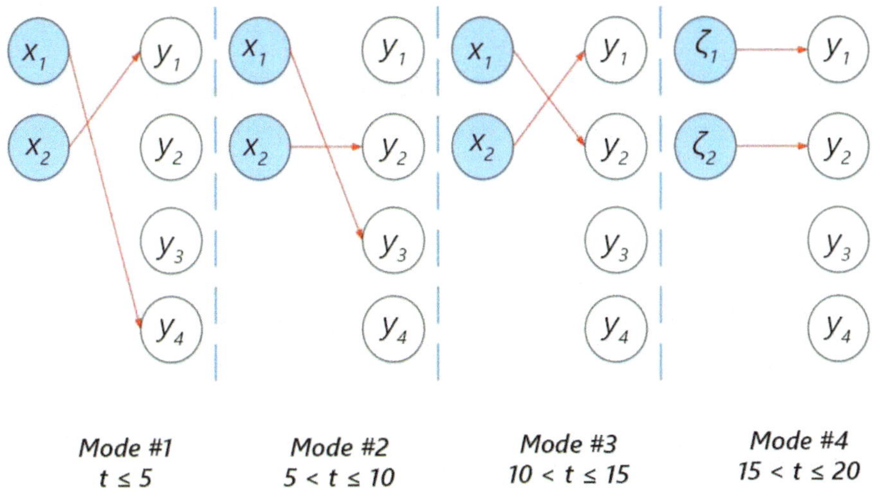

Fig. 3.10 The switching modes

Table 3.2 Simulation condition

$o_{\Lambda_i^1}, i = 1, 2$	–1
$o_{\Lambda_i^2}, i = 1, 2$	1
$\underline{\varsigma}_1$	1
$\underline{\varsigma}_2$	0.9
$\underline{\varsigma}_1$	0.4
$\underline{\varsigma}_2$	0.6
μ	0.5
π	0.5
γ	0.1

(RMSE) results, our proposed controller outperforms other control approaches that are similar. The suggested controller demonstrates enhanced resilience against mode switching and uncertainties in dynamics. Furthermore, it demonstrates substantial progress in performance when functioning in extremely noisy environments. The performance of other controllers significantly deteriorates with increasing measurement noise, but the accuracy of the proposed method remains unaffected by measurement errors. As the level of noise variation goes from 0 to 0.1, the suggested approach's root mean square error (RMSE) experiences just a small increase. In contrast, some other FLSs encounter a more than 50% increase in RMSE [36].

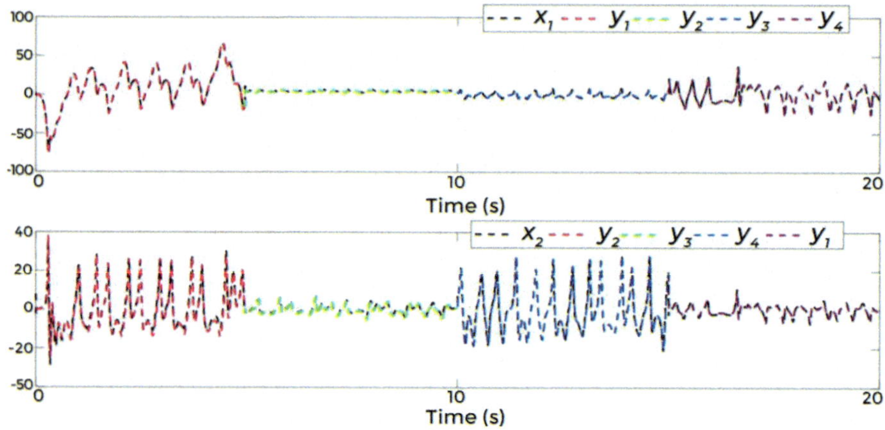

Fig. 3.11 Output signals

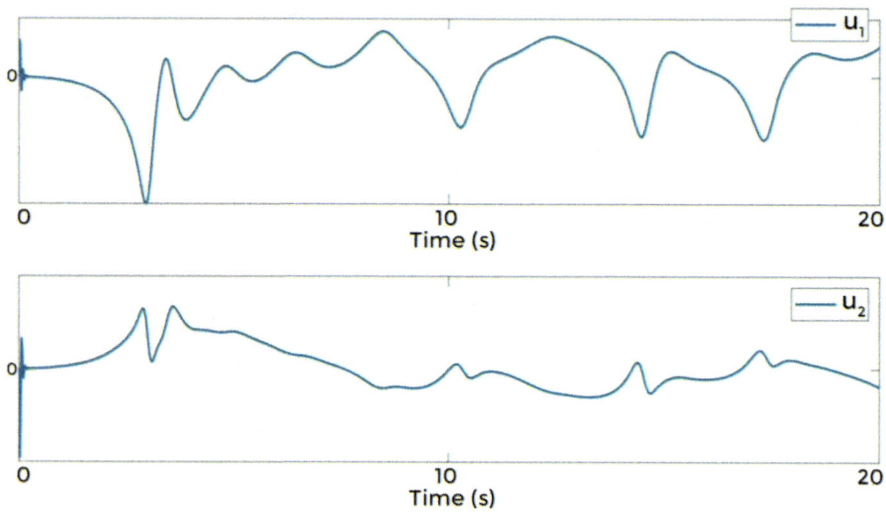

Fig. 3.12 Control signals

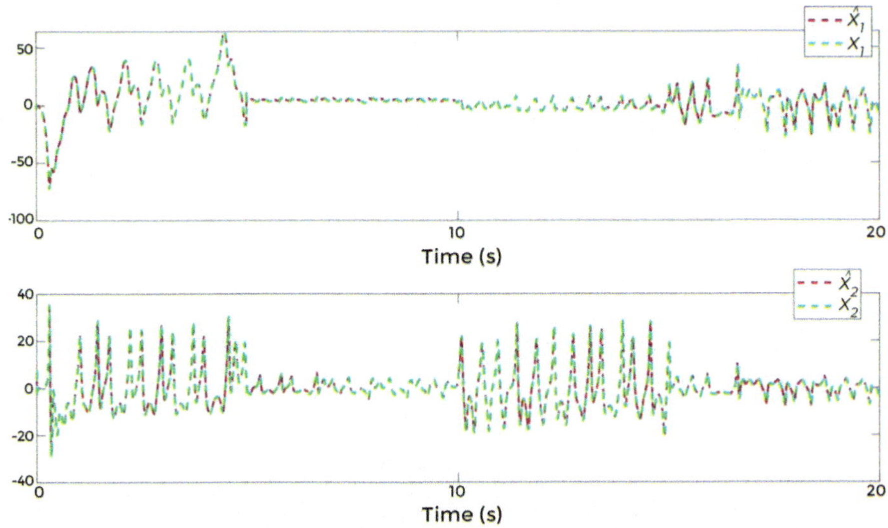

Fig. 3.13 Estimated signals along with the targets

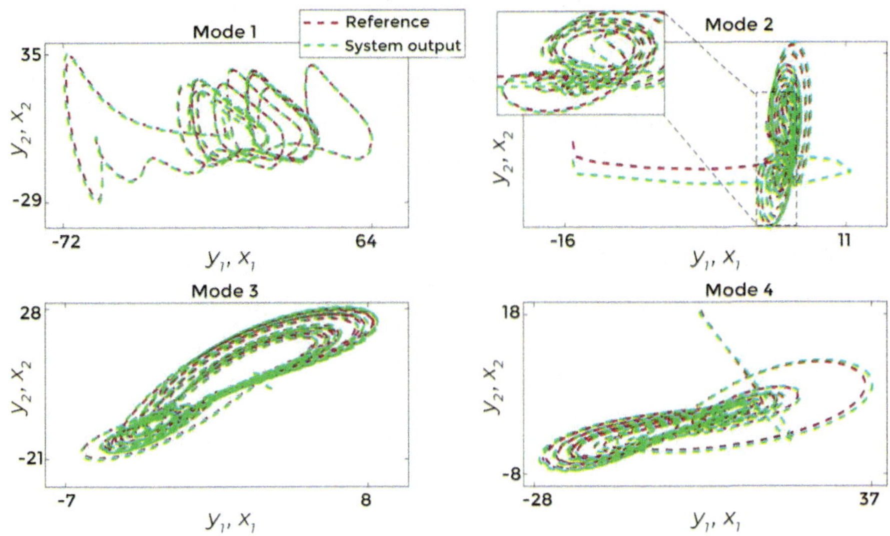

Fig. 3.14 Phase portraits

Table 3.3 The RMSE comparisons

Variance	FLS	$\hat{\chi}_1 - \chi_1$	$\hat{\chi}_2 - \chi_2$
*0	IT2-FLS	5.2150	5.0214
	NT3FS	2.4124	3.0741
*0.05	IT2-FLS	2.2514	4.2415
	NT3FS	1.0171	1.1748
*0.1	IT2-FLS	4.3414	7.2430
	NT3FS	1.7291	1.6729

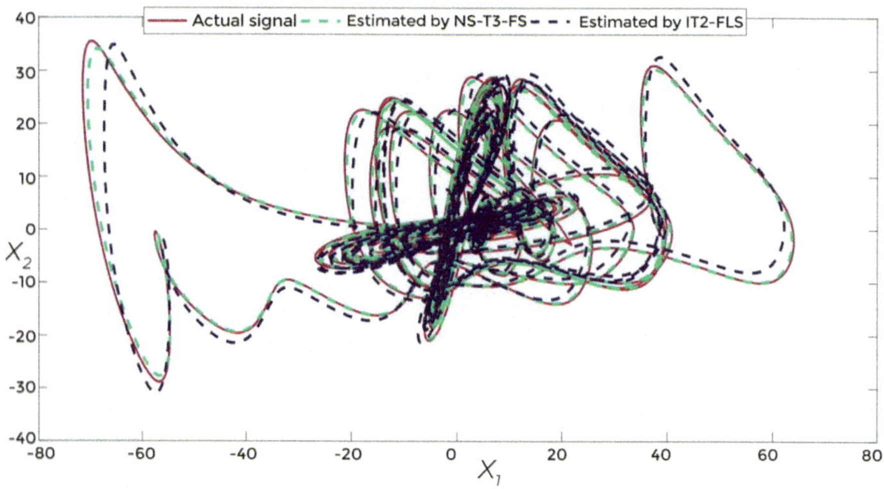

Fig. 3.15 The estimation comparison of FLSs

3.7 Event-Triggered Type-3 Fuzzy Control for Multi-Agent Systems

3.7.1 Problem Formulation

A Multi-Agent System (MAG) consists of multiple interactive agents. Multi-agent systems can be utilized to address complex challenges that are challenging or unattainable for an individual agent or a unified system. MASs have proven to be effective in addressing many challenges across diverse fields, including information technology, communications, industry, defense, education, and military. Examples of MASs include several aircraft, multiple submarines, various robots and satellites. MAGs possess enhanced system performance, flexibility, and reliability, resulting in the capability to execute more intricate activities. Additionally, MAGs offer decreased operational costs and the possibility to generate new capacities. The study of MAGs involves analyzing the network topology that describes the link between agents [37]. We will now examine the following case of MASs:

$$\dot{y}_i^1 = y_i^2 \vdots \dot{y}_i^n = f_i(y_i) + u_i + d_i \tag{3.1}$$

where, the agent numbers are shown by N for $i = 1, ..., N$, we also have $y_i = \left[y_i^1, ..., y_i^n\right]^T \in R^n$, d_i designates the disturbance, and $f_i(y_i)$ is an undetermined function. The leader system is considered as follows:

$$\dot{y}_0^1 = y_0^2 \dots \dot{y}_0^n = f_0(y_0) \tag{3.2}$$

where, $y_0 = \left[y_0^1, ..., y_0^n\right]^T \in R^n$. The primary goal is to create u_i in such a way that the outputs of the agent follow the leader system. The control block diagram can be seen in Fig. 3.1, while the detailed scheme is shown in Fig. 3.2. In Fig. 3.3, the dynamics of each agent are represented by the suggested T3-FLS model, and the main controller is designed using this model in which a compensator ensures stability. The control signal generated is not immediately utilizied by the agents at every reading time. Instead, the actuators are activated using a designed event-triggered arrangement. Further details can be found in the subsequent sections.

3.7.2 Type-3 FLS

The high-order fuzzy logic systems (FLSs) exhibit superior efficiency in diverse noisy applications. The superior efficacy of T3-FLSs has been demonstrated in numerous investigations. The primary factor is that T3-FLSs possess a greater number

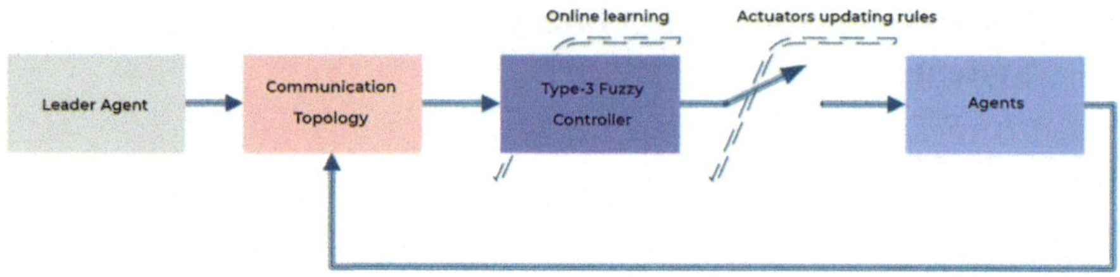

Fig. 3.1 The schematic view of the proposed controller

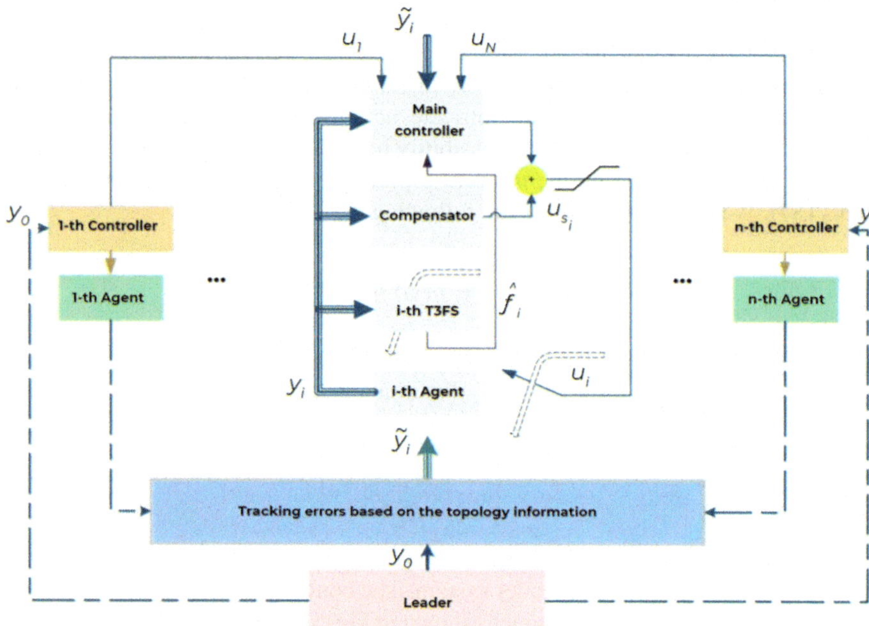

Fig. 3.2 Schematic diagram of control block

of degrees of freedom for representing uncertainty [27]. Furthermore, the secondary fuzzy set in T3-FLSs lacks a unique numerical value, and the footprint-of-uncertainty is also not a unique value, but rather a fuzzy set. When controlling MASs, the agent dynamics and topology exhibit a significant level of nonlinearity. Subsequently, a novel control system is created utilizing the T3-FLS framework for MASs.

Figure 3.3 illustrates the proposed T3-FLS. The membership functions (MFs) are of type-3 as shown in Fig. 3.3. Figure 3.4 provides a more detailed illustration of the type-3 MF. Figure 3.4 illustrates that the secondary membership is a type-2 membership function, rather than a unique value. The intricate calculation is elucidated in a systematic manner below.

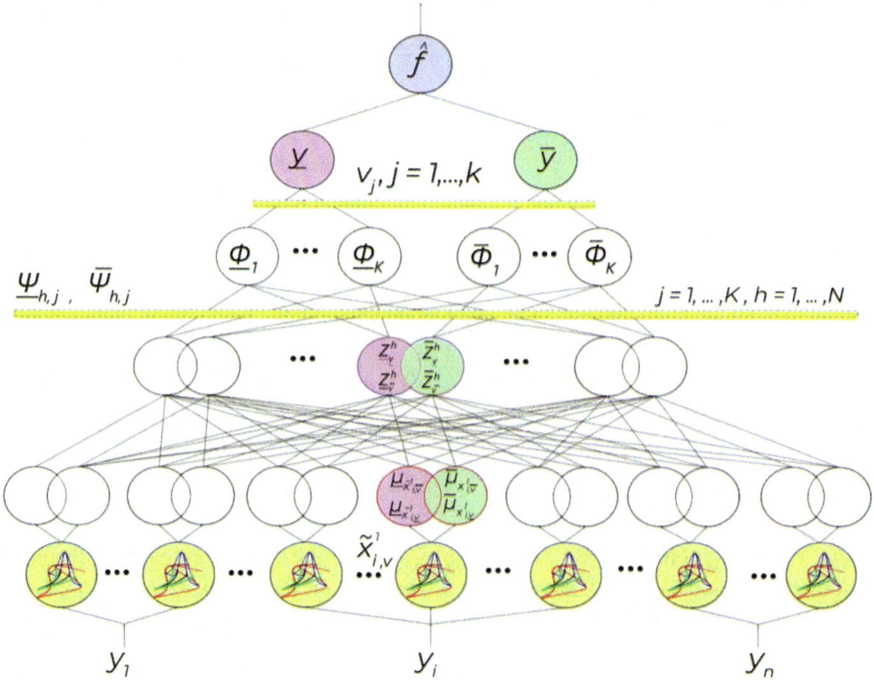

Fig. 3.3 Suggested T3-FLS

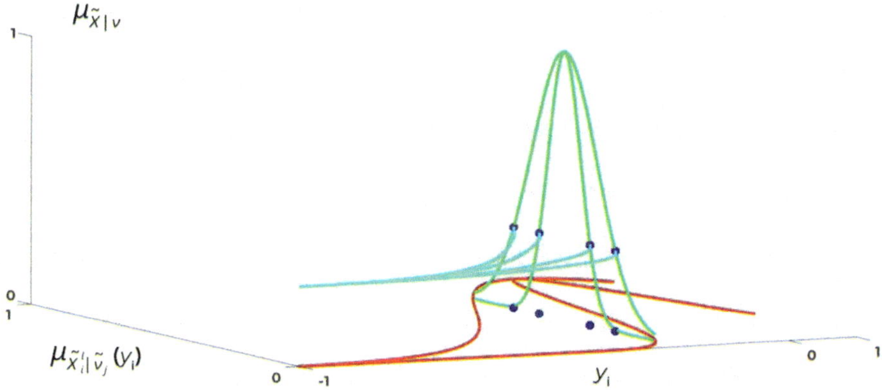

Fig. 3.4 Suggested MF

(1) The T3-FLS inputs are shown by y_i, $i = 1, ..., n$ as the states of agents

(2) As it is demonstrated in Fig. 3.3, every input of y_i has a number of M MFs as $\tilde{\chi}_i^j$. For the Gaussian MFs the memberships are calculated to be:

$$\underline{\mu}_{\tilde{\chi}_i^l|\underline{\nu}_j}(y_i) = exp\left(-\frac{\left(y_i - c_{\tilde{\chi}_i^l|\underline{\nu}_j}\right)^2}{\underline{\sigma}^2_{\tilde{\chi}_i^l|\underline{\nu}_j}}\right) \tag{3.3}$$

$$\underline{\mu}_{\tilde{\chi}_i^l|\underline{\nu}_j}(y_i) = exp\left(-\frac{\left(y_i - c_{\tilde{\chi}_i^l|\underline{\nu}_j}\right)^2}{\underline{\sigma}^2_{\tilde{\chi}_i^l|\underline{\nu}_j}}\right) \tag{3.4}$$

$$\underline{\mu}_{\tilde{\chi}_i^l|\underline{\nu}_j}(y_i) = exp\left(-\frac{\left(y_i - c_{\tilde{\chi}_i^l|\underline{\nu}_j}\right)^2}{\underline{\sigma}^2_{\tilde{\chi}_i^l|\underline{\nu}_j}}\right) \tag{3.5}$$

$$\underline{\mu}_{\tilde{\chi}_i^l|\underline{\nu}_j}(y_i) = exp\left(-\frac{\left(y_i - c_{\tilde{\chi}_i^l|\underline{\nu}_j}\right)^2}{\underline{\sigma}^2_{\tilde{\chi}_i^l|\underline{\nu}_j}}\right) \tag{3.6}$$

where, $c_{\tilde{\chi}_i^l}$ is the center of MF. $\tilde{\chi}_i^l$, $\underline{\sigma}_{\tilde{\chi}_i^l}$, and $\underline{\sigma}_{\tilde{\chi}_i^l}$ are the typical categories for upper bounds (UBs) and lower bounds (LBs) of the MF $\tilde{\chi}_i^l$. The value of horizontal slice are shown by $\nu_j j = 1, ..., K$. To obtain more information regarding the calculation of $\underline{\sigma}_{\tilde{\chi}_i^l}$ and $\underline{\sigma}_{\tilde{\chi}_i^l}$, please refer to reference [15].

(3) By taking into account the memberships of MFs $\tilde{\chi}_i^l$, the rule firings may be derived as shown in Fig. 3.4.

$$\underline{z}^h_{\underline{\nu}_j} = \underline{\mu}_{\tilde{\chi}_i^{h_1}|\underline{\nu}_j}(y_1)\underline{\mu}_{\tilde{\chi}_i^{h_2}|\underline{\nu}_j}(y_2)....\underline{\mu}_{\tilde{\chi}_i^{hn}|\underline{\nu}_j}(y_n) \tag{3.7}$$

$$\underline{z}^h_{\underline{\nu}_j} = \underline{\mu}_{\tilde{\chi}_i^{h_1}|\underline{\nu}_j}(y_1)\underline{\mu}_{\tilde{\chi}_i^{h_2}|\underline{\nu}_j}(y_2)....\underline{\mu}_{\tilde{\chi}_i^{hn}|\underline{\nu}_j}(y_n) \tag{3.8}$$

$$\underline{z}^h_{\underline{\nu}_j} = \underline{\mu}_{\tilde{\chi}_i^{h_1}|\underline{\nu}_j}(y_1)\underline{\mu}_{\tilde{\chi}_i^{h_2}|\underline{\nu}_j}(y_2)....\underline{\mu}_{\tilde{\chi}_i^{hn}|\underline{\nu}_j}(y_n) \tag{3.9}$$

$$\underline{z}^h_{\underline{\nu}_j} = \underline{\mu}_{\tilde{\chi}_i^{h_1}|\underline{\nu}_j}(y_1)\underline{\mu}_{\tilde{\chi}_i^{h_2}|\underline{\nu}_j}(y_2)....\underline{\mu}_{\tilde{\chi}_i^{hn}|\underline{\nu}_j}(y_n) \tag{3.10}$$

where, $j = 1, ..., K$, $h = 1, ..., N$, $\underline{\mu}_{\tilde{\chi}_i^{h_i}|\underline{\nu}_j}(y_i)$ shows the membership for MF $\tilde{\chi}_i^{h_i}|\underline{\nu}_j$. It is worth mentioning that the $i - th$ rule is stated as follows:

$$if\ y_1\ is\ \tilde{\chi}_1^{h_1}\Big|\underline{\nu}_j, \cdots y_i\ is\ \tilde{\chi}_i^{h_i}\Big|\underline{\nu}_j \cdots and\ y_n\ is\ \tilde{\chi}_n^{h_n}|\underline{\nu}_j\ Then \hat{f} \in \left[\underline{\Psi}_{h,j}, \underline{\Psi}_{h,j}\right] \tag{3.11}$$

Next, the rule firings are obtained using the T-norm product, as shown in Eqs. (3.7–3.10).

(4) For the first type-reduction, using Nie-Tan type-reduction [38], we have:

$$\underline{\phi}_j = \frac{\sum_{h=1}^{N}\left(\underline{z}_{\underline{\nu}_j}^{h} + \underline{z}_{\underline{\nu}_j}^{h}\right)\underline{\Psi}_{h,j}}{\sum_{h=1}^{N} \underline{z}_{\underline{\nu}_j}^{h} + \underline{z}_{\underline{\nu}_j}^{h}} \tag{3.12}$$

$$\underline{\phi}_j = \frac{\sum_{h=1}^{N}\left(\underline{z}_{\underline{\nu}_j}^{h} + \underline{z}_{\underline{\nu}_j}^{h}\right)\underline{\Psi}_{h,j}}{\sum_{h=1}^{N} \underline{z}_{\underline{\nu}_j}^{h} + \underline{z}_{\underline{\nu}_j}^{h}} \tag{3.13}$$

(5) For the second type-reduction, we obtain:

$$\underline{y} = \frac{\sum_{j=1}^{K} \underline{\phi}_j \nu_j}{\sum_{j=1}^{K} \nu_j} \tag{3.14}$$

$$\underline{y} = \frac{\sum_{j=1}^{K} \underline{\phi}_j \nu_j}{\sum_{j=1}^{K} \nu_j} \tag{3.15}$$

(6) The output is:

$$\hat{f} = \frac{1}{2}\underline{y} + \frac{1}{2}\underline{y} \tag{3.16}$$

Equation (3.16) is stated as:

$$\hat{f}(y|w) = w^T \psi(y) \tag{3.17}$$

where,

$$w^T = \left[\underline{\Psi}_{1,1}, \ldots, \underline{\Psi}_{N,1}, \ldots, \underline{\Psi}_{1,K}, \ldots, \underline{\Psi}_{N,K}\underline{\Psi}_{1,1}, \ldots, \underline{\Psi}_{N,1}, \ldots, \underline{\Psi}_{1,K}, \ldots, \underline{\Psi}_{N,K}\right] \tag{3.18}$$

$$\begin{aligned} \psi(y) = {} & \frac{\nu_j}{2\sum_{j=1}^{K} \nu_j}\Big[\frac{1}{\sum_{h=1}^{N} \underline{z}_{\underline{\nu}_1}^{h} + \underline{z}_{\underline{\nu}_1}^{h}}\left[\underline{z}_{\underline{\nu}_1}^{1} + \underline{z}_{\underline{\nu}_1}^{1}, \ldots, \underline{z}_{\underline{\nu}_1}^{N} + \underline{z}_{\underline{\nu}_1}^{N}\right] \\ & \cdots \frac{1}{\sum_{h=1}^{N} \underline{z}_{\underline{\nu}_K}^{h} + \underline{z}_{\underline{\nu}_K}^{h}}\left[\underline{z}_{\underline{\nu}_K}^{1} + \underline{z}_{\underline{\nu}_K}^{1}, \ldots, \underline{z}_{\underline{\nu}_K}^{N} + \underline{z}_{\underline{\nu}_K}^{N}\right] \\ & \frac{1}{\sum_{h=1}^{N} \underline{z}_{\underline{\nu}_1}^{h} + \underline{z}_{\underline{\nu}_1}^{h}}\left[\underline{z}_{\underline{\nu}_1}^{1} + \underline{z}_{\underline{\nu}_1}^{1}, \ldots, \underline{z}_{\underline{\nu}_1}^{N} + \underline{z}_{\underline{\nu}_1}^{N}\right] \\ & \cdots \frac{1}{\sum_{h=1}^{N} \underline{z}_{\underline{\nu}_K}^{h} + \underline{z}_{\underline{\nu}_K}^{h}}\left[\underline{z}_{\underline{\nu}_K}^{1} + \underline{z}_{\underline{\nu}_K}^{1}, \ldots, \underline{z}_{\underline{\nu}_K}^{N} + \underline{z}_{\underline{\nu}_K}^{N}\right]\Big]^T \end{aligned} \tag{3.19}$$

3.7.3 Control Without Restriction on on Sampling Time

This part presents the design and analysis of the controller without any restrictions on the sampling time. The primary relationships are outlined in Theorem 1.

Theorem 1 It is stated that the agents Eq. (3.1) will track the leader Eq. (3.2) if the controllers, adaptive compensator systems, and learning rules are defined as Eqs. (3.20–3.22).

$$m_{ii}u_i = -\sum_{j=1}^{N} m_{ij}\left(\hat{F}_j(y_i) - \dot{y}_0^n\right) - \sum_{j=1,j\neq i}^{N} m_{ij}u_j - \lambda_i^1\tilde{y}_i^1 - \lambda_i^2\tilde{y}_i^2 - \cdots - \lambda_i^n\tilde{y}_i^n + u_{s_i} \tag{3.20}$$

$$u_{s_i} = -sign\left(\tilde{y}_i^T \pi_i \zeta\right) K_i \tag{3.21}$$

$$\dot{w}_i = \gamma \tilde{y}_i^T \pi_i \zeta m_{ii} \psi_i(y_i) \tag{3.22}$$

where, $\lambda_i^j, j = 1, ..., n$ are considered to be positive constants. These values are determined in a way that the matrix eigenvalues to be negative:

$$\psi_i = \left[0\vdots 0 - \lambda_i^1 1 \vdots 0 - \lambda_i^2 \cdots \ddots \cdots\cdots 0 \vdots 1 - \lambda_i^n\right] \tag{3.23}$$

The tracking errors are considered to be:

$$\tilde{y}_i^1 = \sum_{j\in N_i} \alpha_{ij}\left(y_j^1 - y_i^1\right) + \beta_i\left(y_0^1 - y_i^1\right) \vdots \tilde{y}_i^n = \sum_{j\in N_i} \alpha_{ij}\left(y_j^n - y_i^n\right) + \beta_i\left(y_0^n - y_i^n\right) \tag{3.24}$$

where, $\alpha_{ij} = 1$ if agent i and j are connected, and $\beta_i = 1$ if agent i is connected to the leader. Expressed in matrix notation, the equation may be written as:

$$\tilde{y}^1 = -(\Lambda + B)\left(y^1 - \underline{1}y_0^1\right) \vdots \tilde{y}^n = -(\Lambda + B)\left(y^n - \underline{1}y_0^n\right) \tag{3.25}$$

where, $\underline{1} \in R^N$ is matrix with unit elements, $\tilde{y}^j = \left[\tilde{y}_1^j, ..., \tilde{y}_N^j\right]^T \in R^N$, $B = diag\{\beta_i\}$ and Λ is determined as:

$$\Lambda_{ij} = \left\{\sum_{j\in N_i} \alpha_{i,j} i = j - \alpha_{i,j},\ i\ and\ j\ are\ connceted\ 0,\ Otherwise\right. \tag{3.26}$$

Time-derivative of the errors, culminates in:

$$\dot{\tilde{y}}^1 = \tilde{y}^2 \vdots \dot{\tilde{y}}^n = M\left(F + u - \underline{1}\dot{y}_0^n\right) \tag{3.27}$$

where, $M = -(\Lambda + B), F = f + d, F = [F_1, ..., F_N]^T$. The tracking error dynamics for $i - th$ agent are stated as:

$$\dot{\tilde{y}}_i^1 = \tilde{y}_i^2 \dot{\tilde{y}}_i^n = \sum_{j=1}^{N} m_{ij}\left(F_j - \dot{y}_0^n\right) + \sum_{j=1, j \neq i}^{N} m_{ij}u_j + m_{ii}u_i \tag{3.28}$$

where, m_{ij} shows the element of (i, j) of matrix M. Equation (3.28), is formulated as:

$$\dot{\tilde{y}}_i = A\tilde{y}_i + \zeta\left(\sum_{j=1}^{N} m_{ij}\left(F_j - \dot{y}_0^n\right) + \sum_{j=1, j \neq i}^{N} m_{ij}u_j + m_{ii}u_i\right) \tag{3.29}$$

where,

$$A = \left[0 \vdots 0 0 1 \vdots 0 0 \cdots \ddots \cdots \cdots 0 \vdots 1 0\right], \zeta = \left[0 \vdots 0 1\right] \tag{3.30}$$

Utilizing the control signal Eq. (3.20), the racking error Eq. (3.29) is calculated to be:

$$\dot{\tilde{y}}_i = \psi_i \tilde{y}_i + \zeta u_{s_i} + \zeta m_{ii}\left(F_i - \hat{F}_i\right) + \zeta \sum_{j=1, j \neq i}^{N} m_{ij}\left(F_j - \hat{F}_j\right) \tag{3.31}$$

By performing addition and subtraction operations on the optimal T3-FLS $F_i^*(y_i, w_i^*)$, the value of $\dot{\tilde{y}}_i$ in Eq. (3.31) changes to:

$$\begin{aligned}
\dot{\tilde{y}}_i &= \psi_i \tilde{y}_i + \zeta u_{s_i} + \zeta m_{ii}\left(F_i^*(y_i, w_i^*) - \hat{F}_i(y_i, w_i)\right) \\
&+ \zeta \sum_{j=1, j \neq i}^{N} m_{ij}\left(\hat{F}_j^*\left(y_j, w_j^*\right) - \hat{F}_j(y_j, w_j)\right) \\
&+ \zeta \sum_{j=1}^{N} m_{ij}\left(F_j - \hat{F}_j^*\left(y_j, w_j^*\right)\right)
\end{aligned} \tag{3.32}$$

where the approximation error is determined to be:

$$v_i = \sum_{j=1, j \neq i}^{N} m_{ij}\left(\hat{F}_j^*\left(y_j, w_j^*\right) - \hat{F}_j(y_j, w_j)\right) + \sum_{j=1}^{N} m_{ij}\left(F_j - \hat{F}_j^*\left(y_j, w_j^*\right)\right) \tag{3.33}$$

Next, Eq. (3.32), is stated as:

$$\dot{\tilde{y}}_i = \psi_i \tilde{y}_i + \zeta u_{s_i} + \zeta v_i + \zeta m_{ii} \tilde{w}_i^T \psi_i(y_i) \tag{3.34}$$

Now, the Lyapunov function Eq. (3.35) is formulated as follows:

$$V = \frac{1}{2} \sum_{i=1}^{N} \tilde{y}_i^T \pi_i \tilde{y}_i + \frac{1}{2\gamma} \sum_{i=1}^{N} \tilde{w}_i^T \tilde{w}_i \tag{3.35}$$

where, $\tilde{w}_i = w_i^* - w_i$ and $\psi_i^T \pi_i + \pi_i \psi_i = -Q_i$. Time derivative of V in Eq. (3.35), gives:

$$\dot{V} = \sum_{i=1}^{N} \tilde{y}_i^T (\psi_i^T \pi_i + \pi_i \psi_i) \tilde{y}_i + \sum_{i=1}^{N} \tilde{y}_i^T \pi_i \zeta u_{s_i} + \sum_{i=1}^{N} \tilde{y}_i^T \pi_i \zeta v_i$$
$$+ \sum_{i=1}^{N} \tilde{y}_i^T \pi_i \zeta m_{ii} \tilde{w}_i^T \psi_i(y_i) - \frac{1}{\gamma} \sum_{i=1}^{N} \tilde{w}_i^T \dot{w}_i \tag{3.36}$$

Replacing the adaptation laws $\dot{w}_i = \gamma \tilde{y}_i^T \pi_i \zeta m_{ii} \psi_i(y_i)$ in Eq. (3.22), we can write:

$$\dot{V} = \sum_{i=1}^{N} \tilde{y}_i^T (\psi_i^T \pi_i + \pi_i \psi_i) \tilde{y}_i + \sum_{i=1}^{N} \tilde{y}_i^T \pi_i \zeta u_{s_i} + \sum_{i=1}^{N} \tilde{y}_i^T \pi_i \zeta v_i \tag{3.37}$$

Using Eq. (3.37), we have:

$$\dot{V} \leq -\sum_{i=1}^{N} \tilde{y}_i^T Q_i \tilde{y}_i + \sum_{i=1}^{N} \tilde{y}_i^T \pi_i \zeta u_{s_i} + \sum_{i=1}^{N} \left| \tilde{y}_i^T \pi_i \zeta \right| \underline{v}_i \tag{3.38}$$

Using the compensator $u_{s_i} = -sign\left(\tilde{y}_i^T \pi_i \zeta\right) K$ in Eq. (3.21), the inequality (3.38) is reformulated as:

$$\dot{V} \leq -\sum_{i=1}^{N} \tilde{y}_i^T Q_i \tilde{y}_i - \sum_{i=1}^{N} \tilde{y}_i^T \pi_i \zeta sign\left(\tilde{y}_i^T \pi_i \zeta\right) K_i + \sum_{i=1}^{N} \left| \tilde{y}_i^T \pi_i \zeta \right| \underline{v}_i \tag{3.39}$$

Given the equation $\tilde{y}_i^T \pi_i \zeta sign\left(\tilde{y}_i^T \pi_i \zeta\right) = \left| \tilde{y}_i^T \pi_i \zeta \right|$, it can be deduced form Eq. (3.39) that $\dot{V} < 0$, thereby concluding the proof.

Note 1 *In order to prevent the occurrence of chattering in the control signal, the expression* $sign\left(\tilde{y}_i^T \pi_i \zeta\right)$ *is substituted by* $tanh\left(\tilde{y}_i^T \pi_i \zeta\right)$.

3.7.4 Event-Triggered Scheme

Currently, we are addressing the issue of sampling and making updates to the control signal. The digital devices are used to implement the planned controller. Subsequently, it is updated at the sample times $t_1, ..., t_h, ...$ The control signal remains constant for the time span $t \in [t_h, t_{h+1}]$. Put simply, throughout the time interval T_h, which is equal to $T_h = t_{h+1} - t_h$, the control signals remain unchanged. During time period T_h, the system experiences the following error:

$$e_i(t) = \tilde{y}_i(t) - \tilde{y}_i(t_h) \tag{3.40}$$

Theorem 2 when the sample time decreases, the digital implementation approaches the analogue implementation, resulting in a reduced error e_i. In this section, the controller is updated using the event-triggered technique, and the stability is demonstrated. The issue at hand is determining the maximum allowable increase in size, denoted as T_h, that will still maintain system stability. The closed-loop system will stay stable if the control signal, compensator, compensator gain, and adaption laws are as specified in Eqs. (3.42–3.45), respectively, with respect to the sampling time period defined in Eq. (3.41).

$$T_h \geq ln\left[\frac{\sigma\alpha_i}{\Gamma(\tilde{y}_i(t_h)) + \beta} + 1\right] \tag{3.41}$$

$$\begin{aligned} m_{ii}u_i(t) = &-\sum_{j=1}^{N} m_{ij}\left(\hat{F}_j(t_h) - y_0^n\right) - \sum_{j=1, j\neq i}^{N} m_{ij}u_j(t) - \lambda_i^1\tilde{y}_i^1(t_h) \\ &- \lambda_i^2\tilde{y}_i^2(t_h) - \cdots - \lambda_i^n\tilde{y}_i^n(t_h) + u_{s_i}(t_h) \end{aligned} \tag{3.42}$$

$$u_{s_i} = -Ksign\left(\tilde{y}_i^T(t_h)\pi_i\zeta\right) \tag{3.43}$$

$$K > \underline{E}_i + \alpha_i \tag{3.44}$$

$$\dot{w}_i = \gamma\tilde{y}_i^T(t)\pi_i\zeta m_{ii}\mu(y_i) \tag{3.45}$$

where,

$$\Gamma_i(\tilde{y}_i(t_h)) = |\psi_i\tilde{y}_i(t_h)|\beta_i = \underline{E}_i + K \tag{3.46}$$

$$\psi_i^T\pi_i + \pi_i\psi_i = -Q_i \tag{3.47}$$

The proof is given in Sect. 3.7.6.

3.7.5 Simulations

This section provides instances to validate the efficacy of the devised scheme. To properly assess the effectiveness of the suggested control mechanism, we analyze a chaotic system for each agent. The first example uses a chaotic gyro, while the second example uses Chua's chaotic system. These systems exhibit complex dynamics.

The vertical axis of simulation graphics (y) displays the values of the signals represented in the legends. The voltage is the universal unit for all signals.

Example 1 *The first example examines the following multi-agent systems:*

$$\dot{y}_i^1 = y_i^2 \dot{y}_i^2 = \frac{-100\left(1 - cosy_i^1\right)^2}{sin^3 y_i^1} - 0.5y_i^2 - 0.05\left(y_i^2\right)^3$$
$$+ sin\left(y_i^1\right) + 35.5sin(25t)siny_i^1 + d(t) + u(t) \quad (3.48)$$

where, $d(t) = 0.15cos(t)y_i^2 + 0.1sin(2t)$. The remaining control settings are displayed in Table 3.1. Figure 3.5 depicts the communication structure and illustrates that all agents are interconnected with their neighboring agents. The primary agent has only connected with the initial agent. The other agents obtain the leader's information indirectly.

Figure 3.6 displays the output paths. The output signals of all agents ($y_i^1, i = 1, \ldots, 5$) closely follow the leader's output (y_0^1), as observed. The tracking faults are displayed in Fig. 3.7. The tracking errors ($\tilde{y}_i^1, i = 1, \ldots, 5$) converge to zero within a finite time. The efficacy of the planned event-triggered scenario is confirmed in Fig. 3.8. All of the actuators are mostly old compared to traditional methods. The control signals ($u_i, i = 1, \ldots, 5$) are illustrated in Fig. 3.9. The obtained control signals exhibit smooth profiles without any chattering. In order to demonstrate the superiority of T3-FLSs, the simulations were conducted using alternative types of FLSs instead of T3-FLSs. The resulting RMSE values are presented in Table 3.2

Table 3.1 Simulation condition in Ex. 1

Parameter	Equation
$\lambda_1 = 100, \lambda_2 = 20$	(3.42)
K	(3.43)
$\underline{E}_i = 10$	(3.44)
$\alpha_i = 0.2$	(3.44)
$\sigma = 0.1$	(3.44)
$\gamma = 0.1$	(3.45)
$\pi_i = [2.62500.00500.00500.0253]$	(3.43)
$c_{\tilde{\chi}_i^1} = -1, c_{\tilde{\chi}_i^2} = 0, c_{\tilde{\chi}_i^3} = 1$	(3-6)
$\underline{\nu}_1 = 0.7, \underline{\nu}_2 = 0.8, \underline{\nu}_3 = 0.9$	(3-6)
$\underline{\nu}_1 = 0.1, \underline{\nu}_2 = 0.2, \underline{\nu}_3 = 0.3$	(3-6)

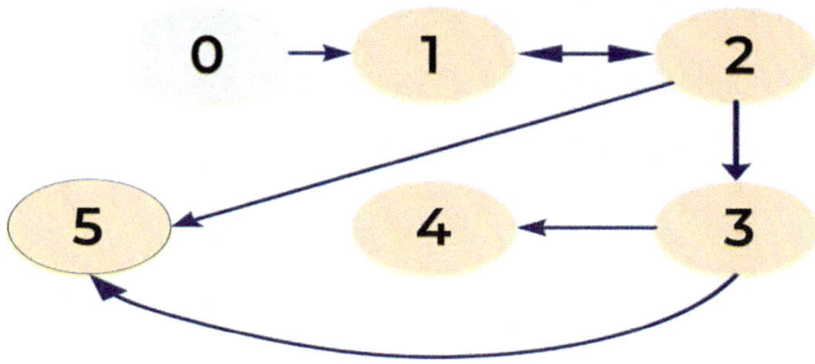

Fig. 3.5 Topology in Ex. 1

wherein utilizing the recommended T3-FLS leads to improved leader following performance.

Example 2 *The second example focuses on the synchronization of the following agents:*

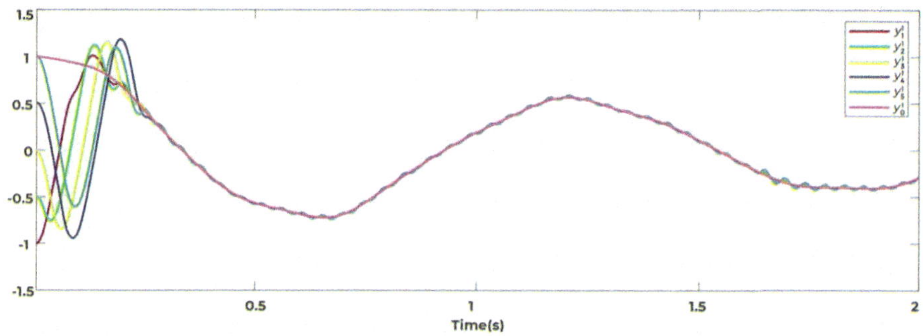

Fig. 3.6 Output trajectories in Ex. 1

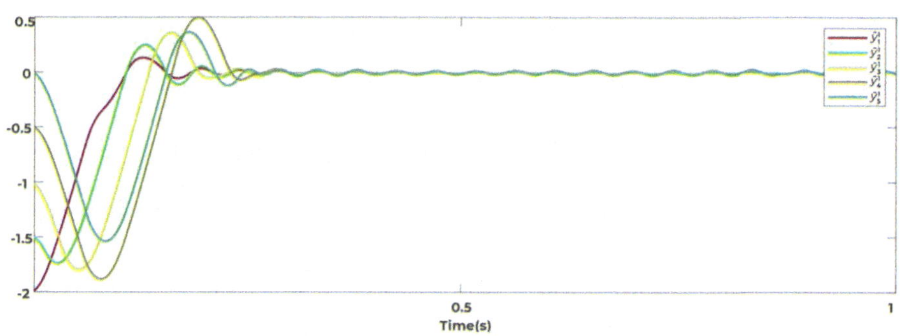

Fig. 3.7 Trajectories of the tracking error in Ex. 1

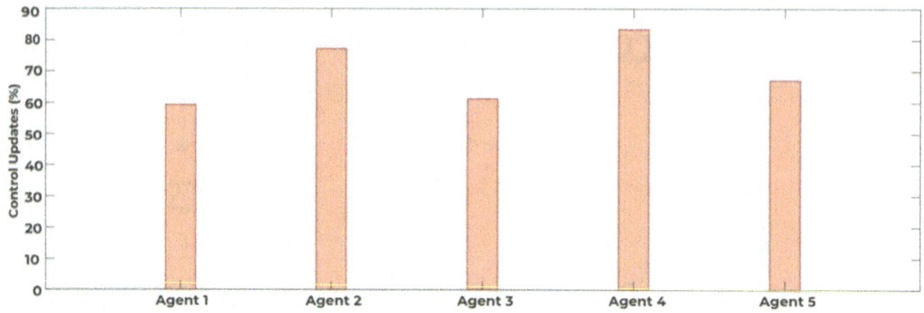

Fig. 3.8 Percentage of the control system updates in Ex. 1

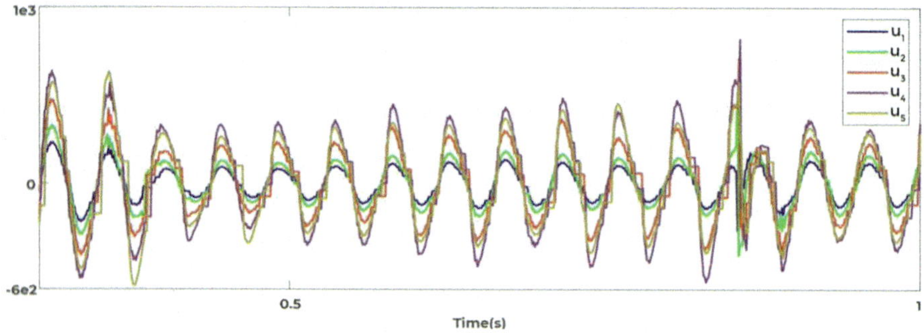

Fig. 3.9 Example 1: Control signals

Table 3.2 Example 2: Comparison results

	$\tilde{y}_1^1$	$\tilde{y}_1^2$	$\tilde{y}_1^3$	$\tilde{y}_1^4$	$\tilde{y}_1^5$
T1-FLS	0.3712	0.4048	0.1071	0.2012	1.4807
T2-FLS	0.2014	0.1241	0.0894	0.1387	0.1079
T3-FLS	0.1315	0.0857	0.0897	0.1494	0.0954

$$\dot{y}_i^1 = y_i^2 \dot{y}_i^2 = y_i^3 \dot{y}_i^3 = -y_i^1 - 1.1y_i^2 - 0.45y_i^3 + \left(y_i^2\right)^2 + d_i(t) + u_i(t) \tag{3.49}$$

$$\dot{y}_0^1 = y_0^2 \dot{y}_0^2 = y_0^3 \dot{y}_0^3 = 0.8y_0^1 - 1.1y_0^2 - 0.45y_0^3 - \left(y_0^1\right)^3 \tag{3.50}$$

The conditions are identical to those in Example 1, except $\lambda_i = [120000, 7400, 150]^T$ and

$$\pi_{1i} = \pi_{2i} = [625.187718.75060.000018.75060.75010.00060.00000.00060.0001] \tag{3.51}$$

The result trajectories are displayed in Fig. 3.10. A high level of expertise in tracking has been attained as well as that obtained in Example 1. Figure 3.10 demonstrates that the output signals of all agents ($y_i^1, i = 1, \ldots, 5$) have successfully reached a state of convergence with the leader's output (y_0^1). Figure 3.11 displays the deviations in tracking. The tracking errors ($\tilde{y}_i^1, i = 1, \ldots, 5$) are minimized to zero within a desired timeframe. Figure 3.12 illustrates that all control systems are less updated compared to conventional procedures. Observations indicate that the actuators are updated throughout 15–40% of all sample times. The control signals, denoted as $u_i, i = 1, \ldots, 5$, are illustrated in Fig. 3.13. The obtained control signals exhibit smooth and continuous patterns without any oscillations or abrupt changes. The phase plan in the absence of control is illustrated in Fig. 3.14. The trajectories depicted in Fig. 3.14 demonstrate that the agents are moving away from the leader and there is a lack of synchronization. The phase diagram after implementing the suggested controller is depicted in Fig. 3.15 in which the trajectories demonstrate that all agents closely converge to the chaotic leader system. In a manner similar to Example 1, the simulations were conducted again using different types of FLSs instead of T3-FLSs to demonstrate the superiority of T3-FLSs. The resulting values of RMSEs are presented in Table 3.3 in which findings indicate that employing the recommended T3-FLS leads to enhanced performance in the performance of leader following.

Note 2 The primary distinction and benefits of the developed controller are as follows:

- The whole dynamics of the agents as well as the leaders are uncertain. The developed controller is independent of the agents mathematical dynamics.
- The actuators are not triggered at every instance of sampling. However, a proposed optimal activation strategy is demonstrated to enhance the longevity of devices.
- It is proven that in the presence of noise, the recommended type-3 FLS-based scheme achieves superior accuracy.

Note 3 It is important to highlight that the control efficiency is independent of the topology. The sole requirement is that the communication graph must be a spanning

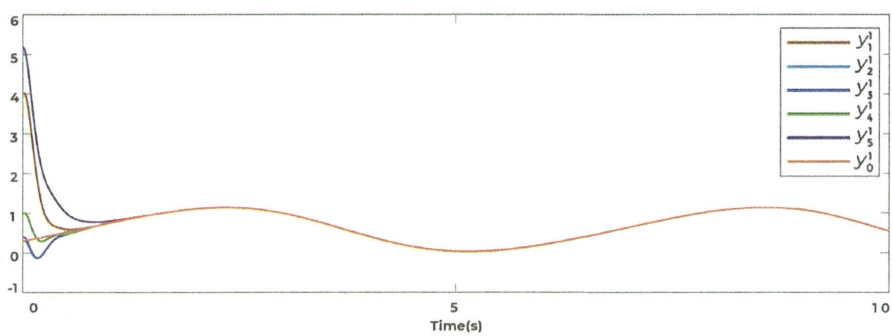

Fig. 3.10 Output trajectories in Ex. 2

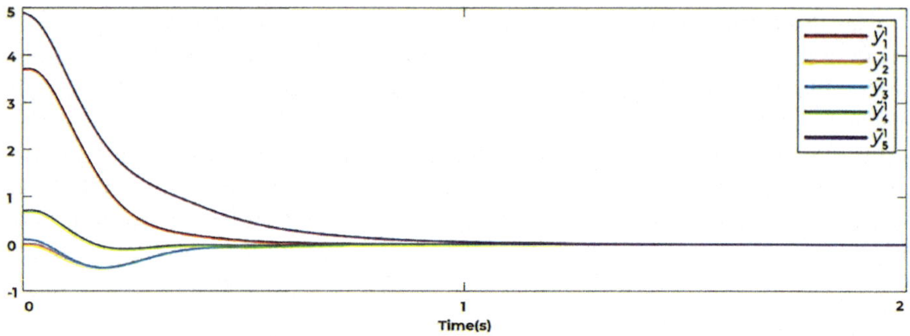

Fig. 3.11 Trajectories of the tracking error in Ex. 2

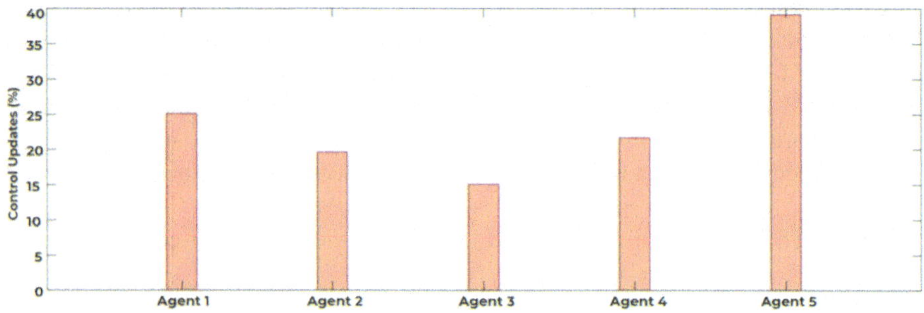

Fig. 3.12 Percentage of control system updates in Ex. 2

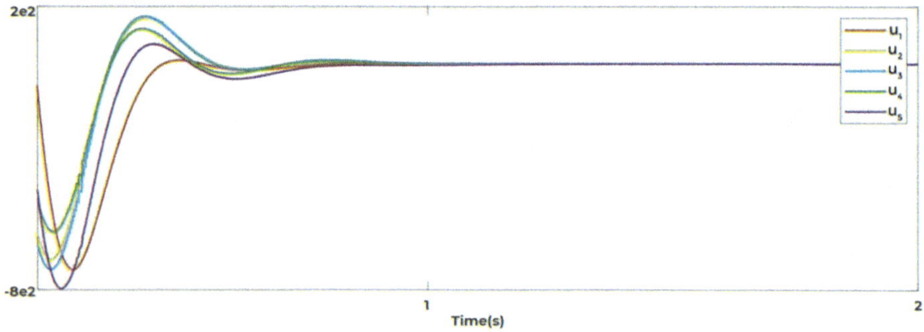

Fig. 3.13 Control signals in Ex. 2

tree, and all agents must be directly or indirectly connected to the leader agent. Put simply, there needs to be a direct connection between every agent and the leader agent.

Note 4 The controller is implemented in Matlab Simulink. Figure 3.16 provides an overview of the simulation.

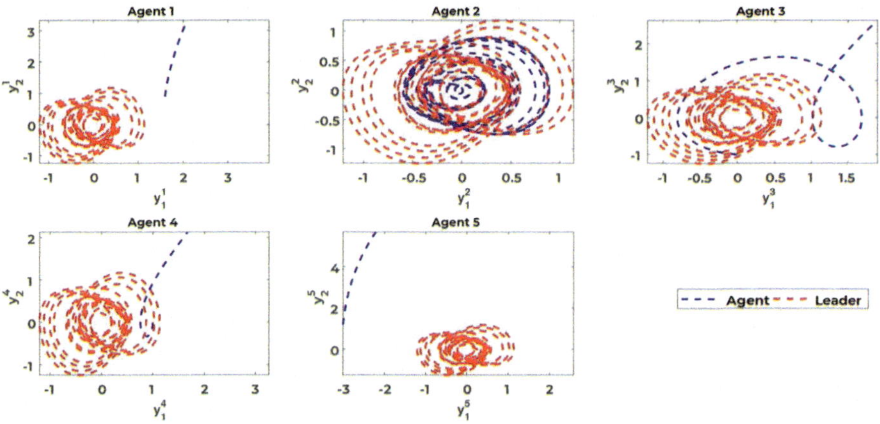

Fig. 3.14 The phase plan; without control in Ex. 2

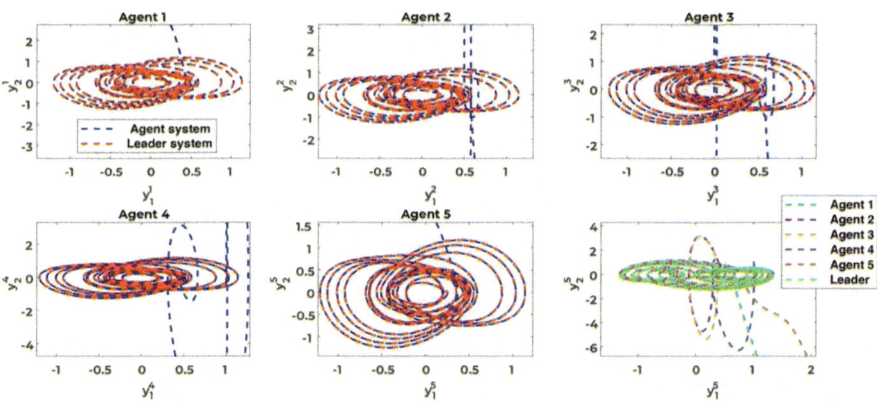

Fig. 3.15 The phase plan with control in Ex. 2

Table 3.3 Comparison results for Ex. 2

	$\tilde{y}_1^1$	$\tilde{y}_1^2$	$\tilde{y}_1^3$	$\tilde{y}_1^4$	$\tilde{y}_1^5$
T1-FLS	0.7231	0.2317	0.2479	0.3472	1.2180
T2-FLS	0.4274	0.1247	0.2017	0.1047	0.8713
T3-FLS	0.3952	0.0649	0.0625	0.0626	0.5568

3.7.6 Proof of Theorem 2

Proof Taking into account $e_i(t) = \tilde{y}_i(t) - \tilde{y}_i(t_h)$ and using the contrller obtained in Eq. (3.42), the dynamics of the tracking error are changed to:

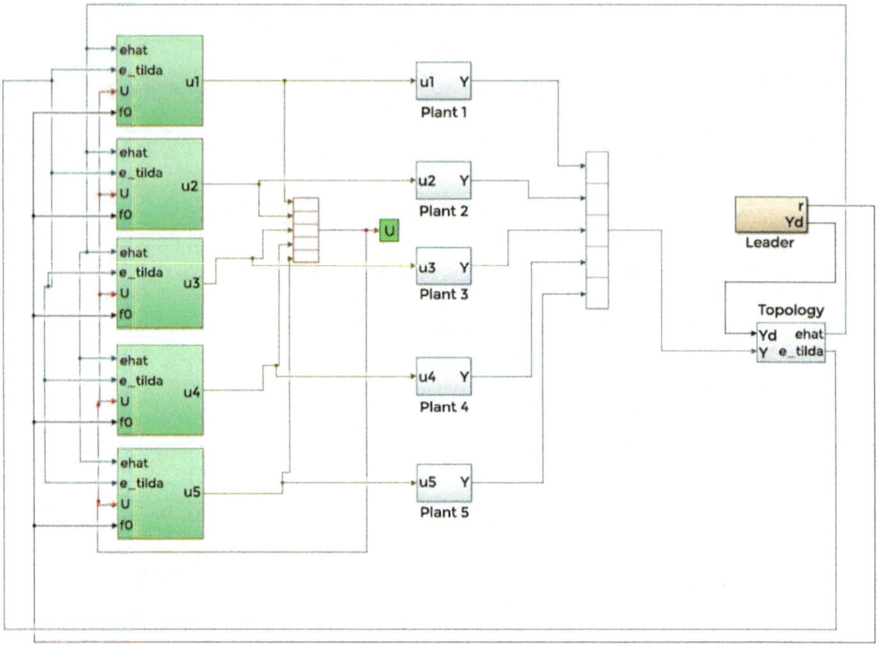

Fig. 3.16 Matlab implementation scheme

$$\dot{\tilde{y}}_i = A\tilde{y}_i(t) + \zeta \sum_{j=1}^{N} m_{ij}\big(F_j(t) - \dot{y}_0^n(t)\big) - \zeta \sum_{j=1}^{N} m_{ij}\big(\hat{F}_j(t_h) - \dot{y}_0^n(t_h)\big) + \zeta \sum_{j=1, j\neq i}^{N} m_{ij}\big(u_j(t) - u_j(t_h)\big) + \zeta u_{s_i}(t_h) + \zeta\big[-\lambda_i^1\tilde{y}_i^1(t_h) - \lambda_i^2\tilde{y}_i^2(t_h) - \cdots - \lambda_i^n\tilde{y}_i^n(t_h)\big] \tag{3.52}$$

From Eq. (3.23), $\dot{\tilde{y}}_i$ comes into:

$$\dot{\tilde{y}}_i = \psi_i\tilde{y}_i(t_h) + Ae_i(t) + \zeta \sum_{j=1}^{N} m_{ij}\big(F_j(t) - \dot{y}_0^n(t)\big) - \zeta \sum_{j=1}^{N} m_{ij}\big(\hat{F}_j(t_h) - \dot{y}_0^n(t_h)\big) + \zeta \sum_{j=1, j\neq i}^{N} m_{ij}\big(u_j(t) - u_j(t_h)\big) + \zeta u_{s_i}(t_h) \tag{3.53}$$

By performing the operations of addition and subtraction on the optimal T3-FLS $\hat{F}_j^*(t_h)$, one obtains:

$$\dot{\tilde{y}}_i = \psi_i \tilde{y}_i(t_h) + Ae_i(t) + \zeta \sum_{j=1}^{N} m_{ij}\big(F_j(t) - \dot{y}_0^n(t)\big) - \zeta \sum_{j=1}^{N} m_{ij}\big(\hat{F}_j(t_h) - \dot{y}_0^n(t_h)\big) -$$
$$\zeta \sum_{j=1}^{N} m_{ij}\Big(\hat{F}_j^*(t_h) - \dot{y}_0^n(t_h)\Big) + \zeta \sum_{j=1}^{N} m_{ij}\Big(\hat{F}_j^*(t_h) - \dot{y}_0^n(t_h)\Big)$$
$$+\zeta \sum_{j=1, j\neq i}^{N} m_{ij}\big(u_j(t) - u_j(t_h)\big) + \zeta u_{s_i}(t_h) \tag{3.54}$$

After simplification, Eq. (3.54) can be stated as:

$$\dot{\tilde{y}}_i = \psi_i \tilde{y}_i(t_h) + Ae_i(t) + \zeta \sum_{j=1}^{N} m_{ij}\big(F_j(t) - \dot{y}_0^n(t)\big) - \zeta \sum_{j=1}^{N} m_{ij}\Big(\hat{F}_j^*(t_h) - \dot{y}_0^n(t_h)\Big) +$$
$$m_{ii}\big(\hat{F}_i^*(t_h) - \hat{F}_i(t_h)\big) + \zeta \sum_{j=1, j\neq i}^{N} m_{ij}\Big(\hat{F}_j^*(t_h) - \hat{F}_j(t_h)\Big)$$
$$+\zeta \sum_{j=1, j\neq i}^{N} m_{ij}\big(u_j(t) - u_j(t_h)\big) + \zeta u_{s_i}(t_h) \tag{3.55}$$

From Eq. (3.55), we have:

$$\dot{\tilde{y}}_i = \psi_i \tilde{y}_i(t_h) + Ae_i(t) + \zeta m_{ii} \tilde{w}_i^T \mu(y_i) + \zeta u_{s_i}(t_h) + \zeta \sum_{j=1}^{N} m_{ij}\big(F_j(t) - \dot{y}_0^n(t)\big) -$$
$$\zeta \sum_{j=1}^{N} m_{ij}\Big(\hat{F}_j^*(t_h) - \dot{y}_0^n(t_h)\Big) + \zeta \sum_{j=1, j\neq i}^{N} m_{ij}\Big(\hat{F}_j^*(t_h) - \hat{F}_j(t_h)\Big)$$
$$+\zeta \sum_{j=1, j\neq i}^{N} m_{ij}\big(u_j(t) - u_j(t_h)\big) \tag{3.56}$$

Consider the following definition:

$$E_i(t) = \sum_{j=1}^{N} m_{ij}\big(F_j(t) - \dot{y}_0^n(t)\big) - \sum_{j=1}^{N} m_{ij}\Big(\hat{F}_j^*(t_h) - \dot{y}_0^n(t_h)\Big)$$
$$+ \sum_{j=1, j\neq i}^{N} m_{ij}\Big(\hat{F}_j^*(t_h) - \hat{F}_j(t_h)\Big) + \sum_{j=1, j\neq i}^{N} m_{ij}\big(u_j(t) - u_j(t_h)\big) \tag{3.57}$$

Then from Eqs. (3.56, 3.57), we get:

$$\dot{\tilde{y}}_i(t) = \psi_i \tilde{y}_i(t_h) + Ae_i(t) + \zeta m_{ii} \tilde{w}_i^T \mu(y_i) + \zeta E_i(t) + \zeta u_{s_i}(t_h) \tag{3.58}$$

In order to conduct stability analysis, it is necessary to take into account the Lyapunov function as follows:

$$V = \frac{1}{2} \sum_{i=1}^{N} \tilde{y}_i^T \pi_i \tilde{y}_i + \frac{1}{2\gamma} \sum_{i=1}^{N} \tilde{w}_i^T \tilde{w}_i \tag{3.59}$$

Taking time derivative of Eq. (3.59), yileds:

$$\dot{V}(t) = \sum_{i=1}^{N} \tilde{y}_i^T(t)\big(\psi_i^T \pi_i + \pi_i \psi_i\big)\tilde{y}_i(t) + \sum_{i=1}^{N} \tilde{y}_i^T(t)\pi_i A_i e_i(t) + \sum_{i=1}^{N} \tilde{y}_i^T(t)\pi_i \zeta u_{s_i}(t_h) + \sum_{i=1}^{N} \tilde{y}_i^T \pi_i \zeta E_i(t) + \sum_{i=1}^{N} \tilde{y}_i^T(t)\pi_i \zeta m_{ii} \tilde{w}_i^T \mu(y_i) - \frac{1}{2\gamma} \sum_{i=1}^{N} \tilde{w}_i^T \dot{w}_i \tag{3.60}$$

By substituting the $\dot{w}_i = \gamma \tilde{y}_i^T(t)\pi_i \zeta m_{ii} \mu(y_i)$ Eq. (3.45), we have:

$$\begin{aligned} \dot{V}(t) = &\sum_{i=1}^{N} \tilde{y}_i^T(t)\big(\psi_i^T \pi_i + \pi_i \psi_i\big)\tilde{y}_i(t) + \sum_{i=1}^{N} \tilde{y}_i^T(t)\pi_i A_i e_i(t) \\ &+ \sum_{i=1}^{N} \tilde{y}_i^T(t)\pi_i \zeta u_{s_i}(t_h) + \sum_{i=1}^{N} \tilde{y}_i^T \pi_i \zeta E_i(t) \end{aligned} \tag{3.61}$$

Using Eq. (3.61), we have:

$$\begin{aligned} \dot{V}(t) \le &\sum_{i=1}^{N} -\tilde{y}_i^T(t) Q_i \tilde{y}_i(t) + \sum_{i=1}^{N} \tilde{y}_i^T(t)\pi_i A_i e_i(t) \\ &+ \sum_{i=1}^{N} \tilde{y}_i^T(t)\pi_i \zeta u_{s_i}(t_h) + \sum_{i=1}^{N} \big|\tilde{y}_i^T \pi_i \zeta\big| |E_i(t)| \end{aligned} \tag{3.62}$$

If $sign\big(\tilde{y}_i^T(t_h)\pi_i \zeta\big) = sign\big(\tilde{y}_i^T(t)\pi_i \zeta\big)$, and knowing the fact that

$$\|A_i\| \|e_i(t)\| \prec \alpha_i \|\zeta\| \tag{3.63}$$

we obtain:

$$\dot{V}(t) \leq \sum_{i=1}^{N} -\tilde{y}_i^T(t)Q_i\tilde{x}_i(t) + \alpha_i\left|\tilde{y}_i^T(t)\pi_i\zeta\right| - \sum_{i=1}^{N} \tilde{y}_i^T(t)\pi_i\zeta K_i sign\left(\tilde{y}_i^T(t_h)\pi_i\zeta\right)$$
$$+\sum_{i=1}^{N}\left|\tilde{y}_i^T\pi_i\zeta\right||E_i(t)| \tag{3.64}$$

The inequality (3.64) can be rewritten as:

$$\dot{V}(t) \leq \sum_{i=1}^{N} -\tilde{y}_i^T(t)Q_i\tilde{y}_i(t) + \sum_{i=1}^{N}\left|\tilde{y}_i^T\pi_i\zeta\right|(|E_i(t)| + \alpha_i - K) \tag{3.65}$$

Until the condition $sign\left(\tilde{y}_i^T(t_h)\pi_i\zeta\right) = sign\left(\tilde{y}_i^T(t)\pi_i\zeta\right)$ is met, and when $K > \underline{E}_i + \alpha_i$, the closed-loop system will continue to be stable. $\underline{E}_i$ represents the maximum value that $|E_i(t)|$ can reach. Now, we will examine the failure to satisfy the relationship $sign\left(\tilde{y}_i^T(t_h)\pi_i\zeta\right) = sign\left(\tilde{y}_i^T(t)\pi_i\zeta\right)$ and study the appropriateness of the sampling time. The discrepancy between $\tilde{y}_i^T(t)\pi_i\zeta$ and $\tilde{y}_i^T(t_h)\pi_i\zeta$ is equivalent to [39].

$$\left|\tilde{y}_i^T(t_h)\pi_i\zeta - \tilde{y}_i^T(t)\pi_i\zeta\right| = \left|e_i^T(t)\pi_i\zeta\right| \tag{3.66}$$

By taking account of the following inequality:

$$\left|e_i^T(t)\pi_i\zeta\right| \leq \|e_i(t)\|\|\pi_i\|\|\zeta\| \tag{3.67}$$

Then, we have:

$$\left|\tilde{y}_i^T(t_h)\pi_i\zeta - \tilde{y}_i^T(t)\pi_i\zeta\right| \leq \|e_i(t)\|\|\pi_i\|\|\zeta\| \tag{3.68}$$

From the fact that $\|\zeta\| = 1$ and $\|A_i\| = \sqrt{\lambda_{max}\left(A_i^T A_i\right)}$, we calculate:

$$\left|\tilde{y}_i^T(t_h)\pi_i\zeta - \tilde{y}_i^T(t)\pi_i\zeta\right| \leq \alpha_i\|\pi_i\| \tag{3.69}$$

where, $\lambda_{max}\left(A_i^T A_i\right)$ shows the utmost eigenvalue value of $A_i^T A_i$. Inequality (3.69), demonstrates a bound for the change of $\tilde{y}_i^T(t)\pi_i\zeta$ at time interval $[t - t_h]$. The control signal is thus updated if $\pi_i^T(t)\pi_i\zeta > \alpha_i\|\pi_i\|$. Now, take into account the following inequalty: [39]

$$\frac{d}{dt}\|e_i(t)\| \leq \left\|\frac{d}{dt}e_i(t)\right\| \tag{3.70}$$

Then

$$\left\|\frac{d}{dt}e_i(t)\right\| = \left\|\frac{d}{dt}(\tilde{y}_i(t) - \tilde{y}_i(t_h))\right\| \tag{3.71}$$

Using Eq. (71), we calculate:

$$\left\| \frac{d}{dt}(\tilde{y}_i(t) - \tilde{y}_i(t_h)) \right\| \le \left\| \frac{d}{dt}\tilde{y}_i(t) \right\| + \left\| \frac{d}{dt}\tilde{y}_i(t_h) \right\| \le \left\| \frac{d}{dt}\tilde{y}_i(t) \right\| \tag{3.72}$$

Then,

$$\frac{d}{dt}\|e_i(t)\| \le \left\| \frac{d}{dt}\tilde{y}_i(t) \right\| \tag{3.73}$$

Using Eqs. (3.58, 3.73) infers:

$$\frac{d}{dt}\|e(t)\| \le \left\| \psi_i \tilde{y}_i(t_h) + Ae_i(t) + \zeta E_i(t) + \zeta u_{s_i}(t_h) \right\| \tag{3.74}$$

By replacing the compensator Eq. (3.43), the inequality (3.74), becomes:

$$\frac{d}{dt}\|e_i(t)\| \le \left\| \psi_i \tilde{y}_i(t_h) + Ae_i(t) + \zeta E_i(t) - Ksign\left(\tilde{y}_i^T(t_h) P_i \zeta\right)\zeta \right\| \tag{3.75}$$

From Eq. (3.75), one has:

$$\begin{aligned} \frac{d}{dt}\|e_i(t)\| &\le \left\| \psi_i \tilde{y}_i(t_h) + Ae_i(t) + \zeta E_i(t) - Ksign\left(\tilde{y}_i^T(t_h) \pi_i \zeta\right)\zeta \right\| \\ &\le \|Ae_i(t)\| + \|\psi_i \tilde{y}_i(t_h)\| + \|\zeta E_i(t)\| + \|K\zeta\| \end{aligned} \tag{3.76}$$

Considering the upper bound of $\underline{E}_i$ as $\underline{E}_i$, we have:

$$\frac{d}{dt}\|e_i(t)\| \le \|Ae_i(t)\| + \|\psi_i \tilde{y}_i(t_h)\| + \|\zeta_i\| \underline{E}_i + \|K\zeta\| \tag{3.77}$$

Then it can written:

$$\frac{d}{dt}\|e_i(t)\| \le \|A\| \|e_i(t)\| + \|\psi_i \tilde{y}_i(t_h)\| + \underline{E}_i + K \tag{3.78}$$

By replacing Γ form Eq. (3.76), one has:

$$\frac{d}{dt}\|e_i(t)\| \le \|A\| \|e_i(t)\| + \Gamma(\tilde{y}_i(t_h)) + \beta \tag{3.79}$$

From the fact that $\|A\| = \sqrt{\lambda_{max}\left(A^T A\right)} = 1$, one can write:

$$\frac{d}{dt}\|e_i(t)\| \le \|e_i(t)\| + \Gamma(\tilde{y}_i(t_h)) + \beta \tag{3.80}$$

Given the initial condition $e_i(t) = 0$, according to Eq. (3.80), we can conclude:

$$\|e_i(t)\| \leq (\Gamma(\tilde{y}_i(t_h)) + \beta)(exp(t - t_h) - 1) \tag{3.81}$$

Then considering $\|e_i(t_{h+1})\| = \sigma\alpha_i$, one has:

$$\sigma\alpha_i \leq (\Gamma(\tilde{y}_i(t_h)) + \beta)(exp(t - t_h) - 1) \tag{3.82}$$

Considering the relation $T_h = t_{h+1} - t_h$, the inequality (3.82) is stated as:

$$\sigma\alpha_i \leq (\Gamma(\tilde{x}_i(t_h)) + \beta)(exp(T_h) - 1) \tag{3.83}$$

Then a lower bound for T_h is calculated as:

$$\frac{\sigma\alpha_i}{\Gamma(\tilde{y}_i(t_h)) + \beta} + 1 \leq exp(T_h) \Rightarrow T_h \geq ln\left[\frac{\sigma\alpha_i}{\Gamma(\tilde{y}_i(t_h)) + \beta} + 1\right] \tag{3.84}$$

This ends the proof.

3.8 Type-3 Fuzzy Stabilization and Synchronization of Chaotic Systems

3.8.1 Problem Formulation

The following financial CSs are taken into account [40]:

$$\dot{\chi}_1 = (-a + \chi_2)\chi_1 + \chi_3\dot{\chi}_2 = -b\chi_2 - \chi_1^2 + 1\dot{\chi}_3 = -\chi_1 - c\chi_3 \tag{3.1}$$

where b is the investment cost, a is the saving amount, c is the commercial demands elasticity, $\dot{\chi}_1$ is the interest rate, $\dot{\chi}_2$ is the investment demand, and $\dot{\chi}_3$ is the price exponent. The analysis regarding bifurcation and Lyapunov spectrum have been investigated in [41, 42]. Figure 3.1 depicts phase pictures of chaotic attractors.

The investment cost, saving amount, and elasticity of business demands are assumed to be both unknown and dynamic. The proposed T3FLS is utilized to handle fluctuations in parameters as well as dynamic uncertainties and disturbances. Figure 3.2 displays the proposed control diagram. T3-FLSs are designed to stabilize all states and ensure that the outputs accurately follow the reader system. The compensators handle the disturbances.

Note 1 *This research focuses on the case-study CS 1, which is widely recognized in the field of economic plants due to its high popularity and its demonstration of*

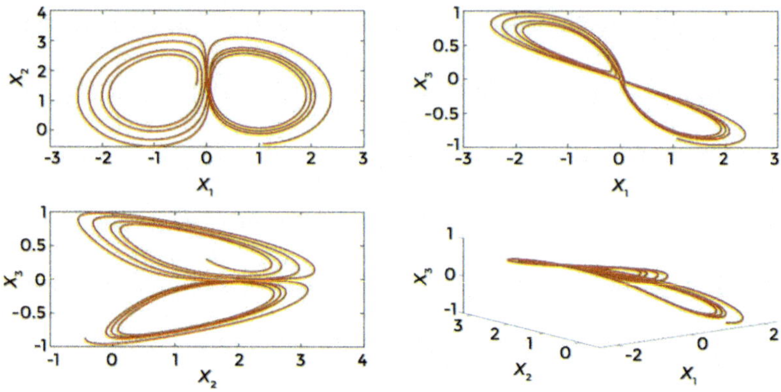

Fig. 3.1 Phase portrait

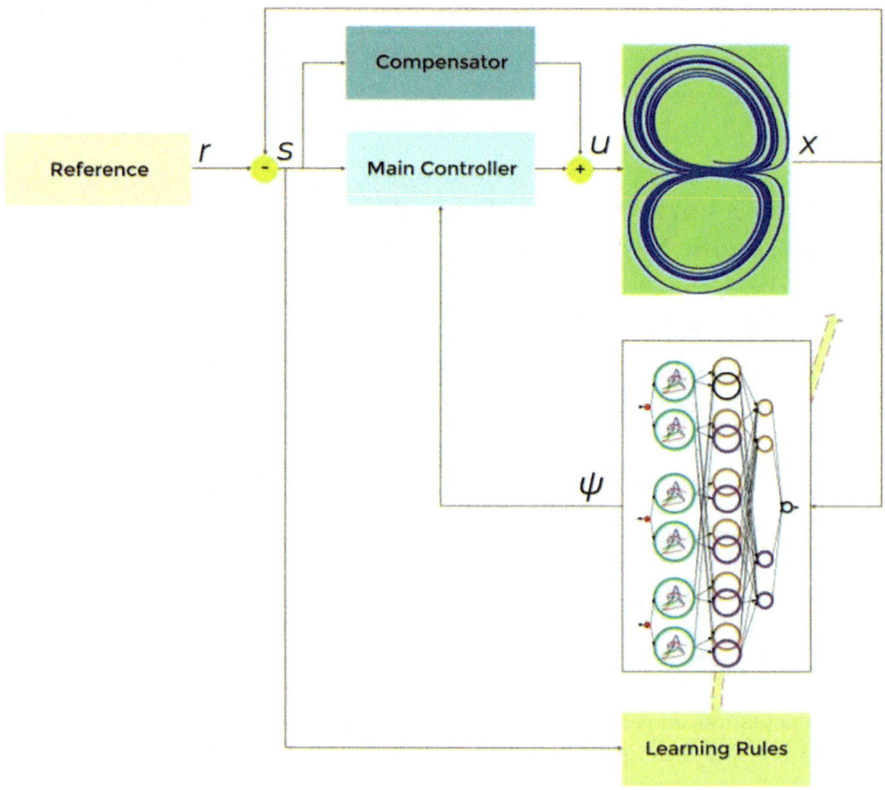

Fig. 3.2 Schematic view of control

intricate chaotic behavior. Nevertheless, the recommended controller is independent of the dynamics and may be readily implemented for different cases of CSs.

3.8.2 Type-3 FLS

Neuro-fuzzy systems and learning approaches are commonly employed to address uncertainties. This work presents a novel and robust strategy to address uncertainty in the intricate dynamics of complex systems (CSs). The proposed FLS scheme is illustrated in Fig. 3.3. Optimized T3-FLSs are employed to address all uncertainty. The calculations are presented as follows.

(1) The inputs of T3-FLS ψ_i are χ_1, χ_2 and χ_3.
(2) For inputs χ_1, χ_2 and χ_3, the Gaussian membership functions depicted in Fig. 3.4 are being taken into account. In contrast to type-2 membership functions, type-3 membership functions consist of four membership values for each horizontal slice. Two of them are linked to the lower incision, whereas the other two are associated with the upper incision. The MFs for χ_i are determined as $\underline{\Omega}_i$ and $\underline{\Omega}_i$ (see Fig. 3.4). The centers of $\underline{\Omega}_i$ and $\underline{\Omega}_i$ are defined by taking into account the upper/lower bounds $\underline{\Omega}_i$. For χ_i, the upper memberships of $\underline{\Omega}_{\chi_i}$ and $\underline{\Omega}_{\chi_i}$, taking account of upper/lower slice $\underline{\nu}_l/\underline{\nu}_l$ are stated as:

$$\mu_{\underline{\Omega}_{i,\underline{\nu}_l}}(\chi_i) = exp\left(-\frac{\left(\chi_i - C_{\underline{\Omega}_i}\right)^2}{\sigma^2_{\underline{\Omega}_{i,\underline{\nu}_l}}}\right) \tag{3.2}$$

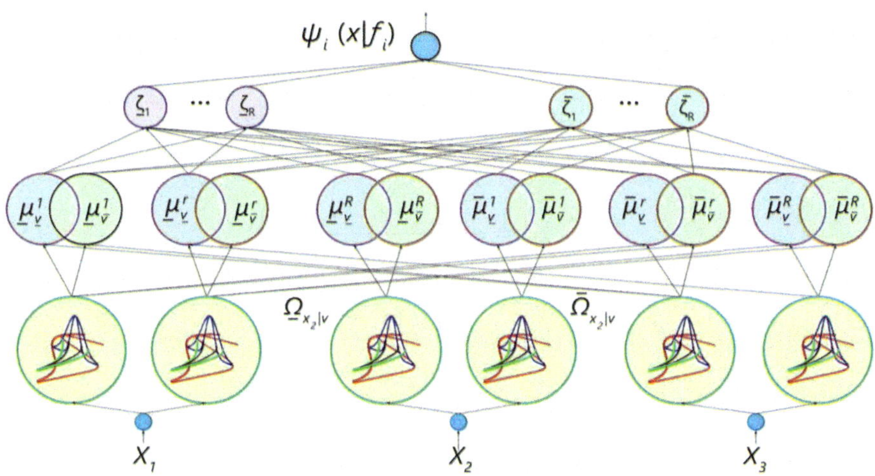

Fig. 3.3 Schematic view of T3-FLS

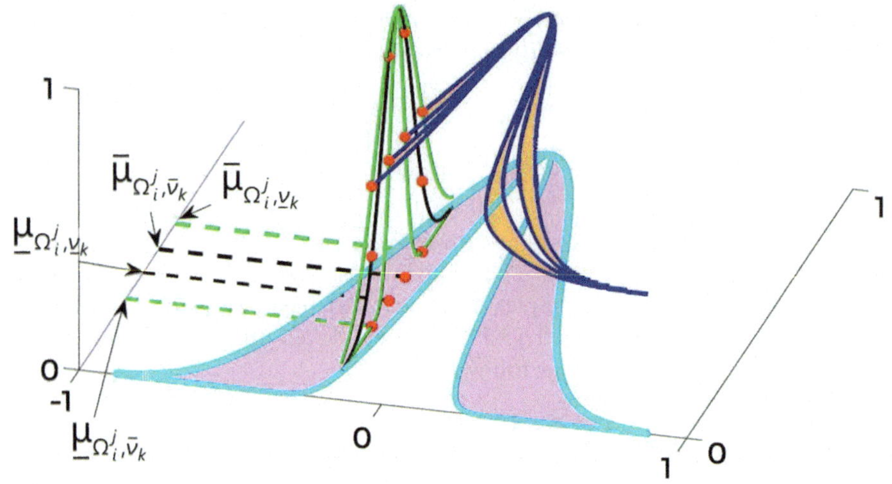

Fig. 3.4 Schematic view of type-3 MF

$$\underline{\mu}_{\underline{\Omega}_{i,\underline{\nu}_l}}(\chi_i) = exp\left(-\frac{\left(\chi_i - C_{\underline{\Omega}_i}\right)^2}{\underline{\sigma}^2_{\underline{\Omega}_{i,\underline{\nu}_l}}}\right) \tag{3.3}$$

$$\underline{\mu}_{\underline{\Omega}_{i,\underline{\nu}_l}}(\chi_i) = exp\left(-\frac{\left(\chi_i - C_{\underline{\Omega}_i}\right)^2}{\underline{\sigma}^2_{\underline{\Omega}_{i,\underline{\nu}_l}}}\right) \tag{3.4}$$

$$\underline{\mu}_{\underline{\Omega}_{i,\underline{\nu}_l}}(\chi_i) = exp\left(-\frac{\left(\chi_i - C_{\underline{\Omega}_i}\right)^2}{\underline{\sigma}^2_{\underline{\Omega}_{i,\underline{\nu}_l}}}\right) \tag{3.5}$$

Similarly, for $\underline{\Omega}_i$, we have:

$$\underline{\mu}_{\underline{\Omega}_{i,\underline{\nu}_l}}(\chi_i) = exp\left(-\frac{\left(\chi_i - C_{\underline{\Omega}_i}\right)^2}{\underline{\sigma}^2_{\underline{\Omega}_{i,\underline{\nu}_l}}}\right) \tag{3.6}$$

$$\underline{\mu}_{\underline{\Omega}_{i,\underline{\nu}_l}}(\chi_i) = exp\left(-\frac{\left(\chi_i - C_{\underline{\Omega}_i}\right)^2}{\underline{\sigma}^2_{\underline{\Omega}_{i,\underline{\nu}_l}}}\right) \tag{3.7}$$

$$\underline{\mu}_{\underline{\Omega}_{i,\underline{\nu}_l}}(\chi_i) = exp\left(-\frac{\left(\chi_i - C_{\underline{\Omega}_i}\right)^2}{\underline{\sigma}^2_{\underline{\Omega}_{i,\underline{\nu}_l}}}\right) \tag{3.8}$$

$$\underline{\mu}_{\underline{\Omega}_{i,\underline{\nu}_l}}(\chi_i) = exp\left(-\frac{\left(\chi_i - C_{\underline{\Omega}_i}\right)^2}{\underline{\sigma}^2_{\underline{\Omega}_{i,\underline{\nu}_l}}}\right) \tag{3.9}$$

where, C_{Ω_i} and $C_{\underline{\Omega}_i}$ show, respectively, centers of $\underline{\Omega}_{\chi_i}$ and $\underline{\Omega}_{\chi_i}$. $\underline{\sigma}_{\Omega_{i,\nu_l}}$ and $\underline{\sigma}^2_{\Omega_{i,\nu_l}}$ denote the upper width of $\underline{\Omega}_{\chi_i}$ at upper and lower slices of $\underline{\nu}_l$ and $\underline{\nu}_l$.

(3) In this step, by taking account of the memberships in the previous step, the rule firings are calculated. R number of rules are available. The $r-th$ rule is stated as:

$$Rule\ \#r:\ If\ \chi_1\ is\ \Omega_{\chi_1,\nu_j}\ and\ \chi_2\ is\ \Omega_{\chi_2,\nu_j}\ and\ \chi_3\ is\ \Omega_{\chi_3,\nu_j}\ Then\ \psi\ \in\left[\underline{x}_r,\ \underline{x}_r\right] \tag{3.10}$$

where, $\underline{x}_r, \underline{x}_r$ are the rule parameters.

(4) Using the product of T-norm and simple type-reduction of [1], the ψ_i output is stated as:

$$\psi_i = \underline{f}_i^T \varsigma_i \tag{3.11}$$

where, $\underline{f}_i$ and ς_i are:

$$f_i = \left[\underline{x}_{i1},\ldots,\underline{x}_{iR},\underline{x}_{i1},\ldots,\underline{x}_{iR}\right]^T$$

$$\varsigma_i = \left[\underline{\varsigma}_{i1},\ldots,\underline{\varsigma}_{iR},\underline{\varsigma}_{i1},\ldots,\underline{\varsigma}_{iR}\right]^T \tag{3.12}$$

where, R is the numbers of rules, $\underline{x}_{ir}$ and $\underline{x}_{ir}$ show the parameters of the $r-th$ rule. Moreover, $\underline{\varsigma}_{ir}$ and $\underline{\varsigma}_{il}$ are:

$$\underline{\varsigma}_r = \frac{\sum_{j=1}^{n_\nu}\underline{\nu}_j\frac{\underline{\theta}^r_{\underline{\nu}_j}}{\sum_{r=1}^{R}\left(\underline{\theta}^r_{\underline{\nu}_j}+\underline{\theta}^r_{\underline{\nu}_j}\right)}}{\sum_{j=1}^{n_\nu}\left(\underline{\nu}_j+\underline{\nu}_j\right)} + \frac{\sum_{j=1}^{n_\nu}\underline{\nu}_j\frac{\underline{\theta}^r_{\underline{\nu}_j}}{\sum_{r=1}^{R}\left(\underline{\theta}^r_{\underline{\nu}_j}+\underline{\theta}^r_{\underline{\nu}_j}\right)}}{\sum_{j=1}^{n_\nu}\left(\underline{\nu}_j+\underline{\nu}_j\right)},\ r=1,\ldots,R \tag{3.13}$$

$$\underline{\varsigma}_r = \frac{\sum_{j=1}^{n_\nu}\underline{\nu}_j\frac{\underline{\theta}^r_{\underline{\nu}_j}}{\sum_{r=1}^{R}\left(\underline{\theta}^r_{\underline{\nu}_j}+\underline{\theta}^r_{\underline{\nu}_j}\right)}}{\sum_{j=1}^{n_\nu}\left(\underline{\nu}_j+\underline{\nu}_j\right)} + \frac{\sum_{j=1}^{n_\nu}\underline{\nu}_j\frac{\underline{\theta}^r_{\underline{\nu}_j}}{\sum_{r=1}^{R}\left(\underline{\theta}^r_{\underline{\nu}_j}+\underline{\theta}^r_{\underline{\nu}_j}\right)}}{\sum_{j=1}^{n_\nu}\left(\underline{\nu}_j+\underline{\nu}_j\right)},\ r=1,\ldots,R \tag{3.14}$$

where, n_ν shows slices numbers and:

$$\underline{\varsigma}_r = \frac{\sum_{j=1}^{n_\nu}\underline{\nu}_j\frac{\underline{\theta}^r_{\underline{\nu}_j}}{\sum_{r=1}^{R}\left(\underline{\theta}^r_{\underline{\nu}_j}+\underline{\theta}^r_{\underline{\nu}_j}\right)}}{\sum_{j=1}^{n_\nu}\left(\underline{\nu}_j+\underline{\nu}_j\right)} + \frac{\sum_{j=1}^{n_\nu}\underline{\nu}_j\frac{\underline{\theta}^r_{\underline{\nu}_j}}{\sum_{r=1}^{R}\left(\underline{\theta}^r_{\underline{\nu}_j}+\underline{\theta}^r_{\underline{\nu}_j}\right)}}{\sum_{j=1}^{n_\nu}\left(\underline{\nu}_j+\underline{\nu}_j\right)},\ r=1,\ldots,R \tag{3.15}$$

$$\underline{\theta}^r_{\underline{\nu}_j} = \underline{\mu}_{\Omega_{\chi_1}|\underline{\nu}_j}(\chi_1)\underline{\mu}_{\Omega_{\chi_2}|\underline{\nu}_j}(\chi_2)\underline{\mu}_{\Omega_{\chi_3}|\underline{\nu}_j}(\chi_3)$$

$$
\begin{aligned}
\underline{\theta}^r_{\underline{\nu}_j} &= \underline{\mu}_{\Omega_{\chi_1}|\underline{\nu}_j}(\chi_1)\underline{\mu}_{\Omega_{\chi_2}|\underline{\nu}_j}(\chi_2)\underline{\mu}_{\Omega_{\chi_3}|\underline{\nu}_j}(\chi_3) \\
\underline{\theta}^r_{\underline{\nu}_j} &= \underline{\mu}_{\Omega_{\chi_1}|\underline{\nu}_j}(\chi_1)\underline{\mu}_{\Omega_{\chi_2}|\underline{\nu}_j}(\chi_2)\underline{\mu}_{\Omega_{\chi_3}|\underline{\nu}_j}(\chi_3) \\
\underline{\theta}^r_{\underline{\nu}_j} &= \underline{\mu}_{\Omega_{\chi_1}|\underline{\nu}_j}(\chi_1)\underline{\mu}_{\Omega_{\chi_2}|\underline{\nu}_j}(\chi_2)\underline{\mu}_{\Omega_{\chi_3}|\underline{\nu}_j}(\chi_3)
\end{aligned}
\tag{3.16}
$$

3.8.3 Stabilization

This part focuses on the design of the stabilizing controller and the analysis of its stability conditions. The primary findings are outlined in Theorem 1.

Theorem 1 *The stability of the system* (3.1) *is achieved by the use of a controller (3.17), adaptation rules* (3.18)*, and a compensator* (3.19)*.*

$$u_i = -\psi_i(\chi|f_i) - k_i\chi_i - u_{ci} \tag{3.17}$$

$$\dot{f}_i = \iota\varsigma_i\chi_i \tag{3.18}$$

$$u_{ci} = E_i sign(\chi_i) \tag{3.19}$$

where, ι and k are considered to be constants, ψ_i shows the T3FLS, χ and f are, respectively, the input vector and rule parameters vector of ψ_i, and the upper bound of AE is denoted by E_i.

Proof Using the controllers u_i, $i = 1, 2, 3$ Eq. (3.17) into Eq. (3.1) infers:

$$
\begin{aligned}
\dot{\chi}_1 &= (-a + \chi_2)\chi_1 + \chi_3 - k_1\chi_1 - \psi_1(\chi|f_1) - u_{c1}\dot{\chi}_2 \\
&= -b\chi_2 - \chi_1^2 + 1 - k_2\chi_2 - \psi_2(\chi|f_2) - u_{c2}\dot{\chi}_3 \\
&= -\chi_1 - c\chi_3 - k_3\chi_3 - \psi_3(\chi|f_3) - u_{c3}
\end{aligned}
\tag{3.20}
$$

From Eq. (3.43), $\dot{\chi}$ is calculated as follows:

$$
\begin{aligned}
\dot{\chi}_1 &= (-a + \chi_2)\chi_1 + \chi_3 - k_1\chi_1 - \psi_1(\chi|f_1) - u_{c1}\dot{\chi}_2 \\
&= -b\chi_2 - \chi_1^2 + 1 - k_2\chi_2 - \psi_2(\chi|f_2) - u_{c2}\dot{\chi}_3 \\
&= -\chi_1 - c\chi_3 - k_3\chi_3 - \psi_3(\chi|f_3) - u_{c3}
\end{aligned}
\tag{3.21}
$$

The ideal Type-3 Fuzzy Logic Systems (T3FLSs) of $\psi_i^*(\chi|f_i^*)$ are T3FLSs that have been defined with optimal parameters of f_i^*. After incorporating the optimal T3FLSs into Eq. (3.44) and using addition and subtraction, we obtain:

$$\dot{\chi}_1 = (-a + \chi_2)\chi_1 + \chi_3 - k_1\chi_1 - \psi_1(\chi|f_1) - u_{c1} + \psi_1^*(\chi|f_1^*) - \psi_1^*(\chi|f_1^*)\dot{\chi}_2 \\ = -b\chi_2 - \chi_1^2 + 1 - k_2\chi_2 - \psi_2(\chi|f_2) - u_{c2} + \psi_2^*(\chi|f_2^*) - \psi_2^*(\chi|f_2^*)\dot{\chi}_3 \\ = -\chi_1 - c\chi_3 - k_3\chi_3 - \psi_3(\chi|f_3) - u_{c3} + \psi_3^*(\chi|f_3^*) - \psi_3^*(\chi|f_3^*) \tag{3.22}$$

After simplification, $\chi_i, i = 1, 2, 3$ dynamics in Eq. (3.45) are reformulated as:

$$\dot{\chi}_1 = \psi_1^*(\chi|f_1^*) - \psi_1(\chi|f_1) - k_1\chi_1 - u_{c1} + (-a + \chi_2)\chi_1 + \chi_3 - \psi_1^*(\chi|f_1^*) \tag{3.23}$$

$$\dot{\chi}_2 = \psi_2^*(\chi|f_2^*) - \psi_2(\chi|f_2) - k_2\chi_2 - u_{c2} - b\chi_2 - \chi_1^2 + 1 - \psi_2^*(\chi|f_2^*) \tag{3.24}$$

$$\dot{\chi}_3 = \psi_3^*(\chi|f_3^*) - \psi_3(\chi|f_3) - k_3\chi_3 - u_{c3} - \chi_1 - c\chi_3 - \psi_3^*(\chi|f_3^*) \tag{3.25}$$

Using the vector form shown in Eq. (3.11), we obtain:

$$\dot{\chi}_1 = \tilde{f}_1^T \varsigma_1 - k_1\chi_1 - u_{c1} + (-a + \chi_2)\chi_1 + \chi_3 - \psi_1^*(\chi|f_1^*) \tag{3.26}$$

$$\dot{\chi}_2 = \tilde{f}_2^T \varsigma_2 - k_2\chi_2 - u_{c2} - b\chi_2 - \chi_1^2 + 1 - \psi_2^*(\chi|f_2^*) \tag{3.27}$$

$$\dot{\chi}_3 = \tilde{f}_3^T \varsigma_3 - k_3\chi_3 - u_{c3} - \chi_1 - c\chi_3 - \psi_3^*(\chi|f_3^*) \tag{3.28}$$

So that, $\tilde{f}_i = f_i^* - f_i$. In order to investigate stability and validate the conclusions of theorem 1, a Lyapunov candidate is chosen as:

$$V = \frac{1}{2\iota}\chi_1^2 + \frac{1}{2\iota}\chi_2^2 + \frac{1}{2\iota}\chi_3^2 + \frac{1}{2\iota}\tilde{f}_1^T\tilde{f}_1 + \frac{1}{2\iota}\tilde{f}_2^T\tilde{f}_2 + \frac{1}{2\iota}\tilde{f}_3^T\tilde{f}_3 \tag{3.29}$$

The derivative of Eq. (3.52) with respect to time gives:

$$\dot{V} = \frac{1}{\iota}\dot{\chi}_1\chi_1 + \frac{1}{\iota}\dot{\chi}_2\chi_2 + \frac{1}{\iota}\dot{\chi}_3\chi_3 - \frac{1}{\iota}\tilde{f}_1^T\dot{f}_1 - \frac{1}{\iota}\tilde{f}_2^T\dot{f}_2 - \frac{1}{\iota}\tilde{f}_3^T\dot{f}_3 \tag{3.30}$$

Inserting $\dot{\chi}_i, i = 1, 2, 3$ from Eqs. (3.26–3.28) into Eq. (3.53) obtains:

$$\dot{V} = [\tilde{f}_1^T \varsigma_1 - k_1\chi_1 - u_{c1} + (-a + \chi_2)\chi_1 + \chi_3 - \psi_1^*(\chi|f_1^*)]\chi_1 \\ +[\tilde{f}_2^T \varsigma_2 - k_2\chi_2 - u_{c2} - b\chi_2 - \chi_1^2 + 1 - \psi_2^*(\chi|f_2^*)]\chi_2 \\ +[\tilde{f}_3^T \varsigma_3 - k_3\chi_3 - u_{c3} - \chi_1 - c\chi_3 - \psi_3^*(\chi|f_3^*)]\chi_3 \\ -\frac{1}{\iota}\tilde{f}_1^T\dot{f}_1 - \frac{1}{\iota}\tilde{f}_2^T\dot{f}_2 - \frac{1}{\iota}\tilde{f}_3^T\dot{f}_3 \tag{3.31}$$

Equation (3.54), is reformulated as:

$$\dot{V} = -k_1\chi_1^2 + \tilde{f}_1^T\varsigma_1\chi_1 - \frac{1}{\iota}\tilde{f}_1^T\dot{f}_1 - u_{c1}\chi_1 - k_2\chi_2^2 + \tilde{f}_2^T\varsigma_2\chi_2 - \frac{1}{\iota}\tilde{f}_2^T\dot{f}_2 - u_{c2}\chi_2$$
$$-k_3\chi_3^2 + \tilde{f}_3^T\varsigma_3\chi_3 - \frac{1}{\iota}\tilde{f}_3^T\dot{f}_3 - u_{c3}\chi_3 + [(-a+\chi_2)\chi_1 + \chi_3 - \psi_1^*(\chi|f_1^*)]\chi_1 +$$
$$[-b\chi_2 - \chi_1^2 + 1 - \psi_2^*(\chi|f_2^*)]\chi_2 + [-\chi_1 - c\chi_3 - \psi_3^*(\chi|f_3^*)]\chi_3 \tag{3.32}$$

After simplification, Eq. (3.32) is changed to:

$$\dot{V} = -k_1\chi_1^2 + \tilde{f}_1^T\left(\varsigma_1\chi_1 - \frac{1}{\iota}\dot{f}_1\right) - u_{c1}\chi_1 - k_2\chi_2^2 + \tilde{f}_2^T\left(\varsigma_2\chi_2 - \frac{1}{\iota}\dot{f}_2\right) - u_{c2}\chi_2$$
$$-k_3\chi_3^2 + \tilde{f}_3^T\left(\varsigma_3\chi_3 - \frac{1}{\iota}\dot{f}_3\right) - u_{c3}\chi_3 + [(-a+\chi_2)\chi_1 + \chi_3 - \psi_1^*(\chi|f_1^*)]\chi_1$$
$$+[-b\chi_2 - \chi_1^2 + 1 - \psi_2^*(\chi|f_2^*)]\chi_2 + [-\chi_1 - c\chi_3 - \psi_3^*(\chi|f_3^*)]\chi_3 \tag{3.33}$$

By taking account of the tuning rules of Eq. (3.18), we have the following form Eq. (3.56):

$$\dot{V} = -k_1\chi_1^2 - k_2\chi_2^2 - k_3\chi_3^2 - u_{c1}\chi_1 - u_{c2}\chi_2 - u_{c3}\chi_3$$
$$+[(-a+\chi_2)\chi_1 + \chi_3 - \psi_1^*(\chi|f_1^*)]\chi_1 + [-b\chi_2 - \chi_1^2 + 1 - \psi_2^*(\chi|f_2^*)]\chi_2 \tag{3.34}$$
$$+[-\chi_1 - c\chi_3 - \psi_3^*(\chi|f_3^*)]\chi_3$$

Using Eq. (3.57), we have:

$$\dot{V} \le -k_1\chi_1^2 - k_2\chi_2^2 - k_3\chi_3^2 - u_{c1}\chi_1 - u_{c2}\chi_2 - u_{c3}\chi_3$$
$$+|(-a+\chi_2)\chi_1 + \chi_3 - \psi_1^*(\chi|f_1^*)||\chi_1| \tag{3.35}$$
$$+|-b\chi_2 - \chi_1^2 + 1 - \psi_2^*(\chi|f_2^*)||\chi_2| + |-\chi_1 - c\chi_3 - \psi_3^*(\chi|f_3^*)||\chi_3|$$

Utilizing the u_{ci} from Eq. (3.19), we have:

$$\dot{V} \le -k_1\chi_1^2 - k_2\chi_2^2 - k_3\chi_3^2 + \{|(-a+\chi_2)\chi_1 + \chi_3 - \psi_1^*(\chi|f_1^*)| - E_1\}|\chi_1|$$
$$+\{|-b\chi_2 - \chi_1^2 + 1 - \psi_2^*(\chi|f_2^*)| - E_2\}|\chi_2|$$
$$+\{|-\chi_1 - c\chi_3 - \psi_3^*(\chi|f_3^*)| - E_3\}|\chi_3| \tag{3.36}$$

Considering the following fact:

$$|(-a+\chi_2)\chi_1 + \chi_3 - \psi_1^*(\chi|f_1^*)| < E_1 \tag{3.37}$$

$$|-b\chi_2 - \chi_1^2 + 1 - \psi_2^*(\chi|f_2^*)| < E_2 \tag{3.38}$$

$$\left|-\chi_1 - c\chi_3 - \psi_3^*\left(\chi|f_3^*\right)\right| < E_3 \tag{3.39}$$

suggests $\dot{V} < 0$, and thus the proof is finished.

3.8.4 Synchronization

This part focuses on the design of the synchronization controller and the analysis of its stability criteria.

Theorem 2 *The system Eq.* (3.1) *monitors the states of the leader* (r_i) *using a controller Eq.* (3.40)*, adaption rules Eq.* (3.41)*, and a compensator Eq.* (3.42)*.*

$$u_i = \dot{r}_i - \psi_i(s|f_i) - k_i s_i - u_{ci} \tag{3.40}$$

$$\dot{f}_i = \iota \varsigma_i s_i \tag{3.41}$$

$$u_{ci} = E_i sign(s_i) \tag{3.42}$$

where, ι and k are assumed to be constants, ψ_i shows the T3FLS, s_i is determined as $s_i = \chi_i - r_i$, the input vector s is considered to be $s = [s_1, \ldots, s_n]^T$ and n denotes the states number, the rule parameters vector of ψ_i is denoted by f, and E_i is the upper bound of AE.

Proof Using the controllers $u_i, i = 1, 2, 3$ Eq. (3.40), the dynamics shown in Eq. (3.63) are stated as follows:

$$\dot{s}_1 = (-a + \chi_2)\chi_1 + \chi_3 - k_1 s_1 - \psi_1(s|f_1) - u_{c1}$$

$$\dot{s}_2 = -b\chi_2 - \chi_1^2 + 1 - k_2 s_2 - \psi_2(s|f_2) - u_{c2}$$

$$\dot{s}_3 = -\chi_1 - c\chi_3 - k_3 s_3 - \psi_3(s|f_3) - u_{c3} \tag{3.43}$$

From Eq. (3.43), $\dot{s}$ one writes:

$$\dot{s}_1 = (-a + \chi_2)\chi_1 + \chi_3 - k_1 s_1 - \psi_1(s|f_1) - u_{c1}$$

$$\dot{s}_2 = -b\chi_2 - \chi_1^2 + 1 - k_2 s_2 - \psi_2(s|f_2) - u_{c2}$$

$$\dot{s}_3 = -\chi_1 - c\chi_3 - k_3 s_3 - \psi_3(s|f_3) - u_{c3} \tag{3.44}$$

Following the same logic as the proof of Theorem 2, we include the optimal T3FLSs into Eq. (3.44) by addition and subtraction, resulting in:

$$\dot{s}_1 = (-a + \chi_2)\chi_1 + \chi_3 - k_1 s_1 - \psi_1(s|f_1) - u_{c1} + \psi_1^*(s|f_1^*) - \psi_1^*(s|f_1^*)$$

$$\dot{s}_2 = -b\chi_2 - \chi_1^2 + 1 - k_2 s_2 - \psi_2(s|f_2) - u_{c2} + \psi_2^*(s|f_2^*) - \psi_2^*(s|f_2^*)$$

$$\dot{s}_3 = -\chi_1 - c\chi_3 - k_3 s_3 - \psi_3(s|f_3) - u_{c3} + \psi_3^*(s|f_3^*) - \psi_3^*(s|f_3^*) \tag{3.45}$$

The dynamics shown in Eq. (3.45) can be reformulated as:

$$\dot{s}_1 = \psi_1^*(s|f_1^*) - \psi_1(s|f_1) - k_1 s_1 - u_{c1} + (-a + \chi_2)\chi_1 + \chi_3 - \psi_1^*(s|f_1^*) \tag{3.46}$$

$$\dot{s}_2 = \psi_2^*(s|f_2^*) - \psi_2(s|f_2) - k_2 s_2 - u_{c2} - b\chi_2 - \chi_1^2 + 1 - \psi_2^*(s|f_2^*) \tag{3.47}$$

$$\dot{s}_3 = \psi_3^*(s|f_3^*) - \psi_3(s|f_3) - k_3 s_3 - u_{c3} - \chi_1 - c\chi_3 - \psi_3^*(s|f_3^*) \tag{3.48}$$

Applying the vector form Eq. (3.11), we get:

$$\dot{s}_1 = \tilde{f}_1^T \varsigma_1 - k_1 s_1 - u_{c1} + (-a + \chi_2)\chi_1 + \chi_3 - \psi_1^*(s|f_1^*) \tag{3.49}$$

$$\dot{s}_2 = \tilde{f}_2^T \varsigma_2 - k_2 s_2 - u_{c2} - b\chi_2 - \chi_1^2 + 1 - \psi_2^*(s|f_2^*) \tag{3.50}$$

$$\dot{s}_3 = \tilde{f}_3^T \varsigma_3 - k_3 s_3 - u_{c3} - \chi_1 - c\chi_3 - \psi_3^*(s|f_3^*) \tag{3.51}$$

where, $\tilde{f}_i = f_i^* - f_i$. In order to prove the outcomes of Theorem 2, a Lyapunov candidate is taken into account as follows:

$$V = \frac{1}{2\iota}s_1^2 + \frac{1}{2\iota}s_2^2 + \frac{1}{2\iota}s_3^2 + \frac{1}{2\iota}\tilde{f}_1^T\tilde{f}_1 + \frac{1}{2\iota}\tilde{f}_2^T\tilde{f}_2 + \frac{1}{2\iota}\tilde{f}_3^T\tilde{f}_3 \tag{3.52}$$

The derivative of Eq. (3.52) with respect to time, yields:

$$\dot{V} = \frac{1}{\iota}\dot{s}_1 s_1 + \frac{1}{\iota}\dot{s}_2 s_2 + \frac{1}{\iota}\dot{s}_3 s_3 - \frac{1}{\iota}\tilde{f}_1^T\dot{f}_1 - \frac{1}{\iota}\tilde{f}_2^T\dot{f}_2 - \frac{1}{\iota}\tilde{f}_3^T\dot{f}_3 \tag{3.53}$$

Inserting $\dot{s}_i$, $i = 1, 2, 3$ from Eqs. (3.49–3.51), gives:

$$\dot{V} = \left[\tilde{f}_1^T \varsigma_1 - k_1 s_1 - u_{c1} + (-a + \chi_2)\chi_1 + \chi_3 - \psi_1^*\left(s|f_1^*\right)\right]s_1$$
$$+\left[\tilde{f}_2^T \varsigma_2 - k_2 s_2 - u_{c2} - b\chi_2 - \chi_1^2 + 1 - \psi_2^*\left(s|f_2^*\right)\right]s_2$$
$$+\left[\tilde{f}_3^T \varsigma_3 - k_3 s_3 - u_{c3} - \chi_1 - c\chi_3 - \psi_3^*\left(s|f_3^*\right)\right]s_3 - \frac{1}{\iota}\tilde{f}_1^T\dot{f}_1 - \frac{1}{\iota}\tilde{f}_2^T\dot{f}_2 - \frac{1}{\iota}\tilde{f}_3^T\dot{f}_3 \tag{3.54}$$

Equation (3.54) can be restated as:

$$\begin{aligned}\dot{V} &= -k_1 s_1^2 + \tilde{f}_1^T \varsigma_1 s_1 - \frac{1}{\iota}\tilde{f}_1^T\dot{f}_1 - u_{c1}s_1 - k_2 s_2^2 + \tilde{f}_2^T \varsigma_2 s_2 - \frac{1}{\iota}\tilde{f}_2^T\dot{f}_2 - u_{c2}s_2 \\ &-k_3 s_3^2 + \tilde{f}_3^T \varsigma_3 s_3 - \frac{1}{\iota}\tilde{f}_3^T\dot{f}_3 - u_{c3}s_3 + \left[(-a + \chi_2)\chi_1 + \chi_3 - \psi_1^*\left(s|f_1^*\right)\right]s_1 \\ &+\left[-b\chi_2 - \chi_1^2 + 1 - \psi_2^*\left(s|f_2^*\right)\right]s_2 + \left[-\chi_1 - c\chi_3 - \psi_3^*\left(s|f_3^*\right)\right]s_3\end{aligned} \tag{3.55}$$

After simplification, Eq. (3.55) changes to:

$$\begin{aligned}\dot{V} &= -k_1 s_1^2 + \tilde{f}_1^T\left(\varsigma_1\chi_1 - \frac{1}{\iota}\dot{f}_1\right) - u_{c1}s_1 - k_2 s_2^2 + \tilde{f}_2^T\left(\varsigma_2\chi_2 - \frac{1}{\iota}\dot{f}_2\right) - u_{c2}s_2 \\ &-k_3 s_3^2 + \tilde{f}_3^T\left(\varsigma_3\chi_3 - \frac{1}{\iota}\dot{f}_3\right) - u_{c3}s_3 + \left[(-a + \chi_2)\chi_1 + \chi_3 - \psi_1^*\left(s|f_1^*\right)\right]s_1 \\ &+\left[-b\chi_2 - \chi_1^2 + 1 - \psi_2^*\left(s|f_2^*\right)\right]s_2 + \left[-\chi_1 - c\chi_3 - \psi_3^*\left(s|f_3^*\right)\right]s_3\end{aligned} \tag{3.56}$$

Considering the tuning rules (3.41), we obtain:

$$\begin{aligned}\dot{V} &= -k_1 s_1^2 - k_2 s_2^2 - k_3 s_3^2 - u_{c1}s_1 - u_{c2}s_2 - u_{c3}s_3 \\ &+\left[(-a + \chi_2)\chi_1 + \chi_3 - \psi_1^*\left(s|f_1^*\right)\right]s_1 \\ &+\left[-b\chi_2 - \chi_1^2 + 1 - \psi_2^*\left(s|f_2^*\right)\right]s_2 + \left[-\chi_1 - c\chi_3 - \psi_3^*\left(s|f_3^*\right)\right]s_3 s\end{aligned} \tag{3.57}$$

From Eq. (3.57), we have:

$$\begin{aligned}\dot{V} &\le -k_1 s_1^2 - k_2 s_2^2 - k_3 s_3^2 - u_{c1}s_1 - u_{c2}s_2 - u_{c3}s_3 \\ &+\left|(-a + \chi_2)\chi_1 + \chi_3 - \psi_1^*\left(s|f_1^*\right)\right||s_1| \\ &+\left|-b\chi_2 - \chi_1^2 + 1 - \psi_2^*\left(s|f_2^*\right)\right||s_2| + \left|-\chi_1 - c\chi_3 - \psi_3^*\left(s|f_3^*\right)\right||s_3|\end{aligned} \tag{3.58}$$

Substituting u_{ci} from Eq. (3.19), we have:

$$\dot{V} \le -k_1 s_1^2 - k_2 s_2^2 - k_3 s_3^2 + \left\{\left|(-a + \chi_2)\chi_1 + \chi_3 - \psi_1^*\left(s|f_1^*\right)\right| - E_1\right\}|s_1|$$
$$+\left\{\left|-b\chi_2 - \chi_1^2 + 1 - \psi_2^*\left(s|f_2^*\right)\right| - E_2\right\}|s_2| + \left\{\left|-\chi_1 - c\chi_3 - \psi_3^*\left(s|f_3^*\right)\right| - E_3\right\}|s_3| \tag{3.59}$$

By taking account of the following facts:

$$\left|(-a+s_2)s_1+s_3-\psi_1^*\left(s|f_1^*\right)\right|<E_1 \tag{3.60}$$

$$\left|-bs_2-s_1^2+1-\psi_2^*\left(s|f_2^*\right)\right|<E_2 \tag{3.61}$$

$$\left|-s_1-cs_3-\psi_3^*\left(s|f_3^*\right)\right|<E_3 \tag{3.62}$$

we can write, $\dot{V} < 0$, and thus the proof is done.

3.8.5 Simulation

The feasibility and correctness of the stabilizing and synchronization system are confirmed by many simulations.

3.8.5.1 Stabilization

The controller Eq. (3.17), tuning rules Eq. (3.18), and compensator Eq. (3.19) stabilize the dynamics of Eq. (3.1) in this section. The simulation conditions are detailed in Table 3.1. The initial conditions are given by $\chi_1(0) = -0.20$, $\chi_2(0) = 1.50$, and $\chi_3(0) = 0.30$. The dynamics are uncertain. Figure 3.5 provides an overview of the simulation scheme. The paths of the outputs χ_1, χ_2 and χ_3 are shown in Fig. 3.6. It is evident that signals χ_1, χ_2 and χ_3 have reached a state of complete convergence towards zero. The control signals u_1, u_2 and u_3 are found in Fig. 3.7, whereas the outputs of T3-FLSs (ψ_1, ψ_2 and ψ_3) are shown in Fig. 3.8. In order to observe the approach towards the zero point more clearly, the phase portrait is depicted in Fig. 3.9. It is observed that the phase trajectories converge to the origin within a limited period of time.

3.8.5.2 Synchronization

This section focuses on evaluating the precision of synchronization. The chaotic system Eq. (3.1) is regarded as a leader system, whereas the slave financial chaotic system is considered to be the following:

$$\begin{aligned}\dot{r}_1 &= 0.8(r_2(t)-r_1(t))+u_1(t)\dot{r}_2 = 0.2r_1(t)-r_2(t)-r_1(t)r_3(t)\\ &+u_2(t)\dot{r}_3 = r_1(t)r_2(t)-1.9r_3(t)+u_3(t)\end{aligned} \tag{3.63}$$

The controller Eq. (3.40), tuning rules Eq. (3.41) and compensator Eq. (3.42) keep the system Eq. (3.63) in sync with Eq. (3.1). The conditions for the simulation are the same as in the last part. $r_1(0) = -0.20$, $r_2(0) = 1.50$, $r_3(0) = 0.30$, $\chi_1(0) = 1.0$,

Table 3.1 Simulation condition

Parameter	Value	Equation
$C\Omega\ i$ –	–1	(7)
$C_{\Omega}\bar{\ }i$	1	(3)
ν^{-}	0.7, 0.8, 1	(3)
$\nu^{\ -}$	0.1, 0.3, 4	(7)
k_i	100	(17)
ι	0.1	(18)
E_i	100	(19)
Algorithm type	ode3(Bogacki-Shampine)	
Integration step	0.001	
Truncation errors	0.001	

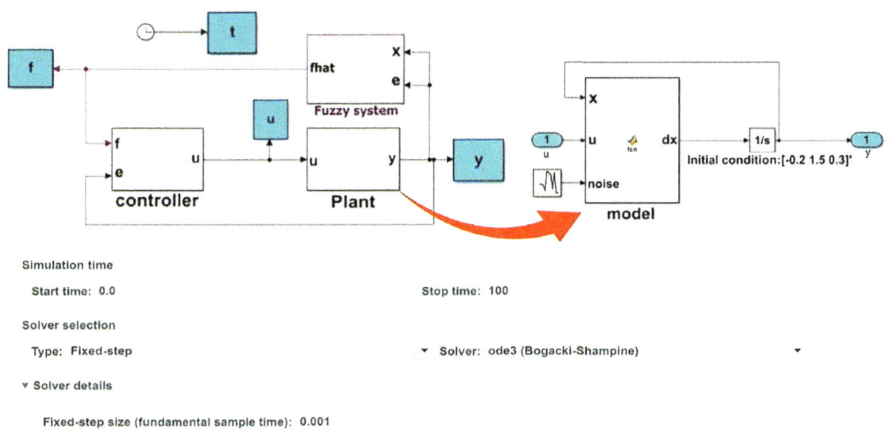

Fig. 3.5 Schematic view of the simulation

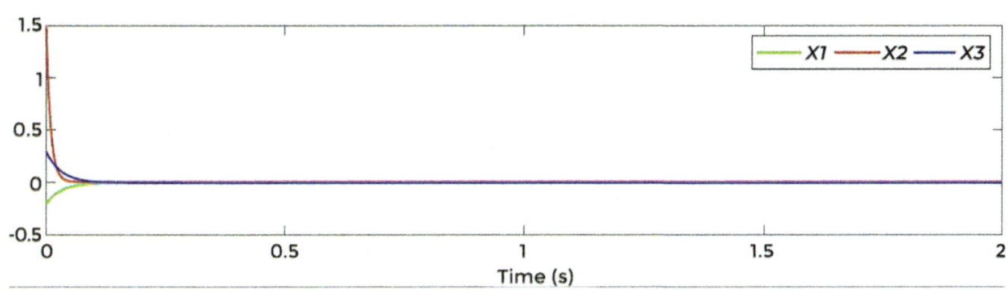

Fig. 3.6 Output signals in Scenario 3.8.5.1

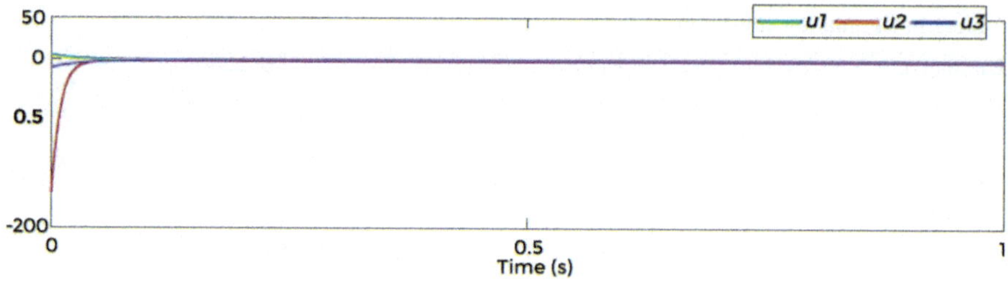

Fig. 3.7 Control signals in Scenario 3.8.5.1

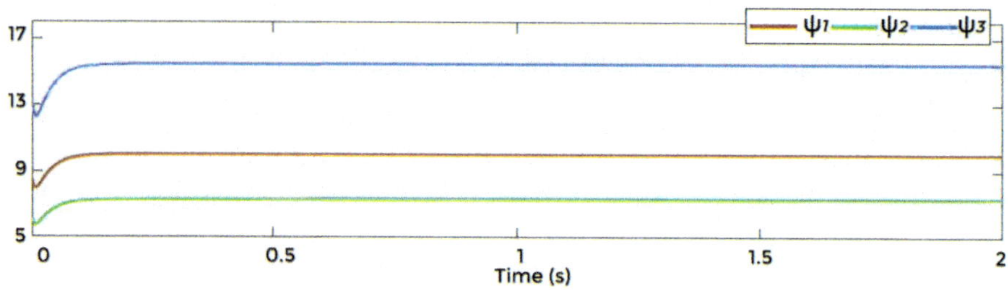

Fig. 3.8 T3-FLS signals in Scenario 3.8.5.1

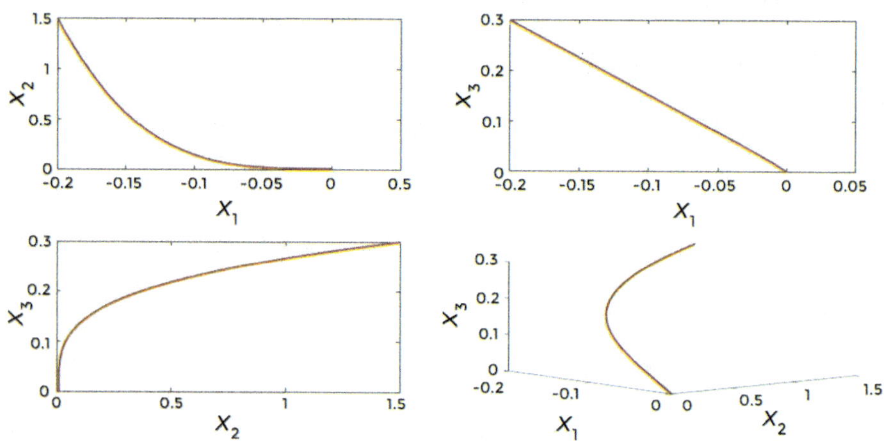

Fig. 3.9 Phase portrait in Scenario 3.8.5.1

$\chi_2(0) = -2.0$, and $\chi_3(0) = 5.0$ are the starting points. The paths of outputs χ_1, r_1, χ_2, r_2 and χ_3, r_3 are shown in Fig. 3.10. It's clear that signals χ_1, χ_2 and χ_3 have joined up well with r_1, r_2 and r_3. The synchronization errors in Fig. 3.11 show that the synchronization worked well. Figure 3.12 shows the control signals u_1, u_2 and u_3, and Fig. 3.13 shows the outputs of the T3-FLSs, which are ψ_1, ψ_2 and ψ_3. The

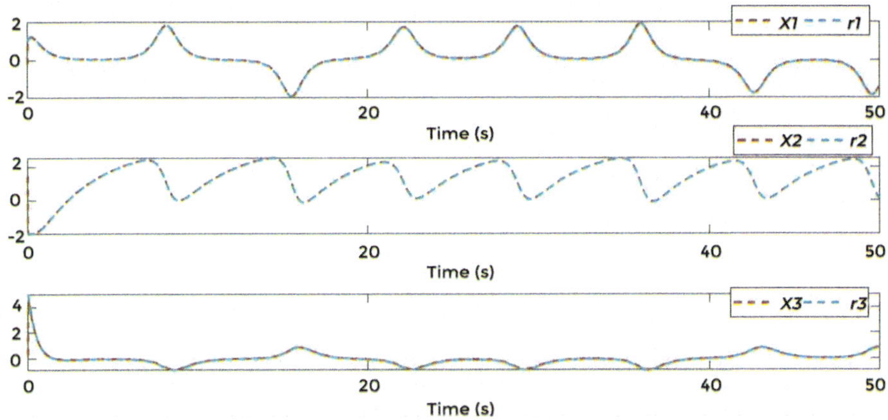

Fig. 3.10 Output signals in Scenario 3.8.5.2

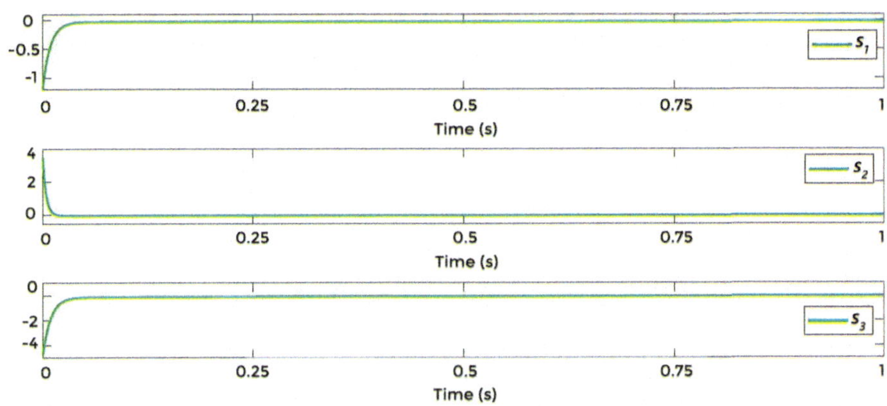

Fig. 3.11 Synchronization error signals in Scenario 3.8.5.2

phase picture in Fig. 3.14 can help you see how well the two signals are in sync. The phase paths of the slave system are very close to those of the leader one at a limited period of time.

3.8.5.3 High Noisy Condition and Comparison

In this part, a large perturbation is put on the system demonstrated in Fig. 3.15. The settings of the simulation are the same as the last test. The paths of outputs χ_1, r_1, χ_2, r_2 and χ_3, r_3 are shown in Fig. 3.16. We can see that the perturbations have been properly dealt with and that the signals χ_1, χ_2 and χ_3 have converged to r_1, r_2 and r_3. In Fig. 3.17, the synchronization errors are shown, along with the synchronization

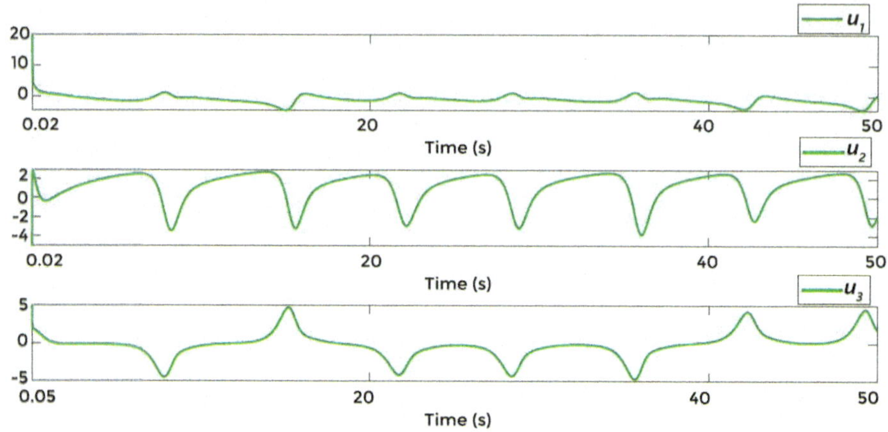

Fig. 3.12 Control signals in Scenario 3.8.5.2

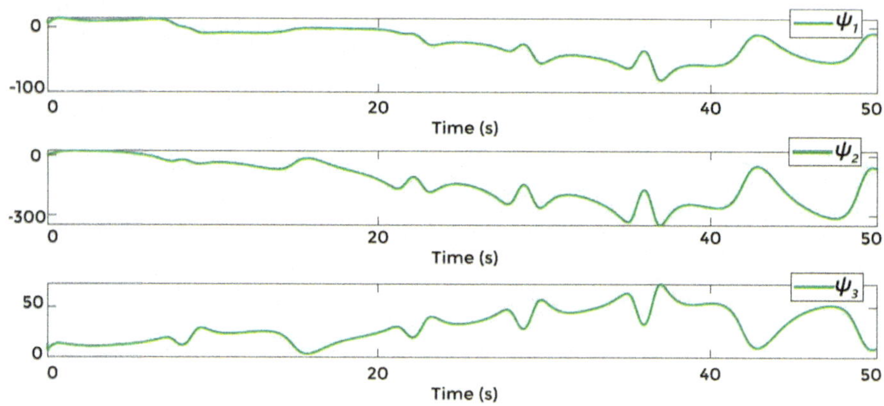

Fig. 3.13 T3-FLS signals in Scenario 3.8.5.2

that should happen when there are a lot of disturbances. In Fig. 3.18, you can see the control signals u_1, u_2 and u_3, and, in Fig. 3.19, you can see the phase picture. As we saw in the last study, the phase paths of the slave system are very close to those of the leader system at a limited period of time when there is noise.

A comparison is provided in order to assess the superiority of the proposed control scenario compared to other similar methods. The root-mean-square (RMS) of synchronization errors for the developed controller is being compared to the type-1 Fuzzy Logic System (T1-FLC) based controller [43], the type-2 Fuzzy Logic System (T2-FLC) based controller [44], and the generalized Fuzzy Logic System (GT2-FLC) based controller [45]. The comparative findings are displayed in Table 3.2. It is evident that the suggested controller achieves superior synchronization precision in highly noisy settings [46].

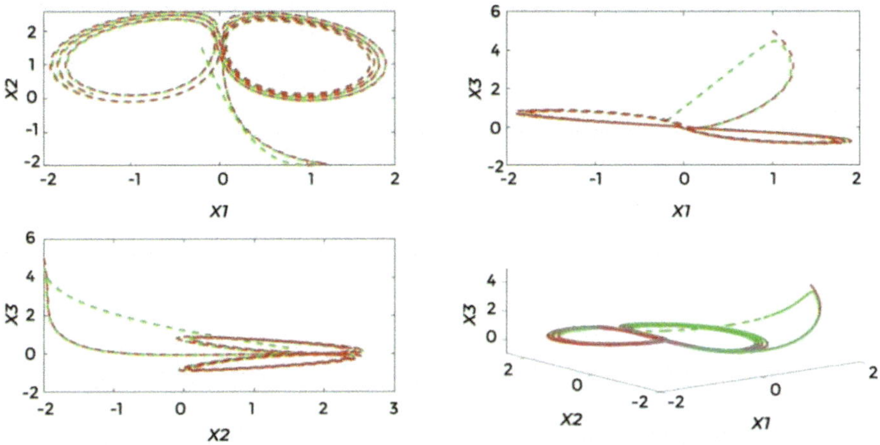

Fig. 3.14 Phase portrait in Scenario 3.8.5.2

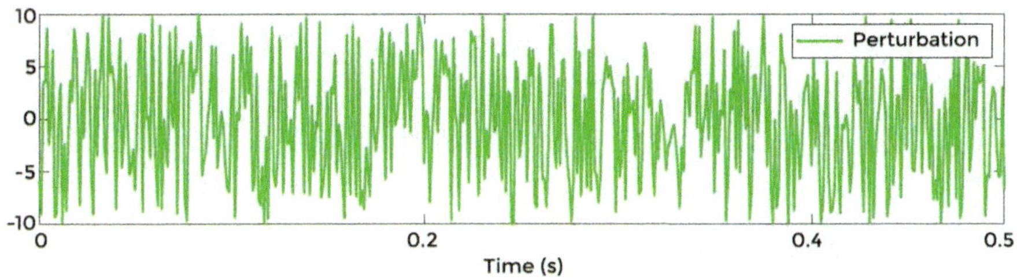

Fig. 3.15 Random dynamic perturbation in Scenario 3.8.5.3

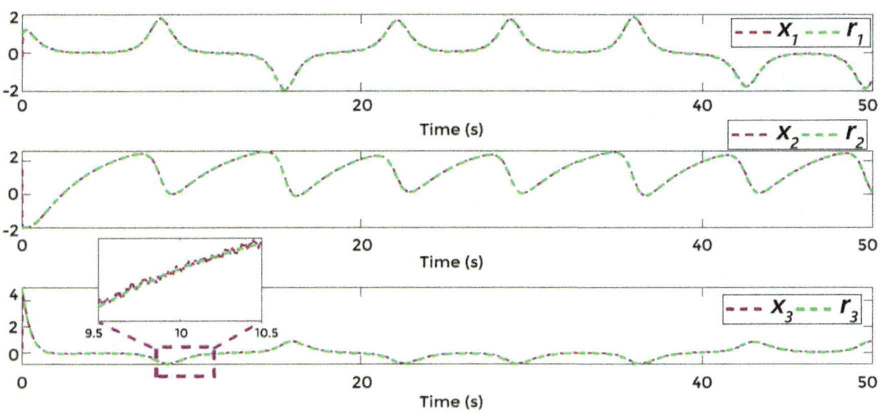

Fig. 3.16 Output signals in Scenario 3.8.5.3

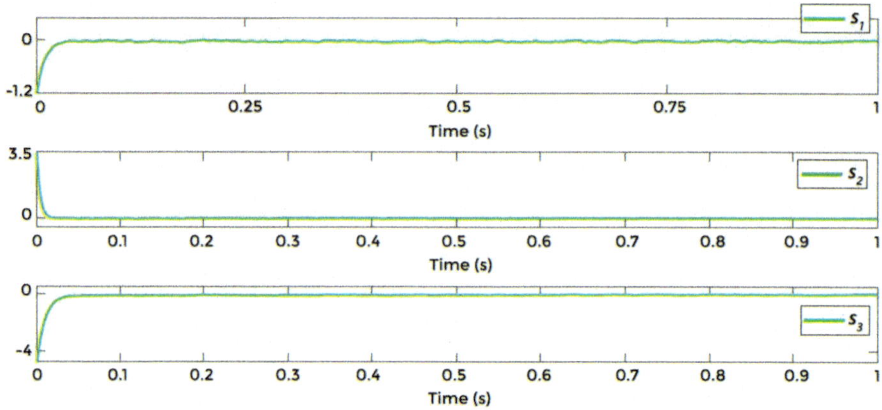

Fig. 3.17 Synchronization error signals in Scenario 3.8.5.3

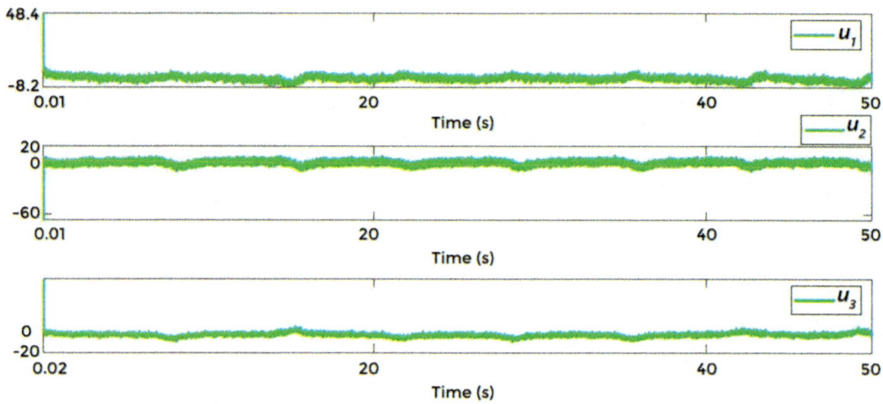

Fig. 3.18 Control signals in Scenario 3.8.5.3

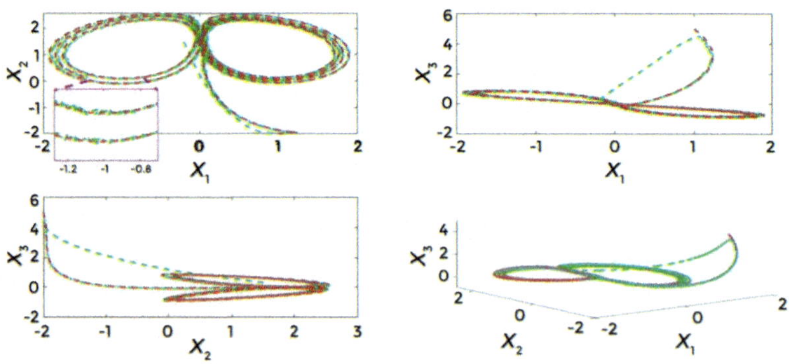

Fig. 3.19 Phase portrait in Scenario 3.8.5.3

Table 3.2 RMSE Comparison

Method	s_1	s_2	s_3
Proposed	0.0217	0.0266	0.0530
T1-FLC [43]	0.8421	1.0871	1.2820
T2-FLC [44]	0.4023	0.9142	1.0621
GT2-FLC [45]	0.4128	0.8071	1.0278

Note 2 This section presents the development of a novel T3-FLS based controller that achieves both stability and synchronization of financial CSs. The simulations conducted under different scenarios demonstrate that the developed controller achieves the intended level of efficiency. Despite the presence of loud noise and unpredictable dynamics, the provided technique effectively controls the target trajectories to match the desired reference. In addition, the resilience of the system is effectively maintained even when subjected to significant disturbances, thanks to the designed compensator. The controller architecture is independent of the specific model of CSs, allowing for easy application to various examples of CSs [46]

3.9 Multiple-Model Type-3 Fuzzy Control for Nonlinear Systems

3.9.1 System Description

The class of NSs with unmodeled and uncertain dynamics is being regarded as:

$$\dot{x} = f\left(x_1, \ldots, x_n, \varsigma_1, \ldots, \varsigma_p, u\right)$$

$$\dot{\varsigma} = g\left(x_1, \ldots, x_n, \varsigma_1, \ldots, \varsigma_p\right)$$

$$y = h\left(x_1, \ldots, x_n, \varsigma_1, \ldots, \varsigma_p, u\right) \tag{3.1}$$

Given $x = [x_1, \ldots, x_n]^T$ represents the vector of states in the modeled dynamics, denoted as $\varsigma = \left[\varsigma_1, \ldots, \varsigma_p\right]^T$ is the vector of states for the unmodeled dynamics, while $y = \left[y_1, \ldots, y_m\right]^T$ denotes the output vector of the system, and the control vector is represented as $u = [u_1, \ldots, u_m]^T$, whereas the unknown functions are represented by f, g, and h, which are nonlinear in nature.

The goal is to build control signals $u_1, \ldots, u_m$ in such a way that the system outputs $y_1, \ldots, y_m$ closely approximate the target signals $r_1, \ldots, r_m$. An overview of the proposed control mechanism is shown in Figs. 3.1 and 3.2. The system's outputs are represented by many first-order fractional-order dynamic fuzzy models, each with

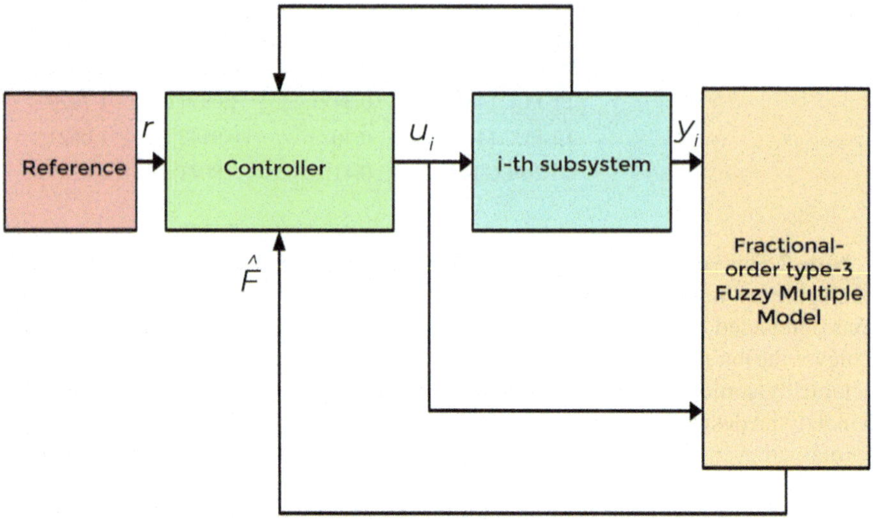

Fig. 3.1 Schematic view of the suggested control system

a distinct fractional-order. At each time interval, the model with the highest estimating capability is selected to construct the controller. Fractional-order calculus is superior to integer-order calculus in the identification of dynamic systems. To approximate a basic fractional-order integral using an integer-order integral, a transfer function of higher order is required. Instead of employing a high-order model for identification using integer-order calculus, a first-order model based on fractional-order calculus is offered. Several first-order dynamic models with varying fractional orders are proposed to demonstrate that the value of FO is not constant. Put simply, the impact of the value of FO is also considered.

The approach to estimate the dynamics of each output using several models is expressed as follows:

$$
\begin{aligned}
&\{D_t^{q_{11}}\hat{y}_{11}(t) = \hat{F}_{11}\left[y(t)|\theta_{11}(t)\right] - \kappa_{11}\hat{e}_{11} + u_{11}(t)\vdots D_t^{q_{1K}}\hat{y}_{1K}(t)\\
&= \hat{F}_{1K}\left[y(t)|\theta_{1K}(t)\right] - \kappa_{1K}\hat{e}_{1K} + u_{1K}(t)\\
&\vdots\{D_t^{q_{m1}}\hat{y}_{m1}(t) = \hat{F}_{m1}\left[y(t)|\theta_{m1}(t)\right] - \kappa_{m1}\hat{e}_{m1} + u_{m1}(t)\vdots D_t^{q_{mK}}\hat{y}_{mK}(t)\\
&= \hat{F}_{mK}\left[y(t)|\theta_{mK}(t)\right] - \kappa_{mK}\hat{e}_{mK} + u_{mK}(t)
\end{aligned}
\tag{3.2}
$$

The parameters vector of the FLS in the $k-th$ model for the $i-th$ subsystem is represented by $\theta_{ik}, i = 1, \ldots, m, k = 1, \ldots, K$, the variable κ_{ik} is determined from stability analysis and $\hat{e}_{ik}$ is considered to be $\hat{e}_{ik} = \hat{y}_{ik} - y_i$ and $D_t^{q_i}\hat{y}_i(t)$ shows the derivative of the fractional-order which is determined as follows:

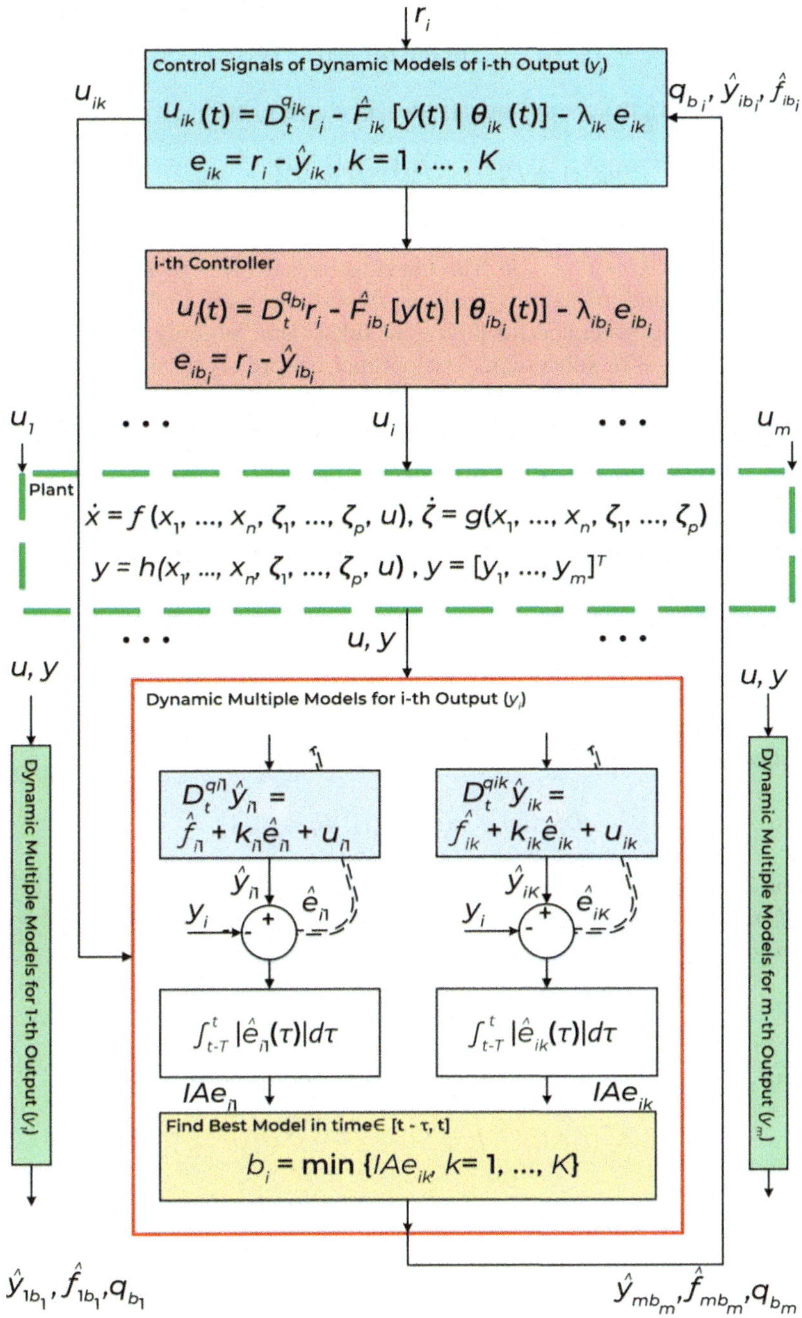

Fig. 3.2 The schematic diagram of the suggested control block

$$D_t^{q_i}\hat{y}_i(t) = \frac{1}{\Gamma(1-q_i)}\int_0^t \frac{\hat{y}_i^{(1)}(\tau)}{(t-\tau)^{q_i}}d\tau \tag{3.3}$$

Using Eq. (3.2), the control signals are calculated to be:

$$u_{ik}(t) = D_t^{q_{ik}} r_i - \hat{F}_{ik}\left[y(t)|\theta_{ik}(t)\right] - \lambda_{ik}e_{ik} \tag{3.4}$$

The control signal for the $k-th$ model in the $i-th$ subsystem is denoted as u_{ik}, $i = 1, \ldots, m$, $k = 1, \ldots, K$. The tracking error is given by $e_{ik} = r_i - \hat{y}_{ik}$. The value of λ_{ik} is found by stability analysis. According to Eq. (3.4), it is seen that K control signals are generated for the $i-th$ subsystem, but only the most effective one is used for $i-th$ subsystem. The optimal control signal is derived from the most accurate dynamic model. In order to determine the optimal dynamic model, the integral of the absolute estimation error is computed for each model ($IA_{\hat{e}_{ik}} = \int_{t-T}^{t} |\hat{e}_{ik}(\tau)|d\tau, k = 1, \ldots, K$) across the time period $[t-T, t]$. Subsequently, the model with the lowest $IA_{\hat{e}_{ik}}$ is selected as the optimal dynamic model (refer to the control block diagram shown in Fig. 3.2). The controller design process relies on the utilization of the following Lemmas and Assumptions.

Assumption 1 *The boundedness of the states of the unmodeled subsystem is assumed.*

Lemma 1 ([47]): *Taking account of the Lyapunov function $x^2(t)$, the fractional derivative of order q of $x^2(t)$ satisfies the inequality below:*

$$D_t^q x^2(t) \le x(t)D_t^q x(t), \forall q \in (0, 1) \tag{3.5}$$

Lemma 2 ([48]): *Taking into account the symmetric positive matrices Π_{11}, Π_{22} and Π_{12}, then the inequalities $\Pi_{11} - \Pi_{12}^T\Pi_{22}^{-1}\Pi_{12} > 0$ and $\Pi_{22} - \Pi_{12}^T\Pi_{11}^{-1}\Pi_{12} > 0$ can be reformulated as:*

$$[\Pi_{11}\Pi_{12}\Pi_{12}\Pi_{22}] < 0 \tag{3.6}$$

Lemma 3 ([49]): *Take account of the real matrices ψ, e and Δ such that:*

$$\Delta^T(t)\Delta(t) \le I, \psi\Delta(t)e + e^T\Delta^T(t)\psi^T < 0 \tag{3.7}$$

Then, there is a positive scaler ς so that the below inequality is valid:

$$\varsigma\psi\psi^T + \varsigma^{-1}\Delta^T(t)\Delta(t) < 0 \tag{3.8}$$

Lemma 4 ([50]): *The system $D_t^q x(t) = Ax(t)$, $0 < q < 1$ is asymptotically stable, if there exist symmetric matrices $Q_{11}, Q_{21} > 0$ and skew symmetric matrices Q_{12}, Q_{22} so that the below LMIs are satisfied:*

$$\sum_{i=1}^{2}\sum_{j=1}^{2} Sym\{\Omega_{ij} \otimes (AQ_{ij})\} <, [Q_{11}Q_{12} - Q_{12}Q_{11}] > 0, [Q_{21}Q_{22} - Q_{22}Q_{21}] > 0 \tag{3.9}$$

where, $Sym\{A\}$ is defined as $Sym\{A\} = A + A^T$, $\otimes$ is the Kronecker product, and Ω_{ij}, $i, j = 1, 2$ is calculated to be:

$$\begin{aligned}
\Omega_{11} &= [sin(q\pi/2) - cos(q\pi/2)cos(q\pi/2)sin(q\pi/2)] \\
\Omega_{12} &= [cos(q\pi/2)sin(q\pi/2) - sin(q\pi/2)cos(q\pi/2)] \\
\Omega_{21} &= [sin(q\pi/2)cos(q\pi/2) - cos(q\pi/2)sin(q\pi/2)] \\
\Omega_{22} &= [-cos(q\pi/2)sin(q\pi/2) - sin(q\pi/2) - cos(q\pi/2)]
\end{aligned} \tag{3.10}$$

3.9.2 The Interval Type-3 FLS

The interval type-3 fuzzy logic systems (IT3FLSs) are an extension of ordinary GT2FLSs that are better equipped to deal with significant levels of uncertainty [51]. This work utilizes IT3FLSs to do online identification of unclear NSs. Interval type-3 fuzzy sets possess an additional degree of freedom compared to regular general type-2 fuzzy sets. The secondary membership in interval type-3 fuzzy sets is represented by an interval type-2 fuzzy set, as opposed to the type-1 secondary membership in general type-2 fuzzy sets. Moreover, the levels of uncertainty in interval type-3 fuzzy sets, both upper and lower, are not predictable. Instead, they are represented as interval type-1 fuzzy sets. These characteristics make IT3FLSs more efficient in dealing with uncertainty and estimating nonlinear dynamic systems. Let's examine the $i - th$ output of the system Eq. (3.1), which is represented by the following dynamic system:

$$D_t^{q_{ik}} \hat{y}_{ik}(t) = \hat{F}_{ik}[y(t)|\theta_{ik}(t)] - \kappa_{ik}\hat{e}_{1k} + u_i(t) \tag{3.11}$$

where, $\hat{F}_{ik}[y(t)|\theta_{ik}(t)]$ is 3 IT3FLS. The other variables are obtained in Eq. (3.2). The details of the output $\hat{F}_{ik}[y(t)|\theta_{1k}(t)]$ calculation are explained in below.

(1) The inputs are considered to be $y_1, y_2, \ldots, y_m$.
(2) The memberships yield:

$$\underline{\vartheta}^{h}_{i,\underline{\alpha}_j}(y_i) = exp\left(-\frac{\left(y_i - c_{\psi_i^h}\right)}{\underline{\sigma}_{\psi^h_{i,\underline{\alpha}_j}}}\right) \tag{3.12}$$

$$\underline{\vartheta}^h_{i,\underline{\alpha}_j}(y_i) = exp\left(-\frac{\left(y_i - c_{\psi_i^h}\right)}{\underline{\sigma}_{\psi^h_{i,\underline{\alpha}_j}}}\right) \tag{3.13}$$

$$\underline{\vartheta}^h_{i,\underline{\alpha}_j}(y_i) = exp\left(-\frac{\left(y_i - c_{\psi_i^h}\right)}{\underline{\sigma}_{\psi^h_{i,\underline{\alpha}_j}}}\right) \tag{3.14}$$

$$\underline{\vartheta}^h_{i,\underline{\alpha}_j}(y_i) = exp\left(-\frac{\left(y_i - c_{\psi_i^h}\right)}{\underline{\sigma}_{\psi^h_{i,\underline{\alpha}_j}}}\right) \tag{3.15}$$

Here, ψ_i^h represents the $h-th$ membership function for the $i-th$ input, and its center is denoted as $c_{\psi_i^h}$. The lower and upper standard deviations for ψ_i^h at the upper horizontal slice $\underline{\alpha}_j$ are representted by $\underline{\sigma}_{\psi^h_{i,\underline{\alpha}_j}}$ and $\underline{\sigma}_{\psi^h_{i,\underline{\alpha}_j}}$ respectively. In a similar manner, $\underline{\sigma}_{\psi^h_{i,\underline{\alpha}_j}}$ and $\underline{\sigma}_{\psi^h_{i,\underline{\alpha}_j}}$ represent the lower and upper standard deviations for ψ_i^h at the upper horizontal slice $\underline{\alpha}_j$. The variables $\underline{\vartheta}^h_{i,\underline{\alpha}_j}$ and $\underline{\vartheta}^h_{i,\underline{\alpha}_j}$ represent the higher memberships for the $j-th$ α-cut. On the other hand, $\underline{\vartheta}^h_{i,\underline{\alpha}_j}$, $\underline{\vartheta}^h_{i,\underline{\alpha}_j}$ represent the lower memberships for the $j-th$ α-cut.

(3) The rule firing are:

$$\underline{z}^l_{\mu_s=\underline{\alpha}_j} = \underline{\vartheta}^l_{1,\underline{\alpha}_j}\underline{\vartheta}^l_{i,\underline{\alpha}_j}\cdots\underline{\vartheta}^l_{i,\underline{\alpha}_j} \tag{3.16}$$

$$\underline{z}^l_{\mu_s=\underline{\alpha}_j} = \underline{\vartheta}^l_{1,\underline{\alpha}_j}\underline{\vartheta}^l_{i,\underline{\alpha}_j}\cdots\underline{\vartheta}^l_{i,\underline{\alpha}_j} \tag{3.17}$$

$$\underline{z}^l_{\mu_s=\underline{\alpha}_j} = \underline{\vartheta}^l_{1,\underline{\alpha}_j}\underline{\vartheta}^l_{i,\underline{\alpha}_j}\cdots\underline{\vartheta}^l_{i,\underline{\alpha}_j} \tag{3.18}$$

$$\underline{z}^l_{\mu_s=\underline{\alpha}_j} = \underline{\vartheta}^l_{1,\underline{\alpha}_j}\underline{\vartheta}^l_{i,\underline{\alpha}_j}\cdots\underline{\vartheta}^l_{i,\underline{\alpha}_j} \tag{3.19}$$

(4) The output of $\hat{F}_{ik}\left[y(t)|\theta_{1k}(t)\right]$ is calculated to be:

$$\hat{F}_{ik} = \theta_{ik}\xi \tag{3.20}$$

where, θ_{ik} shoes the vector of successive parameters and ξ is the vector of the derivative of $\hat{f}_{ik}$ with respect to the consecutive variables. The vectors θ_{ik} and ξ are determined as:

$$\theta_{ik} = \left[\underline{w}_{ik1}, \ldots, \underline{w}_{ikR}, \underline{w}_{ik1}, \ldots, \underline{w}_{ikR}\right]^T, \xi = \left[\underline{\zeta}_1, \ldots, \underline{\zeta}_R, \underline{\zeta}_1, \ldots, \underline{\zeta}_R\right]^T \tag{3.21}$$

where, $\underline{w}_{ikl}$ and $\underline{w}_{ikl}$ show the parameters of the $l-th$ rule. Moreover, R is the number of total rules, and $\underline{\zeta}_l$ and $\underline{\zeta}_l$ are considered to be:

$$\underline{\zeta}_l = \frac{\sum_{j=1}^{n_\alpha} \underline{\alpha}_j \frac{\underline{z}^l_{\mu s=\underline{\alpha}_j}}{\sum_{l=1}^{R}\left(\underline{z}^l_{\mu s=\underline{\alpha}_j}+\underline{z}^l_{\mu s=\underline{\alpha}_j}\right)}}{\sum_{j=1}^{n_\alpha}\left(\underline{\alpha}_j+\underline{\alpha}_j\right)} + \frac{\sum_{j=1}^{n_\alpha} \underline{\alpha}_j \frac{\underline{z}^l_{\mu s=\underline{\alpha}_j}}{\sum_{l=1}^{R}\left(\underline{z}^l_{\mu s=\underline{\alpha}_j}+\underline{z}^l_{\mu s=\underline{\alpha}_j}\right)}}{\sum_{j=1}^{n_\alpha}\left(\underline{\alpha}_j+\underline{\alpha}_j\right)} \tag{3.22}$$

$$\underline{\zeta}_l = \frac{\sum_{j=1}^{n_\alpha} \underline{\alpha}_j \frac{\underline{z}^l_{\mu s=\underline{\alpha}_j}}{\sum_{l=1}^{R}\left(\underline{z}^l_{\mu s=\underline{\alpha}_j}+\underline{z}^l_{\mu s=\underline{\alpha}_j}\right)}}{\sum_{j=1}^{n_\alpha}\left(\underline{\alpha}_j+\underline{\alpha}_j\right)} + \frac{\sum_{j=1}^{n_\alpha} \underline{\alpha}_j \frac{\underline{z}^l_{\mu s=\underline{\alpha}_j}}{\sum_{l=1}^{R}\left(\underline{z}^l_{\mu s=\underline{\alpha}_j}+\underline{z}^l_{\mu s=\underline{\alpha}_j}\right)}}{\sum_{j=1}^{n_\alpha}\left(\underline{\alpha}_j+\underline{\alpha}_j\right)} \tag{3.23}$$

where, n_α is the number of α-cuts α, $\underline{\alpha}_j$ and $\underline{\alpha}_j$ are the upper and lower of the secondary membership in the $j-th$ α-cut (refere to Fig. 3.3), $l=1,\ldots,R$, R is the number of rules, $\underline{z}^l_{\mu_s=\underline{\alpha}_j}$, $\underline{z}^l_{\mu_s=\underline{\alpha}_j}$ and $\underline{z}^l_{\mu_s=\underline{\alpha}_j}$, $\underline{z}^l_{\mu_s=\underline{\alpha}_j}$ are the lower and the upper rule firings. For furthur discussion see [51].

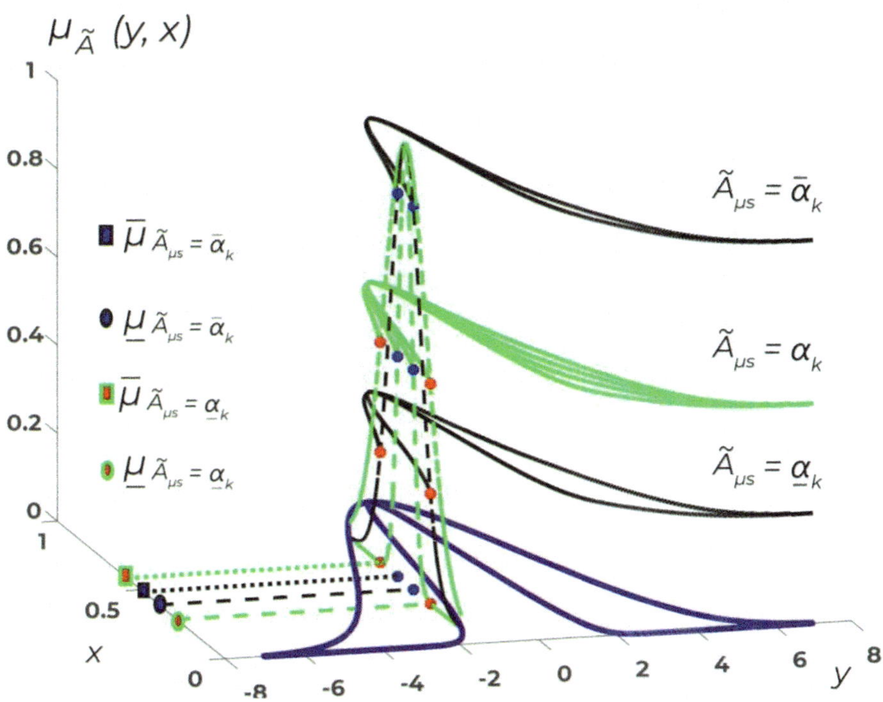

Fig. 3.3 The interval type-3 MF

3.9.3 Stability Analysis

The primary stability findings are succinctly outlined in the subsequent theorem:

Theorem 1 *Take into account the i-th subsystem and its estimated models. If the adaptation laws are as follows, the closed-loop i-th subsystem will be asymptotically stable.*

$$D^{q_{ik}}\theta_{ik} = -\gamma\hat{e}_{ik}\xi \tag{3.24}$$

and the variables of $\Theta_{ik}, k = 1, \ldots, K$, $\chi_{ik}, k = 1, \ldots, K$, $\kappa_{ik}, k = 1, \ldots, K$ exist, and two real scalars $\varsigma_j, j = 1, 2$, so that we have:

$$[\Pi_{11}\Pi_{12}\Pi_{12}\Pi_{22}] < 0 \tag{3.25}$$

where, γ denotes the adaptation rate and

$$\Pi_{11} = \sum_{j=1}^{2} Sym\left\{\Omega_{j1} \otimes \begin{bmatrix} -\kappa_{i1}0\xi^T - \kappa_{i1}\chi_{i1}\underline{0} - \gamma\xi\vdots 00\underline{00}\vdots 00\underline{00}\vdots \underline{000} \\ \cdots 00 \cdots 00 \cdots \vdots \cdots\cdots\cdots \underline{0}\vdots - \kappa_{iK} \\ -\kappa_{iK} - \gamma\xi\underline{0}\vdots 0 - \chi_{iK}\underline{0000}\vdots\xi^T\underline{00} \end{bmatrix}\right\}$$
$$+\sum_{j=1}^{2}\varsigma_j\left\{I_2 \otimes \begin{bmatrix} 0\psi_{i1}\psi_{i1}^T\underline{000}00\vdots 00\underline{00}\vdots 00\underline{00}\vdots\underline{000} \\ \cdots 00 \cdots 00 \cdots \vdots \cdots\cdots\cdots \underline{0}\vdots 00\underline{00}\vdots\psi_{iK}\psi_{iK}^T\underline{000}00\vdots\underline{000} \end{bmatrix}\right\} \tag{3.26}$$

$$\Pi_{22} = \begin{bmatrix} I_2 \otimes \begin{bmatrix} 00\underline{0}0\Theta_{i1}\underline{00}\vdots 00\underline{00}\vdots 00\underline{00}\vdots\underline{000} \cdots 00 \cdots 00 \cdots \vdots \cdots\cdots\cdots \\ \underline{0}\vdots 00\underline{00}\vdots 0\Theta_{iK}\underline{0000}\vdots\underline{000} \end{bmatrix}^T \\ I_2 \otimes \begin{bmatrix} 00\underline{0}0\Theta_{i1}\underline{00}\vdots 00\underline{00}\vdots 00\underline{00}\vdots\underline{000} \cdots 00 \cdots 00 \cdots \vdots \cdots\cdots\cdots \\ \underline{0}\vdots 00\underline{00}\vdots 0\Theta_{iK}\underline{0000}\vdots\underline{000} \end{bmatrix}^T \end{bmatrix} \tag{3.27}$$

$$\Pi_{12} = [-\varsigma_1 00 - \varsigma_1] \otimes I_2 \tag{3.28}$$

Note 1 *The sign 0 represents a scalar, while* $\underline{0}$ *represents a vector with zero entries and the correct dimension.*

Proof By inserting Eq. (3.4) into Eq. (3.2), the dynamics of the tracking errors are calculated as follows:

$$\{D_t^{q_{11}} e_{11}(t) = -\lambda_{11} e_{11} - \kappa_{11}\hat{e}_{11} \vdots D_t^{q_{1K}} e_{1K}(t) = -\lambda_{1K} e_{1K} - \kappa_{1K}\hat{e}_{1K}$$
$$\{D_t^{q_{m1}} e_{m1}(t) = -\lambda_{m1} e_{m1} - \kappa_{m1}\hat{e}_{m1} \vdots D_t^{q_{mK}} e_{mK}(t) = -\lambda_{mK} e_{mK} - \kappa_{mK}\hat{e}_{mK} \tag{3.29}$$

Using Eq. (3.2) and by taking into account the optimal fuzzy systems $\hat{F}_{ik}^{*}\left[y(t)|\theta_{ik}^{*}(t)\right]$, the dynamics of the estimation errors have the following form:

$$\begin{aligned}
&\{D_t^{q_{11}}\hat{e}_{11}(t) = \hat{F}_{11}\left[y(t)|\theta_{11}(t)\right] - \hat{F}_{11}^{*}\left[y(t)|\theta_1^{*}(t)\right] + D_t^{q_1^*} y_1(t) \\
&-D_t^{q_{11}} y_1(t) - \kappa_{11}\hat{e}_{11} \\
&\vdots D_t^{q_{1K}}\hat{e}_{1K}(t) = \hat{F}_{1K}\left[y(t)|\theta_{1K}(t)\right] - \hat{F}_{1K}^{*}\left[y(t)|\theta_1^{*}(t)\right] + D_t^{q_1^*} y_1(t) \\
&-D_t^{q_{1K}} y_1(t) - \kappa_{1K}\hat{e}_{1K} \\
&\vdots\{D_t^{q_{m1}}\hat{e}_{m1}(t) = \hat{F}_{m1}\left[y(t)|\theta_{m1}(t)\right] - \hat{F}_{m1}^{*}\left[y(t)|\theta_m^{*}(t)\right] + D_t^{q_m^*} y_m(t) \\
&-D_t^{q_{m1}} y_m(t) - \kappa_{m1}\hat{e}_{m1} \\
&\vdots D_t^{q_{mK}}\hat{e}_{mK}(t) = \hat{F}_{mK}\left[y(t)|\theta_{mK}(t)\right] - \hat{F}_{mK}^{*}\left[y(t)|\theta_m^{*}(t)\right] + D_t^{q_m^*} y_m(t) \\
&D_t^{q_{mK}} y_m(t) - \kappa_{mK}\hat{e}_{mK}
\end{aligned} \tag{3.30}$$

From Eq. (3.20), we have:

$$\hat{F}_{ik}\left[y(t)|\theta_{ik}(t)\right] - \hat{F}_{ik}^{*}\left[y(t)|\theta_i^{*}(t)\right] = \tilde{\theta}_{ik}^{T}\xi i = 1, \ldots, m, k = 1, \ldots, K \tag{3.31}$$

The approximation error ε_{ik}, $i = 1, \ldots, m, k = 1, \ldots, K$ is demonstrated as follows:

$$\varepsilon_{ik} = D_t^{q_i^*} y_i(t) - D_t^{q_{ik}} y_i(t) \tag{3.32}$$

It is supposed that:

$$\varepsilon_{ik} = \psi_{ik}\Delta_{ik}(t) e_{ik}, \; |\Delta_{ik}(t)| \leq 1 \tag{3.33}$$

where, ψ_{ik} denotes a constant value. This assumption is reasonable, as the impact of the approximation error becomes evident in that of the tracking. Based on Eqs. (3.31, 3.32), the dynamics of the estimating errors are expressed in the following manner:

$$\{D_t^{q_{11}}\hat{e}_{11} = \tilde{\theta}_{11}^{T}\xi + \varepsilon_{11} - \kappa_{11}\hat{e}_{11} \vdots D_t^{q_{1K}}\hat{e}_{1K} = \tilde{\theta}_{1K}^{T}\xi + \varepsilon_{1K} - \kappa_{1K}\hat{e}_{1K}$$
$$\{D_t^{q_{m1}}\hat{e}_{m1} = \tilde{\theta}_{m1}^{T}\xi + \varepsilon_{m1} - \kappa_{m1}\hat{e}_{m1} \vdots D_t^{q_{mK}}\hat{e}_{mK} = \tilde{\theta}_{mK}^{T}\xi + \varepsilon_{mK} - \kappa_{mK}\hat{e}_{mK} \tag{3.34}$$

Let's examine the $i - th$ subsystem and its estimated models. Represented in matrix form, we have:

$$\left[D_t^{q_{i1}}\hat{e}_{i1}D_t^{q_{i1}}e_{i1}D_t^{q_{i1}}\tilde{\theta}_{i1}\vdots D_t^{q_{iK}}\hat{e}_{iK}D_t^{q_{iK}}e_{iK}D_t^{q_{iK}}\tilde{\theta}_{iK}\right] = A_{i1}\left[\hat{e}_{i1}e_{i1}\tilde{\theta}_{i1}\vdots\hat{e}_{iK}e_{iK}\tilde{\theta}_{iK}\right] \quad (3.35)$$

where,

$$A_{i1} = \begin{bmatrix} -\kappa_{i1}\psi_{i1}\Delta_{i1}(t)\xi^T - \kappa_{i1} - \lambda_{i1}\underline{0}_{1\times L} - \gamma\xi\vdots 00\underline{0}_{L\times 1} \\ \underline{0}_{L\times 1}\vdots 00\underline{0}_{L\times 1}\underline{0}_{L\times 1}\vdots\underline{0}_{1\times L}\underline{0}_{1\times L}\underline{0}_{L\times 1} \\ \cdots 00\cdots 00\cdots\vdots\cdots\cdots\cdots\underline{0}_{L\times 1}\vdots - \kappa_{iK} - \kappa_{iK} - \gamma\xi\underline{0}_{L\times 1}\vdots\psi_{iK}\Delta_{iK}(t) \\ -\lambda_{iK}\underline{0}_{L\times 1} \\ \underline{0}_{1\times L}\underline{0}_{1\times L}\underline{0}_{L\times 1}\vdots\xi^T\underline{0}_{1\times L}\underline{0}_{L\times 1} \end{bmatrix} \quad (3.36)$$

Take Lemma 4 into account, and define $Q_{12} = Q_{22} = \underline{0}$ and:

$$Q_{11} = Q_{21} = [1\,0\,\underline{0}\,0\,\Theta_{i1}\,\underline{00}\vdots 0\,0\,\underline{00}\vdots 0\,0\underline{01}\vdots\underline{000}\cdots 0\,0$$
$$\cdots 0\,0\cdots\vdots\cdots\cdots\cdots\underline{0}\vdots 1\,0\,\underline{00}\vdots 0\,\Theta_{iK}\,\underline{0\,0\,0\,0}\vdots\underline{001}] \quad (3.37)$$

Then the system Eq. (3.35) is asymptotically stable if:

$$\sum_{j=1}^{2} Sym\left\{\begin{array}{l} \Omega_{j1}\otimes -\kappa_{i1}\psi_{i1}\Delta_{i1}(t)\xi^T - \kappa_{i1} - \lambda_{i1}\underline{0} \\ -\gamma\xi\vdots 00\underline{00}\vdots 00\underline{00}\vdots\underline{000}\cdots 00\cdots 00\cdots\vdots\cdots\cdots\cdots\underline{0}\vdots \\ -\kappa_{iK} - \kappa_{iK} - \gamma\xi\underline{0}\vdots\psi_{iK}\Delta_{iK}(t) \\ -\lambda_{iK}\underline{0000}\vdots\begin{bmatrix} -\kappa_{i1}\psi_{i1}\Delta_{i1}(t)\xi^T - \kappa_{i1} - \lambda_{i1}\underline{0} - \gamma\xi\vdots 00\underline{00}\vdots 00\underline{00}\vdots\underline{000} \\ \cdots 00\cdots 00\cdots\vdots\cdots\cdots\cdots\underline{0}\vdots - \kappa_{iK} \\ -\kappa_{iK} - \gamma\xi\underline{0}\vdots\psi_{iK}\Delta_{iK}(t) - \lambda_{iK}\underline{0000}\vdots\xi^T\underline{00} \end{bmatrix} \\ \begin{bmatrix} 10\underline{0}0\Theta_{i1}\underline{00}\vdots 00\underline{00}\vdots 00\underline{01}\vdots\underline{000}\cdots 00\cdots \\ 00\cdots\vdots\cdots\cdots\cdots\underline{0}\vdots 10\underline{00}\vdots 0\Theta_{iK}\underline{0000}\vdots\underline{001} \end{bmatrix} \end{array}\right\} \quad (3.38)$$

where, $\underline{0}$ and $\underline{1}$ are, respectively, the zero and unit matrix which have compatible dimensions. Inequality (3.38) can be restated as:

$$\sum_{j=1}^{2} Sym\left\{\Omega_{j1} \otimes \begin{bmatrix} -\kappa_{i1}0\xi^T - \kappa_{i1} - \lambda_{i1}\Theta_{i1}\underline{0} - \gamma\xi \vdots 00\underline{00}\vdots 00\underline{00}\vdots \underline{000} \\ \cdots 00 \cdots 00 \cdots \vdots \cdots\cdots\cdots \underline{0}\vdots - \kappa_{iK} - \kappa_{iK} - \gamma\xi\underline{0}\vdots 0 \\ -\lambda_{iK}\Theta_{iK}\underline{0000}\vdots\xi^T\underline{00} \end{bmatrix}\right\} +$$

$$\sum_{j=1}^{2} Sym\left\{\Omega_{j1} \otimes \begin{bmatrix} 0\psi_{i1}\Delta_{i1}(t)\Theta_{i1}000\underline{00}\vdots 00\underline{00}\vdots 00\underline{00}\vdots\underline{000} \\ \cdots 00 \cdots 00 \cdots \vdots \cdots\cdots\cdots \underline{0}\vdots 00\underline{00}\vdots 00\underline{00}\vdots \psi_{iK}\Delta_{iK}(t) \\ \Theta_{iK}0\underline{0000}\vdots\underline{000} \end{bmatrix}\right\} < 0 \tag{3.39}$$

The second term in Eq. (3.39) can be reformulated as:

$$Sym\left\{\Omega_{j1} \otimes \begin{bmatrix} 0\psi_{i1}\Delta_{i1}(t)\Theta_{i1}\underline{000}\underline{00}\vdots 00\underline{00}\vdots 00\underline{00}\vdots\underline{000} \\ \cdots 00 \cdots 00 \cdots \vdots \cdots\cdots\cdots \underline{0}\vdots 00\underline{00}\vdots \psi_{iK}\Delta_{iK}(t)\Theta_{iK}\underline{00000}\vdots\underline{000} \end{bmatrix}\right\} =$$

$$Sym\left\{\Omega_{j1} \otimes \left(\begin{bmatrix} 0\psi_{i1}\underline{000}\underline{00}\vdots 00\underline{00}\vdots 00\underline{00}\vdots\underline{000} \cdots 00 \cdots 00 \cdots \vdots \cdots\cdots\cdots \\ \underline{0}\vdots 00\underline{00}\vdots \psi_{iK}\underline{00000}\vdots\underline{000} \end{bmatrix} \begin{bmatrix} 0\underline{0}00\Delta_{i1}(t)\underline{00}\vdots 00\underline{00}\vdots 00\underline{00}\vdots \\ \underline{000} \cdots 00 \cdots 00 \cdots \vdots \cdots\cdots\cdots \underline{0}\vdots 00\underline{00}\vdots 0\Delta_{iK}(t)\underline{0000}\vdots\underline{000} \end{bmatrix} . \begin{bmatrix} 0\underline{0}00\Theta_{i1}\underline{00}\vdots 00\underline{00}\vdots 00\underline{00}\vdots\underline{000} \cdots 00 \cdots 00 \cdots \\ \vdots \cdots\cdots\cdots \underline{0}\vdots 00\underline{00}\vdots 0\Theta_{iK}\underline{0} \\ \underline{000}\vdots\underline{000} \end{bmatrix} \right)\right\}$$

$$Sym\left\{\begin{array}{l}\Omega_{j1}\otimes\left[\begin{array}{l}0\psi_{i1}\Delta_{i1}(t)\Theta_{i1}\underline{0000}\underline{0}\vdots 00\underline{00}\vdots 00\underline{00}\vdots \underline{000}\\ \cdots 00\cdots 00\cdots\vdots\cdots\cdots\cdots\underline{0}\vdots 00\underline{00}\vdots\psi_{iK}\Delta_{iK}(t)\Theta_{iK}0\underline{0000}\vdots\underline{000}\end{array}\right]\cdot\\ \left(I_2\otimes\left[\begin{array}{l}00\underline{0}0\Delta_{i1}(t)\underline{00}\vdots 00\underline{00}\vdots 00\underline{00}\vdots\underline{000}\cdots 00\cdots 00\cdots\\ \vdots\cdots\cdots\cdots\underline{0}\vdots 00\underline{00}\vdots 0\\ \Delta_{iK}(t)\underline{0000}\vdots\underline{000}\end{array}\right]\right)\cdot\\ \left(I_2\otimes\left[00\underline{0}0\Theta_{i1}\underline{00}\vdots 00\underline{00}\vdots 00\underline{00}\vdots\underline{000}\cdots 00\cdots 00\cdots\vdots\cdots\cdots\cdots\underline{0}\vdots 00\underline{00}\vdots\right]\right.\\ \left.0\Theta_{iK}\underline{0000}\vdots\underline{000}\right)\end{array}\right\} \tag{3.40}$$

Form Eq. (3.33), we get:

$$\begin{aligned}&\cdot\left(I_2\otimes\left[00\underline{0}0\Delta_{i1}(t)\underline{00}:00\underline{00}:00\underline{00}:\underline{000}\quad\cdots 00\cdots 00\cdots:\cdots\cdots\cdots\right.\right.\\&\left.\left.00\underline{00}:0\Delta_{iK}(t)\underline{0000}:\underline{00}\right]\right)\cdot\\&\qquad\left(I_2\otimes\left[0\underline{00}\Delta_{i1}(t)\underline{00}:00\underline{00}:\underline{000}:\right.\right.\\&\left.\left.\underline{000}\cdots 00\cdots 00\cdots:\cdots\cdots\cdots\underline{0}:00\underline{00}:0\Delta_{iK}(t)\underline{0000}:\underline{000}\right]\right)^T=\\&\left(I_2\otimes\left[\begin{array}{l}00\underline{0}0\Delta_{i1}^T(t)\Delta_{i1}(t)\underline{00}\vdots 00\underline{00}\vdots 00\underline{00}\vdots\underline{000}\\ \cdots 00\cdots 00\cdots\vdots\cdots\cdots\cdots\underline{0}\vdots 00\underline{00}\vdots 0\Delta_{iK}^T(t)\Delta_{iK}(t)\underline{0000}\vdots\underline{000}\end{array}\right]\right)\le I\end{aligned} \tag{3.41}$$

Using Eqs. (3.40, 3.41) and Lemma 3, one has:

$$Sym\left\{\Omega_{j1}\otimes\left[\begin{array}{l}0\psi_{i1}\Delta_{i1}(t)\Theta_{i1}\underline{0000}\underline{0}\vdots 00\underline{00}\vdots 00\underline{00}\vdots\\ \underline{000}\cdots 00\cdots 00\cdots\vdots\cdots\cdots\cdots\underline{0}\vdots 00\underline{00}\vdots\psi_{iK}\Delta_{iK}(t)\Theta_{iK}0\underline{0000}\vdots\underline{000}\end{array}\right]\right\}\le$$

$$\begin{aligned}&\cdot\varsigma_j\left(\Omega_{j1}\otimes\left[0\psi_{i1}\underline{00000}:00\underline{00}:00\underline{00}:\underline{000}\quad\cdots 00\cdots 00\cdots:\cdots\cdots\cdots\right.\right.\\&\left.\left.00\underline{00}:\psi_{iK}\underline{0000}:\underline{000}\right]\right)\cdot\\&\qquad\left(\Omega_{j1}\otimes\left[0\psi_{i1}\underline{0000}:00\underline{00}:00\underline{00}\ :\right.\right.\\&\left.\left.\underline{000}\quad\cdots 00\cdots 00\cdots:\cdots\cdots\cdots\quad\underline{0}:0\underline{00}:\psi_{iK}\underline{00000}:\underline{000}\right]\right)^T+\\&\varsigma_j^{-1}(I_2\otimes[00\underline{0}0\Theta_{i1}\underline{00}:00\underline{00}:00\underline{00}:\underline{000}\quad\cdots 00\cdots 00\cdots\vdots\cdots\cdots\cdots\\&\underline{0}\vdots 00\underline{00}\vdots 0\Theta_{iK}\underline{0000}\vdots\underline{000}])^T\end{aligned}$$

$$\cdot (I_2 \otimes [00\underline{0}0\Theta_{i1}\underline{00}\vdots 00\underline{00}\vdots 00\underline{00}\vdots\underline{000}\cdots 00\cdots 00\cdots\vdots\cdots\cdots\cdots$$

$$\underline{0}\vdots 00\underline{00}\vdots 0\Theta_{iK}\underline{0000}\vdots\underline{000}]) \tag{3.42}$$

Using Eq. (3.42) and considering the fact that $\Omega_{j1} \otimes \Omega_{j1}^T = I_2, j = 1, 2$, the inequality (3.575), changes to:

$$\sum_{j=1}^{2} Sym\{\Omega_{j1} \otimes [-k_{i1}\ 0\ \xi^T\ -k_{i1}\ -\lambda_{i1}\ \Theta_{i1}\ \underline{0}\ -\ \gamma\xi\ \vdots\ 0\,0\,\underline{0}\ \underline{0}\ \vdots\ \underline{0}\,\underline{0}\,\underline{0}\ \cdots$$

$$0\,0\ \cdots\ 0\,0\ \cdots\ \vdots\ \cdots\cdots\cdots\ \underline{0}\ \vdots\ -k_{ik}\ -\ k_{ik}\ -\ \gamma\xi\ \underline{0}\ \vdots\ 0\ -\ \lambda_{ik}\Theta_{i1}\ \underline{0}\ \underline{0}\,\underline{0}\ \underline{0}\ \vdots\ \xi^T\ \underline{0}\,\underline{0}\,]\} +$$

$$\sum_{j=1}^{2}\varsigma_j\left\{\begin{matrix}\left(I_2\otimes\begin{bmatrix}0\psi_{i1}\psi_{i1}^T\underline{0}00\underline{00}\vdots 00\underline{00}\vdots 00\underline{00}\vdots\underline{000}\cdots 00\cdots 00\cdots\vdots\\ \cdots\cdots\cdots\\ \underline{0}\vdots 00\underline{00}\vdots\psi_{iK}\psi_{iK}^T\underline{0}\underline{0000}\vdots\underline{000}\end{bmatrix}\right)+\\ \varsigma_j^{-1}\left(I_2\otimes\begin{bmatrix}00\underline{0}0\Theta_{i1}\underline{00}\vdots 00\underline{00}\vdots 00\underline{00}\vdots\underline{000}\\ \cdots 00\cdots 00\cdots\vdots\cdots\cdots\cdots\underline{0}\vdots 00\underline{00}\vdots 0\Theta_{iK}\underline{0000}\vdots\underline{000}\end{bmatrix}\right)^T\\ \cdot\left(I_2\otimes\begin{bmatrix}0000\Theta_{i1}\underline{00}\vdots 00\underline{00}\vdots 00\underline{00}\vdots\underline{000}\cdots 00\cdots 00\cdots\vdots\cdots\cdots\cdots\underline{0}\\ \vdots 00\underline{00}\vdots 0\Theta_{iK}\underline{0000}\vdots\underline{000}\end{bmatrix}\right)\end{matrix}\right\} < 0 \tag{3.43}$$

The second term in Eq. (3.43) can be restated as below:

$$\sum_{j=1}^{2}\varsigma_j^{-1}\left(I_2\otimes\begin{bmatrix}00\underline{0}0\Theta_{i1}\underline{00}\vdots 00\underline{00}\vdots 00\underline{00}\vdots\underline{000}\cdots 00\cdots 00\\ \cdots\vdots\cdots\cdots\cdots\underline{0}\vdots 00\underline{00}\vdots 0\Theta_{iK}\underline{0000}\vdots\underline{000}\end{bmatrix}\right)$$

$$\cdot\left(I_2\otimes\begin{bmatrix}00\underline{0}0\Theta_{i1}\underline{00}\vdots 00\underline{00}\vdots 00\underline{00}\vdots\underline{000}\cdots 00\cdots 00\cdots\\ \vdots\cdots\cdots\cdots\underline{0}\vdots 00\underline{00}\vdots 0\Theta_{iK}\underline{0000}\vdots\underline{000}\end{bmatrix}\right) =$$

$$\begin{bmatrix}I_2\otimes\begin{bmatrix}00\underline{0}0\Theta_{i1}\underline{00}\vdots 00\underline{00}\vdots 00\underline{00}\vdots\underline{000}\cdots 00\cdots 00\cdots\vdots\\ \cdots\cdots\cdots\underline{0}\vdots 00\underline{00}\vdots 0\Theta_{iK}\underline{0000}\vdots\underline{000}\end{bmatrix}^T\\ I_2\otimes\begin{bmatrix}00\underline{0}0\Theta_{i1}\underline{00}\vdots 00\underline{00}\vdots 00\underline{00}\vdots\underline{000}\cdots 00\cdots 00\\ \cdots\vdots\cdots\cdots\cdots\underline{0}\vdots 00\underline{00}\vdots 0\Theta_{iK}\underline{0000}\vdots\underline{000}\end{bmatrix}^T\end{bmatrix}^T$$

$$.\{[\varsigma_1 00 \varsigma_1]\otimes I_2\}^{-1}$$

$$\begin{bmatrix} I_2 \otimes \begin{bmatrix} 00\underline{0}0\Theta_{i1}\underline{00}\vdots 00\underline{00}\vdots 00\underline{00}\vdots \underline{000}\cdots 00 \cdots 00 \cdots \vdots \\ \cdots\cdots\cdots \underline{0}\vdots 00\underline{00}\vdots 0\Theta_{iK}\underline{0000}\vdots\underline{000} \end{bmatrix}^T \\ I_2 \otimes \begin{bmatrix} 00\underline{0}0\Theta_{i1}\underline{00}\vdots 00\underline{00}\vdots 00\underline{00}\vdots \underline{000}\cdots 00 \cdots 00 \\ \cdots \vdots \cdots\cdots\cdots \underline{0}\vdots 00\underline{00}\vdots 0\Theta_{iK}\underline{0000}\vdots\underline{000} \end{bmatrix}^T \end{bmatrix} \tag{3.44}$$

Now, by substituting $\chi_{ik} = \lambda_{ik}\Theta_{ik}, k = 1, \ldots, K$ and considering Eqs. (3.43, 3.44) and Lemma 2, the inequality (3.25) is obtained which completes the proof.

3.9.4 Simulations

This section presents one simulation using normal conditions, one simulation in real-time, and one experimental verification to show the high performance and efficacy of the proposed control technique.

3.9.4.1 Normal Simulation

Example 1 *The first and second states of the hyperchaotic Lorenz system are regarded as the controlled plant dynamics, while the remaining states are regarded as the unmodeled dynamic subsystem. Next, we have:*

$$D_t^{0.97}x_1 = 10(x_2 - x_1) + \varsigma_2 + u_1 D_t^{0.97}x_2 = 28x_1 - x_2 - x_1\varsigma_1 + u_2 \tag{3.45}$$

where, the unmodeled dynamics can be stated as:

$$D_t^{0.97}\varsigma_1 = x_1x_2 - 8/3\varsigma_1, D_t^{0.97}\varsigma_2 = -x_2\varsigma_1 - \varsigma_2 \tag{3.46}$$

The beginning circumstances are given as follows: $x_1(0) = 0.61$ $x_2(0) = 0.21$ $x_3(0) = 0.61$ and $x_4(0) = 0.21$ 1. Table 3.1 displays the control conditions, whereas Table 3.2 presents the IT3FLS parameters. The reference signals are defined as $r_1 = 0.5sin(2t) + 0.5cos(2.5t)$ and $r_2 = 1/3sin(3t) + cos(t)$. Figure 3.4 illustrates the system's tracking performance and the paths followed by its outputs. Figure 3.5 displays the paths followed by the actual outputs and the estimated ones. The paths followed by the control signals are depicted in Fig. 3.6. From Figs. 3.4–3.6, it is evident that the system's output trajectories (y_1 and y_2) closely follow the reference signals r_1 and r_2. Additionally, the control signals u_1 and u_2 exhibit smooth and desirable forms. The calculated system's output trajectories (y_1 and y_2) closely align with the actual system's output trajectories (y_1 and y_2), indicating strong convergence.

Table 3.3 compares the estimating and tracking errors using various types of fuzzy systems for further analysis. It is evident that the proposed controller utilizing IT3FLS yields superior tracking and estimation performance, particularly in highly noisy conditions. It is important to mention that white noise with varying noise power is introduced into control signals as a disruption.

To assess the impact of fractional-order values on estimation performance, a study was conducted in a control-free mode. The first output, y_1, was estimated using fuzzy dynamic models with various fractional-order values, as well as one model with an integer-order. The root mean-square errors (MSEs) are presented in Table 3.4. The results indicate that the estimation performance varies with the fractional-order value,

Table 3.1 Control parameters

Parameter	Value	Equation	Parameter	Value	Equation
λ_{ik}	500	(4)	κ_{ik}	100	(2)
K	5	(4)	γ	10	(24)
R	3	(22)	$n+\alpha$	3	(22)

Table 3.2 IT3FLS parameters

$\underline{\alpha} \in [0.75, 0.9, 1]$, $\underline{\alpha} \in [0, 0.25, 0.5]$
$c^1_{\psi_i} = -1$, $c^2_{\psi_i} = 0$, $c^3_{\psi_i} = 0$, $i = 1, 2$, $n_\alpha = 3$, $R = 3$
$\underline{\sigma}^h_{\psi_i,0} = 1$, $i = 1, 2$, $h = 1, 2, 3$, $\underline{\sigma}^h_{\psi_i,0} = 1$, $i = 1, 2$, $h = 1, 2, 3$

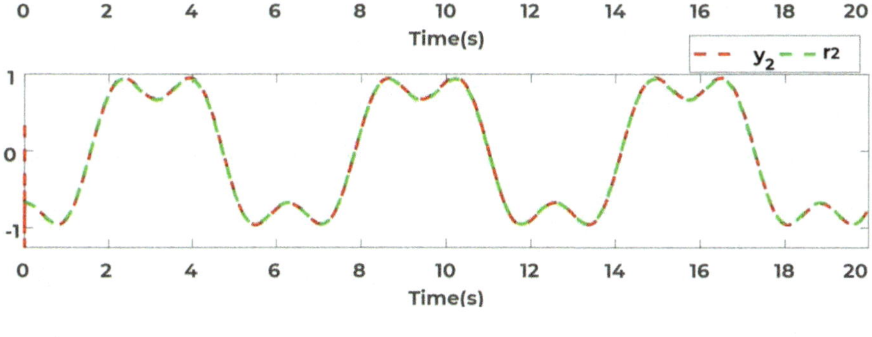

Fig. 3.4 The outputs trajectories and reference signals in Ex. 1

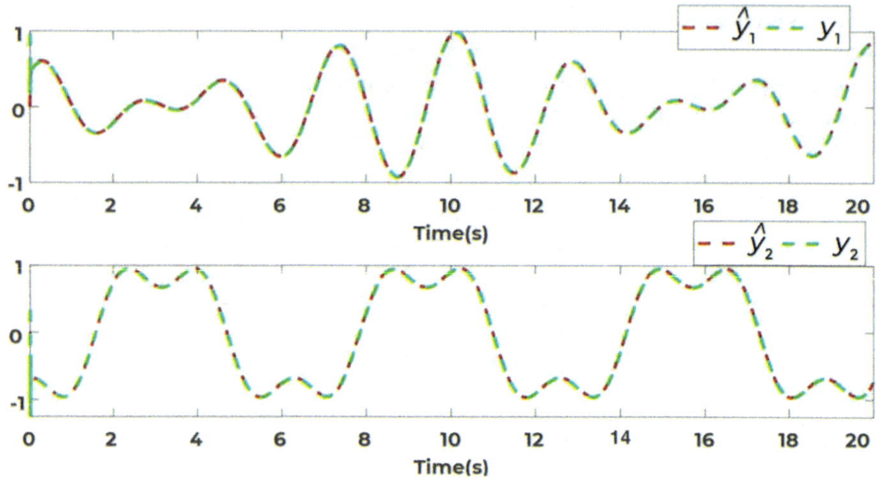

Fig. 3.5 The real and estimated outputs trajectories in Ex. 1

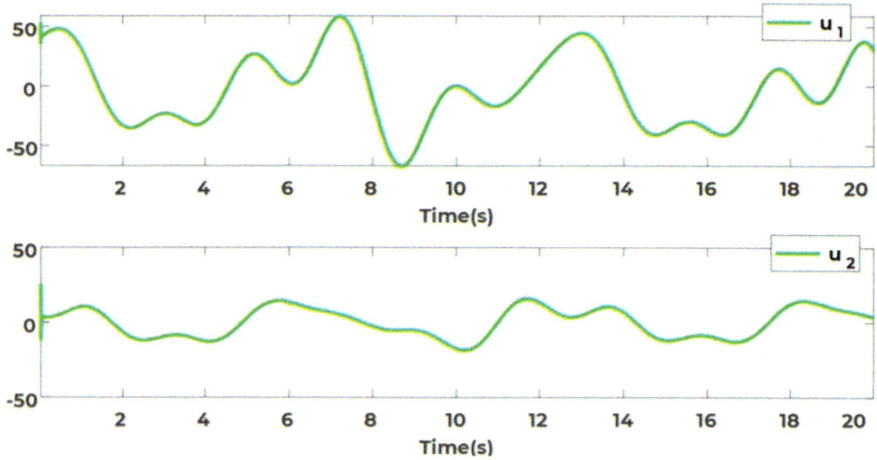

Fig. 3.6 The control signals trajectories in Ex. 1

and the model with a fractional-order value of $q = 0.5$ demonstrates superior accuracy compared to the other models. The estimation error trajectories for the different models are illustrated in Fig. 3.7, revealing that the integer-order dynamic model performs the poorest. To further evaluate the effectiveness of the designed controller, its superiority is demonstrated by a comparison with other recently developed control systems.

The proposed controller is being compared to the fractional-order type-2 fuzzy controller [52] and the fractional-order Takagi–Sugeno (T-S) fuzzy controller [53]. The techniques described in references [52, 53] are implemented on system Eq. (3.45)

Table 3.3 The root mean square magnitude of estimation and tracking errors according to various fuzzy systems and noisy conditions in Ex. 1

		Fuzzy system		
Noise power	Error signal	T2FLS	GT2FLS	IT3FLS
*0	e_1	0.03421	0.0312	0.0301
	e_2	0.0131	0.0132	0.0131
	$\hat{e}_1$	0.0095	0.0094	0.0095
	$\hat{e}_2$	0.0124	0.0119	0.0115
*0.01	e_1	0.0453	0.0335	0.0322
	e_2	0.0273	0.0223	0.0145
	$\hat{e}_1$	0.0267	0.0234	0.0113
	$\hat{e}_2$	0.0314	0.0224	0.0131
*0.1	e_1	0.0981	0.0634	0.0464
	e_2	0.0785	0.0247	0.0236
	$\hat{e}_1$	0.0478	0.0347	0.0215
	$\hat{e}_2$	0.0563	0.0384	0.0227

Table 3.4 The root mean square magnitude of estimation error $y_1 - \hat{y}_1$ according to different values of the FO q in dynamic fuzzy model, (without control mode) in Ex. 1

q	$t \in$			
	$[0, 5s]$	$[0, 10s]$	$[0, 15s]$	$[0, 20s]$
0.9	0.3719	0.3411	0.3374	0.3360
0.8	0.4819	0.4410	0.4345	0.4320
0.7	0.4874	0.4459	0.4388	0.4362
0.6	0.2593	0.2376	0.2336	0.2322
0.5	0.2100	0.1928	0.1896	0.1883
Integer-order	0.6008	0.5495	0.5410	0.5375

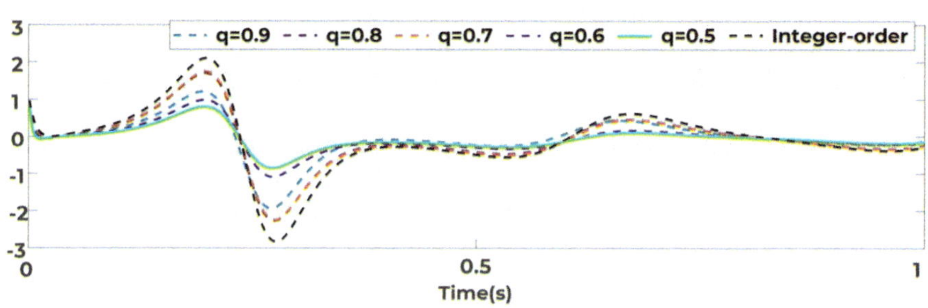

Fig. 3.7 The estimation error $y_1 - \hat{y}_1$ trajectories according to various values of the fractional-order q in dynamic fuzzy model, (without control mode) in Ex. 1

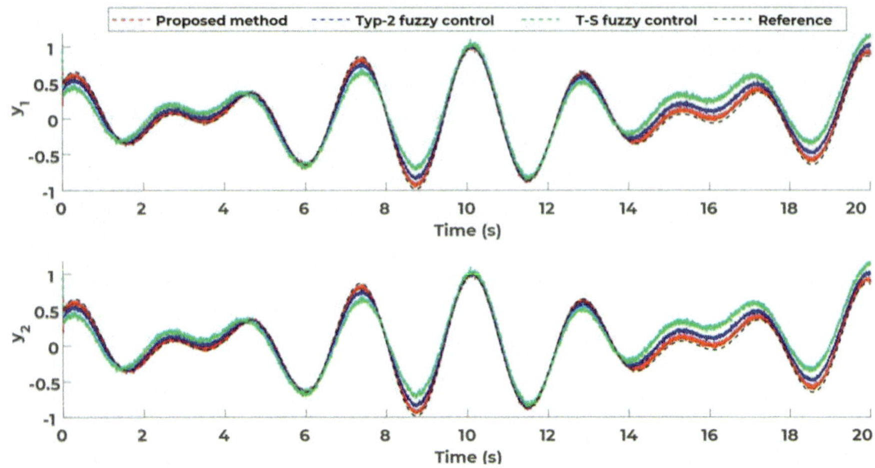

Fig. 3.8 The trajectories as to signals y_1, y_2 with different controllers and reference signals r_1 and r_2 in Ex. 1

Table 3.5 The root mean square magnitude of estimation error $y_i - r_i, i = 1, 2$ by various controllers in Ex. 1

	Signal	
	y_1	y_2
Suggested method	0.0464	0.0236
Type-2 fuzzy control [52]	0.1018	0.0312
T-S fuzzy control [53]	0.1886	0.0472

under identical simulated settings. The paths followed by signals y_1 and y_2 under the influence of several controllers and reference signals are shown in Fig. 3.8 and Table 3.5 provides a numerical comparison. It is evident that the recommended controller exhibits superior control proficiency.

Example 2 *In order to evaluate the proposed control system for integer-order systems, it is implemented on the integer-order chaotic model of brushless DC motors. The dynamics of the first and second states are regarded as the dynamics of the controlled plant, whereas the dynamics of the third state are regarded as the unmodeled dynamics. Subsequently, there exists:*

$$\dot{x}_1 = f_1(x_1, x_2, \varsigma_1) + u_1, \dot{x}_2 = f_2(x_1, x_2, \varsigma_1) + u_2 \tag{3.47}$$

where, the nonlinear f_1 and f_2 are calculated to be:

$$f_1 = 0.16 - x_1 - x_2\varsigma_1 + 58\varsigma_1, f_2 = 21 - 0.23x_2 + x_1\varsigma_1 \tag{3.48}$$

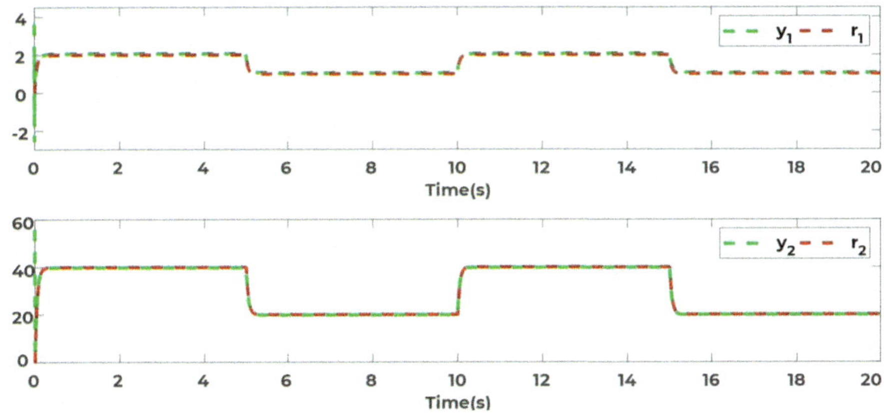

Fig. 3.8 The tracking performance in Ex. 2

The unmodeled subsystem dynamics are considered as:

$$\dot{\varsigma}_1 = 4.4(x_1 - \varsigma_1) + 0.8x_1x_2 - 0.52 \tag{3.49}$$

The initial conditions are as follows: $x_1(0) = 3.65$, $x_2(0) = 56$, $\varsigma(0) = 0.35$. Figure 3.8 illustrates the tracking performance, whereas Figure 3.9 shows the trajectories of the control signals. It has been noted that a high level of precision in tracking is attained. Similarly to the previous example, the estimation performance is demonstrated by varying the fractional-order value in fuzzy dynamic models. The initial output, y_1, is estimated in a mode without control. Additionally, a single dynamic model with integer-order is taken into account. The RMSE values are provided in Table 3.6. Figure 3.10 displays the trajectories of the estimation error $y_1 - \hat{y}_1$, based on various FOs q in the dynamic fuzzy model. It is observed that the use of fractional calculus enhances the accuracy of estimation.

3.9.4.2 Real-Time Simulation

Example 3 *This example evaluates the effectiveness of the proposed controller using a real-time simulation on the Arduino Mega board. Figure 3.11 illustrates the block diagram. The Matlab and Arduino board are connected using serial communication. The control signals created by Matlab are transmitted to the Arduino board, and the system outputs are then returned to Matlab. The proposed method is implemented on the integer-order chaotic model of the brushless DC motors provided in the previous example.*

The reference system dynamics are:

$$\dot{r}_1 = 0.168 - r_1 - r_2\varsigma_1 + 60\chi,\ \dot{r}_2 = 20.66 - 0.26r_2 + r_1\chi\dot{\chi}$$

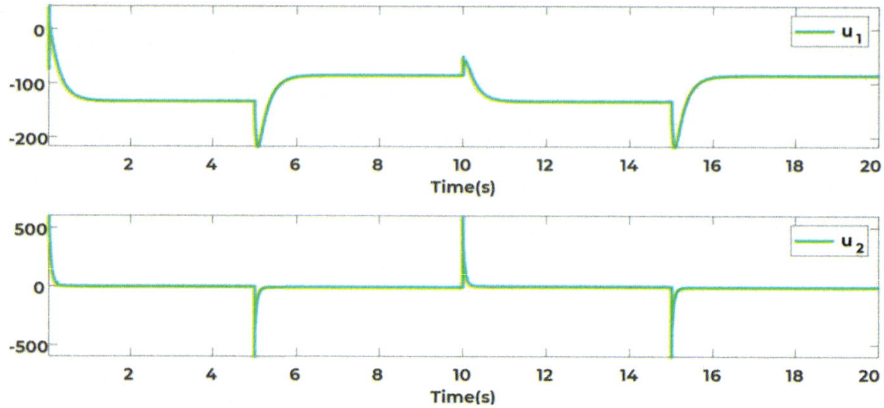

Fig. 3.9 The control signals trajectories in Ex. 2

Table 3.6 The root mean square magnitude of estimation error $y_1 - \hat{y}_1$ according to various values of the FO q in dynamic fuzzy model, (without control mode) in Ex. 2

	$t \in$			
q	$[0, 5s]$	$[0, 10s]$	$[0, 15s]$	$[0, 20s]$
0.9	0.3646	0.3706	0.3685	0.3710
0.8	0.4690	0.4802	0.4783	0.4822
0.7	0.4742	0.4856	0.4837	0.4876
0.6	0.2568	0.2587	0.2564	0.2579
0.5	0.2097	0.2095	0.2071	0.2080
Integer-order	0.5822	0.5985	0.5968	0.6018

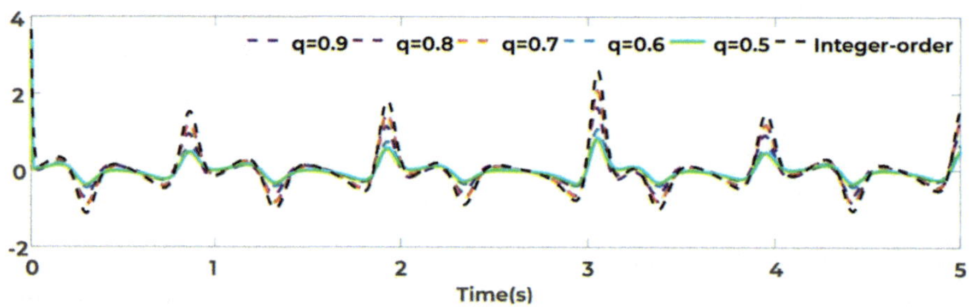

Fig. 3.10 The estimation error $y_1 - \hat{y}_1$ trajectories according to various values of the fractional-order q in dynamic fuzzy model, (without control mode) in Ex. 2

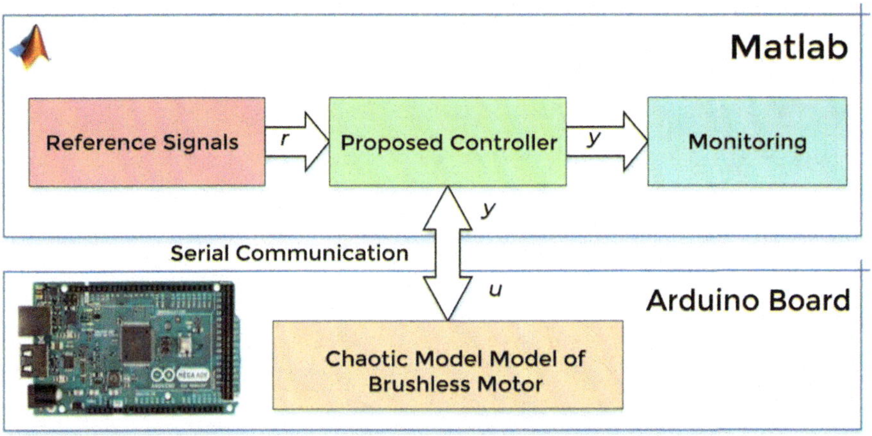

Fig. 3.11 The block diagram of the real-time simulation in Ex. 3

$$= 4.55(r_1 - \chi) + 0.875r_1r_2 - 0.53 \tag{3.50}$$

The initial conditions are as follows:$r_1(0) = 3.63$, $r_2(0) = 56.02$, $\chi(0) = 0.29$, $x_1(0) = 3.65$, $x_2(0) = 56$, $\varsigma(0) = 0.35$. The trajectories of the outputs and reference signals can be seen in Fig. 3.12. It is evident that the output trajectories closely follow the reference signal trajectories. It is important to note that these results were achieved under the following conditions: (1) the system dynamics are completely unknown, (2) the dynamics are affected by unmodeled dynamics, (3) the controller is applied in real-time, and (4) the reference signals are the outputs of a chaotic system. Additionally, similar to Examples 1 and 2, the impact of the fractional-order value on the identification process is examined in Table 3.7, demonstrating the effectiveness of fractional calculus.

3.9.4.3 Experimental Examination

Example 4 *In this example, the recommended control strategy is implemented on a heat transfer system. Figure 3.13 illustrates the block diagram, and the experimental setting that is being suggested is seen in Fig. 3.14. The result obtained is the measurement of temperature, whereas the control signal used to regulate the system is the electrical voltage applied to the heater. The fan's voltage serves as a disturbance and unmodeled dynamic system. Figure 3.15 displays the paths followed by the reference signal and output. It is evident that the output trajectory closely follows the desired signal. It is important to acknowledge that the dynamics of the plant are completely unknown and are disrupted by external disturbances. Figure 3.16 displays both the actual trajectory and the projected trajectory. It is observed that the trajectory of the*

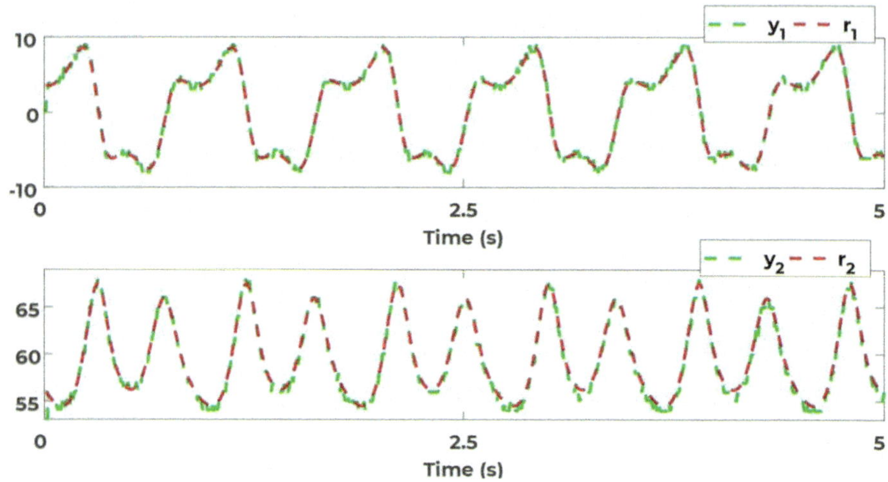

Fig. 3.12 The outputs signals, y_i, $i = 1, 2$ and reference signals r_i for the real-time simulation in Ex. 3

Table 3.7 The root mean square magnitude of estimation error $y_1 - \hat{y}_1$ according to various values of the fractional-order q in dynamic fuzzy model, (without control mode) in Ex. 3

0.9	0.8	0.7	0.6	0.5	Integer-order
0.8741	0.8637	0.8681	0.8572	0.8501	0.8917

estimated temperature closely aligns with the actual temperature. This example illustrates the efficacy of the proposed control system for practical applications. Similar to the previous cases, Table 3.8 demonstrates the impact of the fractional-order value on the identification process [54].

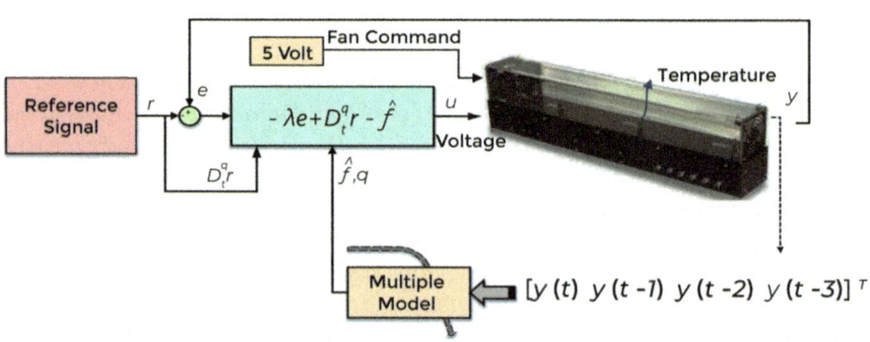

Fig. 3.13 The experimental examination block diagram in Ex. 4

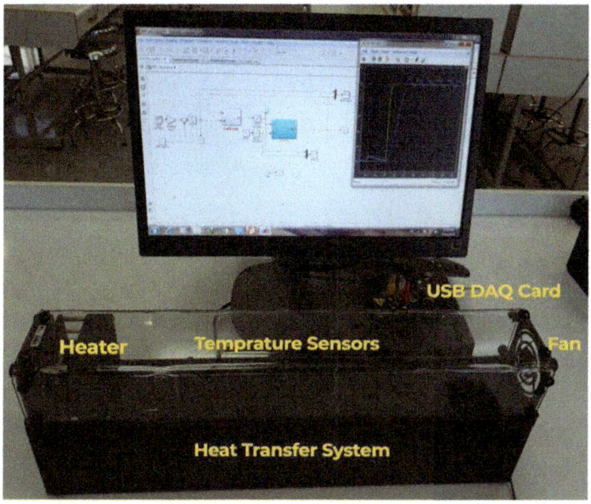

Fig. 3.14 The experimental configuration in Ex. 4

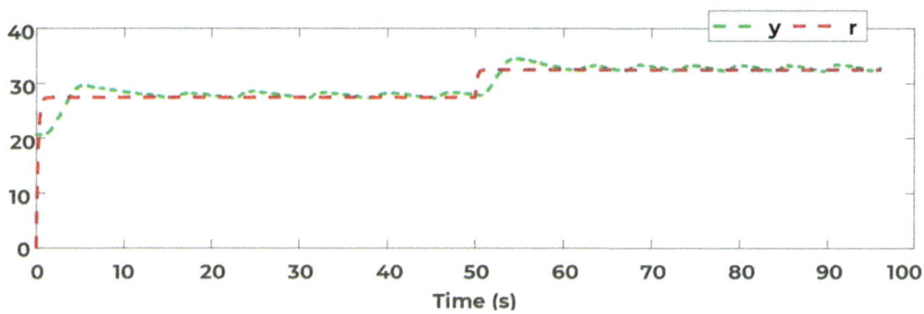

Fig. 3.15 The output and reference signal trajectories in Ex. 4

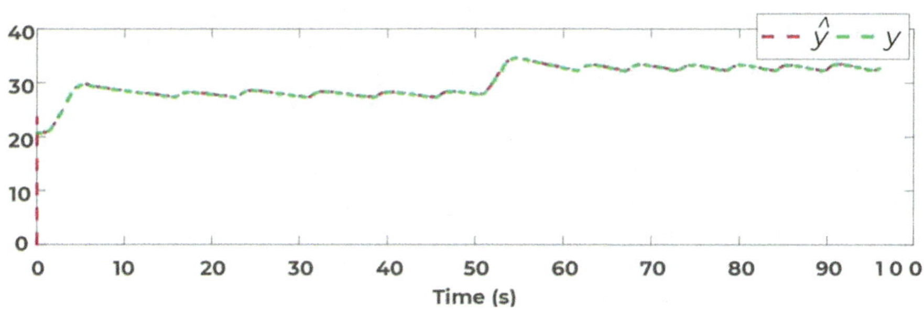

Fig. 3.16 The real and estimated output trajectories in Ex. 4

Table 3.8 The root mean square magnitude of estimation error $y_1 - \hat{y}_1$ according to various values of the FO q in dynamic fuzzy model, (without control mode) in Ex. 4

0.9	0.8	0.7	0.6	0.5	Integer-order
0.4557	0.4509	0.4486	0.4475	0.4472	0.4714

3.9.5 Discussion

The efficacy of fractional-order calculus has been proven in many research for the purpose of simulating physical systems. Inspired by this approach, this section presents a control system for nonlinear systems with uncertain dynamics. Additionally, it proposes a fractional-order fuzzy multiple model for online dynamic identification. A robust controller is constructed based on the specified model to effectively tolerate the impact of perturbations and approximation errors. The excellence and effectiveness of the recommended control system are confirmed by two simulation examples and two experimental evaluations. All examples assume complete uncertainty on the dynamics and also account for the presence of disturbances. The robustness of the planned control system is guaranteed by the effective convergence of output signals to the reference signals, even in the face of dynamic uncertainties and perturbations [54].

3.10 Adaptive Kernel Correntropy Kalman Filter for Type-3 Fuzzy Systems

3.10.1 Problem Definition

Figure 3.1 provides an overview of the primary goal of this investigation. The created correntropy Kalman filters are used to tune both the rule and MF parameters, as demonstrated. The suggested CKF does not have a set kernel size, but instead determines it using a simple fuzzy technique. The proposed framework is illustrated in Fig. 3.2. The depicted membership functions, which include both vertical and horizontal divisions, may be found in Figs. 3.3 and 3.4. When comparing the planned membership function (MF) presented in Fig. 3.6 with the usual general type-2 MFs supplied in Fig. 3.7, it is evident that the upper and lower boundaries of the degree of membership in the proposed MF are not fixed values, but rather interval type-1 MFs. The additional degree of freedom in the suggested FLS allows it to effectively manage a greater amount of uncertainties. Figure 3.5 illustrates the flowchart of the computation process. The specifics of the calculations are provided in the following sections.

3.10.2 Interval Type-3 Fuzzy Logic System

The procedure for acquiring the output of the suggested fuzzy system is elucidated in a comprehensive manner as outlined below.

(1) Obtain the inputs $x_i, i = 1, \ldots, n$.

(2) Calculate the memberships. If we assume that the $j - th$ membership function for the $i - th$ input is denoted as $\tilde{A}_i^j$, then the upper and lower memberships of $\tilde{A}_i^j$ at the secondary levels $u_s = \underline{\alpha}_k$ and $u_s = \underline{\alpha}_k$ are calculated as follows:

$$\underline{\mu}_{\tilde{A}^j_{i|u_s=\underline{\alpha}_k}} = \{1 - \left(\frac{\left|x_i - c_{\tilde{A}_i^j}\right|}{\underline{d}_{\tilde{A}_i^j}}\right)^{\underline{\alpha}_k} if\ c_{\tilde{A}_i^j} - \underline{d}_{\tilde{A}_i^j} < x_i \leq c_{\tilde{A}_i^j} 1 - \left(\frac{\left|x_i - c_{\tilde{A}_i^j}\right|}{\underline{d}_{\tilde{A}_i^j}}\right)^{\underline{\alpha}_k} \tag{3.1}$$

$$if\ c_{\tilde{A}_i^j} < x_i \leq c_{\tilde{A}_i^j} + \underline{d}_{\tilde{A}_i^j} 0$$

$$if\ x_i > c_{\tilde{A}_i^j} + \underline{d}_{\tilde{A}_i^j} or x_i \leq c_{\tilde{A}_i^j} - \underline{d}_{\tilde{A}_i^j}$$

$$\underline{\mu}_{\tilde{A}^j_{i|u_s=\underline{\alpha}_k}} = \{1 - \left(\frac{\left|x_i - c_{\tilde{A}_i^j}\right|}{\underline{d}_{\tilde{A}_i^j}}\right)^{\underline{\alpha}_k} if\ c_{\tilde{A}_i^j} - \underline{d}_{\tilde{A}_i^j} < x_i \leq c_{\tilde{A}_i^j} 1 - \left(\frac{\left|x_i - c_{\tilde{A}_i^j}\right|}{\underline{d}_{\tilde{A}_i^j}}\right)^{\underline{\alpha}_k} \tag{3.2}$$

$$if\ c_{\tilde{A}_i^j} < x_i \leq c_{\tilde{A}_i^j} + \underline{d}_{\tilde{A}_i^j} 0$$

$$if\ x_i > c_{\tilde{A}_i^j} + \underline{d}_{\tilde{A}_i^j}\ or\ x_i \leq c_{\tilde{A}_i^j} - \underline{d}_{\tilde{A}_i^j}$$

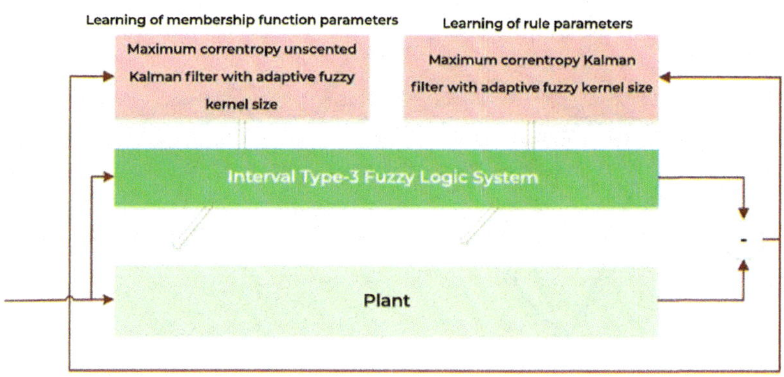

Fig. 3.1 Schematic view of the proposed model

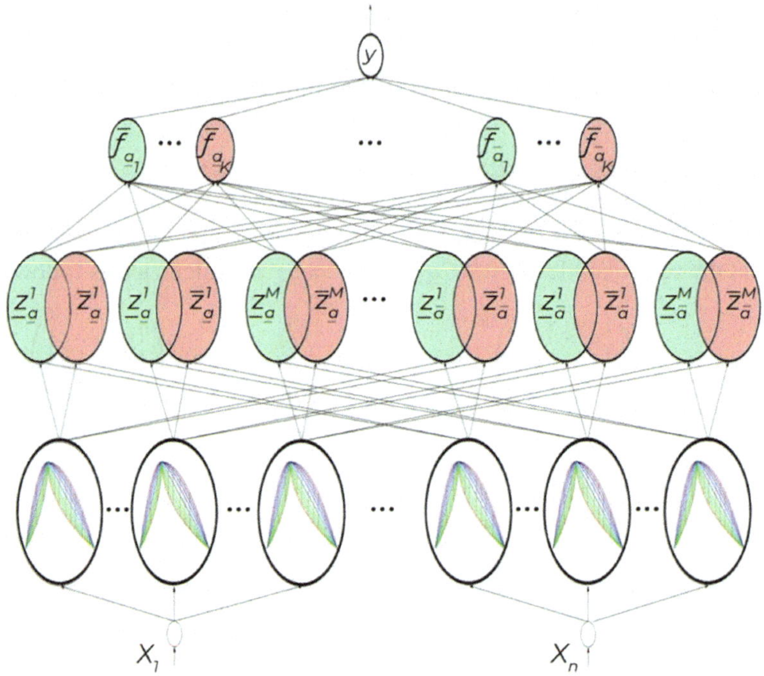

Fig. 3.2 Suggested interval type-3 fuzzy system

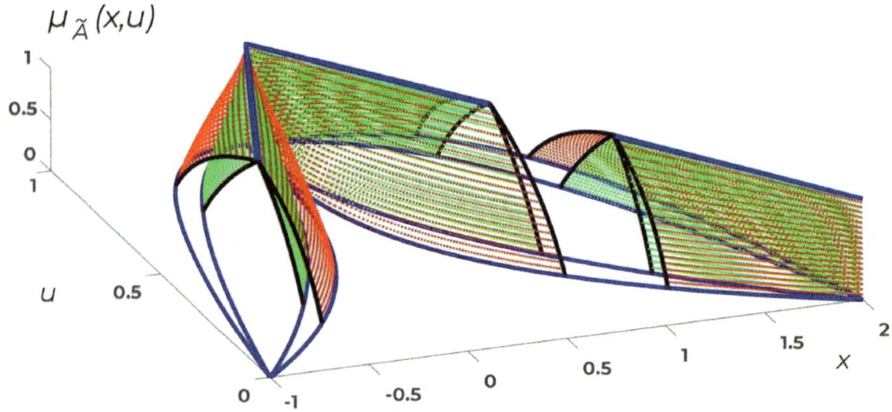

Fig. 3.3 Suggested interval type-3 fuzzy MF

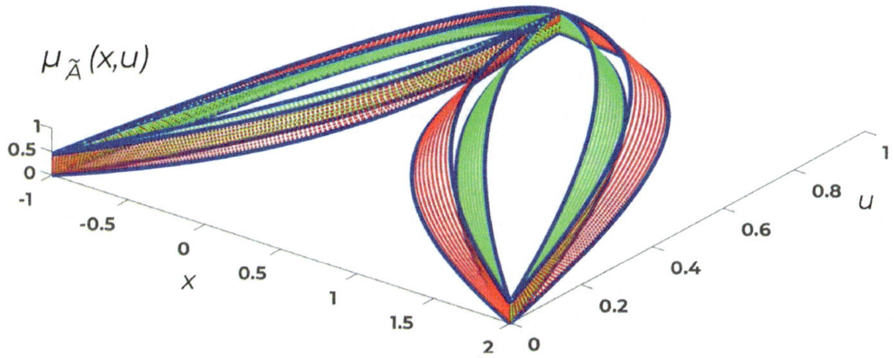

Fig. 3.4 The suggested MF horizontal slice

$$\underline{\mu}_{\tilde{A}^j_{i|u_s=\underline{\alpha}_k}} = \{1 - \left(\frac{\left|x_i - c_{\tilde{A}^j_i}\right|}{\underline{d}_{\tilde{A}^j_i}}\right)^{\frac{1}{\underline{\alpha}_k}} if\ c_{\tilde{A}^j_i} - \underline{d}_{\tilde{A}^j_i} < x_i \leq c_{\tilde{A}^j_i} 1 - \left(\frac{\left|x_i - c_{\tilde{A}^j_i}\right|}{\underline{d}_{\tilde{A}^j_i}}\right)^{\frac{1}{\underline{\alpha}_k}}$$
$$if\ c_{\tilde{A}^j_i} < x_i \leq c_{\tilde{A}^j_i} + \underline{d}_{\tilde{A}^j_i} 0$$
$$if\ x_i > c_{\tilde{A}^j_i} + \underline{d}_{\tilde{A}^j_i}\ or\ x_i \leq c_{\tilde{A}^j_i} - \underline{d}_{\tilde{A}^j_i} \tag{3.3}$$

$$\underline{\mu}_{\tilde{A}^j_{i|u_s=\underline{\alpha}_k}} = \{1 - \left(\frac{\left|x_i - c_{\tilde{A}^j_i}\right|}{\underline{d}_{\tilde{A}^j_i}}\right)^{\frac{1}{\underline{\alpha}_k}} if\ c_{\tilde{A}^j_i} - \underline{d}_{\tilde{A}^j_i} < x_i \leq c_{\tilde{A}^j_i} 1 - \left(\frac{\left|x_i - c_{\tilde{A}^j_i}\right|}{\underline{d}_{\tilde{A}^j_i}}\right)^{\frac{1}{\underline{\alpha}_k}}$$
$$if\ c_{\tilde{A}^j_i} < x_i \leq c_{\tilde{A}^j_i} + \underline{d}_{\tilde{A}^j_i} 0$$
$$if\ x_i > c_{\tilde{A}^j_i} + \underline{d}_{\tilde{A}^j_i}\ or\ x_i \leq c_{\tilde{A}^j_i} - \underline{d}_{\tilde{A}^j_i} \tag{3.4}$$

Here, $\underline{\mu}_{\tilde{A}^j_{i|u_s=\underline{\alpha}_k}} / \underline{\mu}_{\tilde{A}^j_{i|u_s=\underline{\alpha}_k}}$ and $\underline{\mu}_{\tilde{A}^j_{i|u_s=\underline{\alpha}_k}} / \underline{\mu}_{\tilde{A}^j_{i|u_s=\underline{\alpha}_k}}$ are, respectively, the upper and lower memberships for MF $\tilde{A}^j_i$ at the secondary levels $u_s = \underline{\alpha}_k$ and $u_s = \underline{\alpha}_k$. The center of $\tilde{A}^j_i$ is shown by $c_{\tilde{A}^j_i}$, moreover, $\underline{d}_{\tilde{A}^j_i}$ and $\underline{d}_{\tilde{A}^j_i}$ are, respectively, the distances of the center of $\tilde{A}^j_i$ to the left point and right point of $\tilde{A}^j_i$.

(3) Calculate the occurrences of rule firings at secondary levels $u_s = \underline{\alpha}_k, \underline{\alpha}_k$:

$$\underline{z}^l_{\underline{\alpha}_k} = \underline{\mu}_{\tilde{A}^{p_1}_{1|u_s=\underline{\alpha}_k}} \cdot \underline{\mu}_{\tilde{A}^{p_2}_{1|u_s=\underline{\alpha}_k}} \cdots \underline{\mu}_{\tilde{A}^{p_n}_{1|u_s=\underline{\alpha}_k}} \tag{3.5}$$

$$\underline{z}^l_{\underline{\alpha}_k} = \underline{\mu}_{\tilde{A}^{p_1}_{1|u_s=\underline{\alpha}_k}} \cdot \underline{\mu}_{\tilde{A}^{p_2}_{1|u_s=\underline{\alpha}_k}} \cdots \underline{\mu}_{\tilde{A}^{p_n}_{1|u_s=\underline{\alpha}_k}} \tag{3.6}$$

$$\underline{z}^l_{\underline{\alpha}_k} = \underline{\mu}_{\tilde{A}^{p_1}_{1|u_s=\underline{\alpha}_k}} \cdot \underline{\mu}_{\tilde{A}^{p_2}_{1|u_s=\underline{\alpha}_k}} \cdots \underline{\mu}_{\tilde{A}^{p_n}_{1|u_s=\underline{\alpha}_k}} \tag{3.7}$$

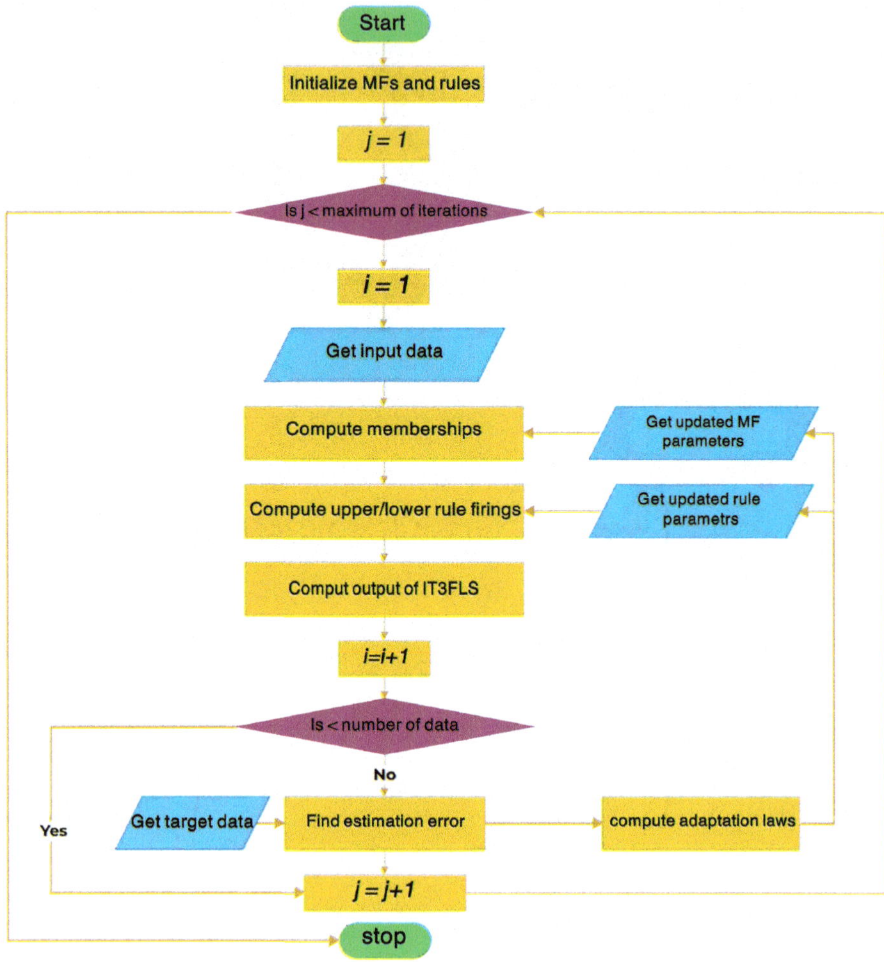

Fig. 3.5 A flowchart for the process of computation

$$\underline{z}^{l}_{\underline{\alpha}_k} = \underline{\mu}_{\tilde{A}^{p_1}_{1|u_s=\underline{\alpha}_k}} \cdot \underline{\mu}_{\tilde{A}^{p_2}_{1|u_s=\underline{\alpha}_k}} \cdots \underline{\mu}_{\tilde{A}^{p_n}_{1|u_s=\underline{\alpha}_k}} \tag{3.8}$$

where $l - th$ rule is stated as below:

$$\begin{aligned} & l - th\,Rule:\ if\ \ x_1\,is\,\tilde{A}_1^{p_1}\ and\,x_2\,is\,\tilde{A}_2^{p_2}\ and\ \cdots x_n\,is\,\tilde{A}_n^{p_n} \\ & Then\,y \in \left[\underline{w}_l,\ \underline{w}_l\right],\ l = 1,\ \ldots,\ M \end{aligned} \tag{3.9}$$

where the $p_j - th$ membership function for input x_i is denoted by $\tilde{A}_i^{p_j}$, and $\underline{w}_l$ and $\underline{w}_l$ show the rule parameters.

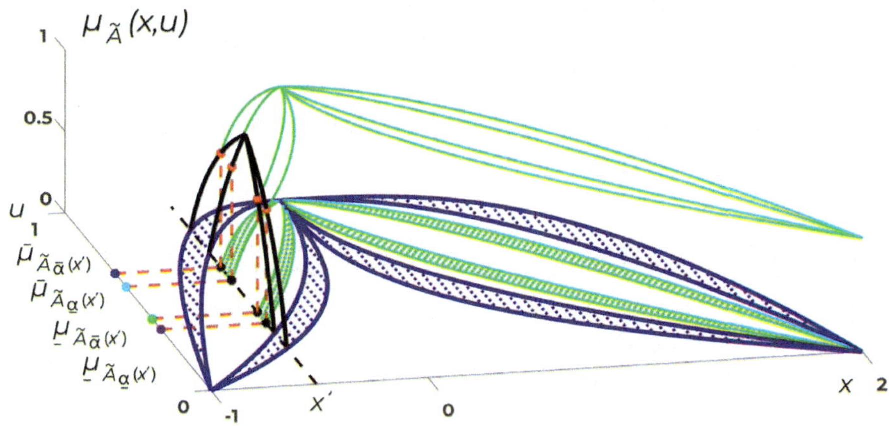

Fig. 3.6 The FOU as to interval type-3 MF

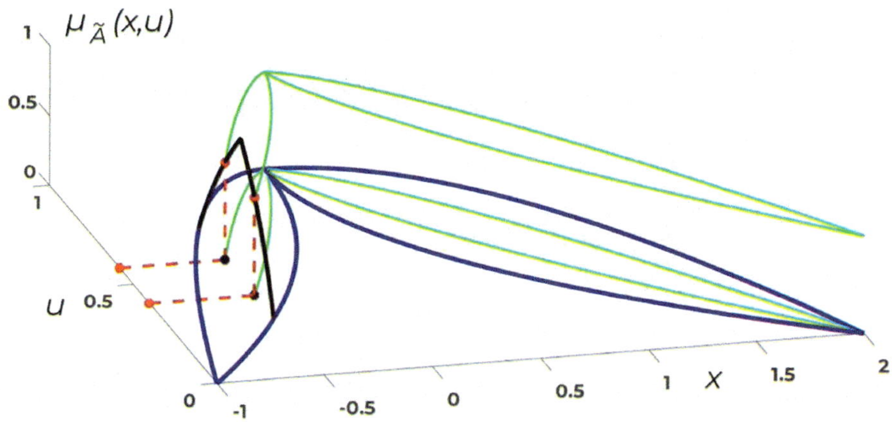

Fig. 3.7 The FOU as to the general type-2 MF

(4) The output is computed to be:

$$y = \sum_{k=1}^{K}\left(\underline{\alpha}_k \underline{f}_k + \underline{\alpha}_k \underline{f}_k\right) / \sum_{k=1}^{K}\left(\underline{\alpha}_k + \underline{\alpha}_k\right) \tag{3.10}$$

where,

$$\underline{f}_k = \frac{\sum_{l=1}^{M}\left(\underline{z}^l_{\alpha_k}\underline{w}_l + \underline{z}^l_{\alpha_k}\underline{w}_l\right)}{\sum_{l=1}^{M}\left(\underline{z}^l_{\alpha_k} + \underline{z}^l_{\alpha_k}\right)} \tag{3.11}$$

$$\underline{f}_k = \frac{\sum_{l=1}^{M} \left(\underline{z}_{\alpha_k}^{l} \underline{w}_l + \underline{z}_{\alpha_k}^{l} \underline{w}_l \right)}{\sum_{l=1}^{M} \left(\underline{z}_{\alpha_k}^{l} + \underline{z}_{\alpha_k}^{l} \right)} \tag{3.12}$$

where, K and M are, respectively, the number of horizontal cuts and rules.

3.10.3 Suggested Adaptive Kernel Size Correntropy Filter

Consider the following system:

$$x(t) = Hx(t-1) + \vartheta(t) y(t) = Gx(t) + \upsilon(t) \tag{3.13}$$

Let x, y, H and G represent matrices of dimensions $n \times 1$, $m \times 1$, $n \times n$ and $m \times n$, respectively. Additionally, ϑ and υ are Gaussian noise with covariance R and Q, respectively. The Kalman filter is the most effective method for accurately estimating the state vector $\hat{x}$, especially when dealing with Gaussian noise. However, the correntropy filter offers an intriguing technique to address the impact of non-Gaussian noise. The primary distinction lies in the fact that this method focuses on minimizing the cost function Eq. (3.14) instead of the traditional cost function used in the Kalman filter Eq. (3.15).

$$J = F_\sigma(\|y(t) - G\hat{x}(t)\|) + F_\sigma(\|\hat{x}(t) - H\hat{x}(t-1)\|) \tag{3.14}$$

$$\begin{aligned} J = {} & \frac{1}{2}(y(t) - G\hat{x}(t))^T R^{-1}(t)(y(t) - G\hat{x}(t)) \\ & + \frac{1}{2}(\hat{x}(t) - H\hat{x}(t-1))^T P^{-1}(t)(\hat{x}(t) - H\hat{x}(t-1)) \end{aligned} \tag{3.15}$$

where, $P(t) = E\left[e(t)e^T(t)\right]$, $e(t) = (x(t) - \hat{x}(t))$. The expectation operation is shown by E, and

$$F_\sigma(x - y) = exp\left(-\|x - y\|^2 / 2\sigma^2\right) \tag{3.16}$$

The primary constraint of this algorithm is that its efficiency is very contingent upon the value of σ. In noisy conditions, a higher value of σ is recommended, while in noise-free scenarios, a lower value of σ is preferred. Nevertheless, in the majority of studies, the parameter σ is assumed to remain constant. In order to address this issue, a novel fuzzy methodology is proposed. The value of σ in the suggested method is decided by a straightforward fuzzy system to ensure optimal performance. The rule database and membership functions for the fuzzy kernel size are, respectively, provided in Table 3.1 and Fig. 3.8.

Table 3.1 The rule database for the fuzzy system σ (kernel size)

Number of rule	If $e \in$	and $\dot{e} \in$	Then $\sigma \in$
1	–	$B_{\dot{e}}$	S_σ
2	–	$S_{\dot{e}}$	S_σ
3	B_e	-	S_σ
4	S_e	-	B_σ
5	M_e	–	M_σ
6	–	$M_{\dot{e}}$	M_σ
7	S_e	$M_{\dot{e}}$	B_σ
8	M_e	$M_{\dot{e}}$	B_σ

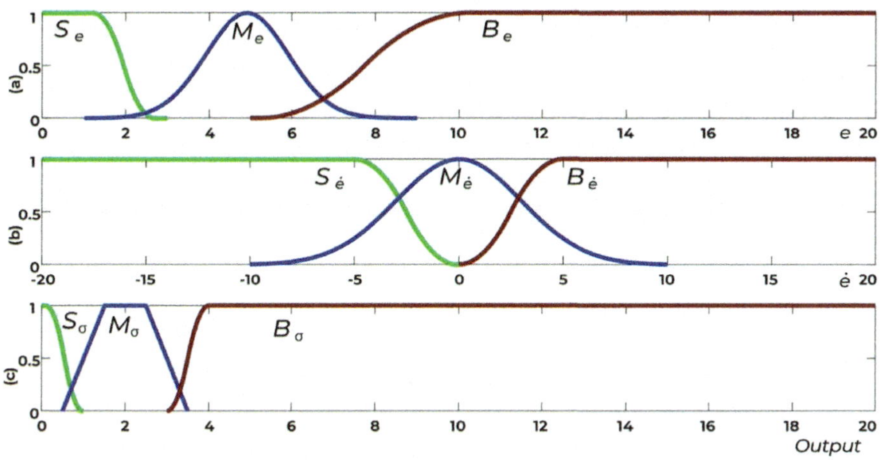

Fig. 3.8 The MFs for the fuzzy kernel size, **a** Membership functions for input e, **b** Membership functions for $\dot{e}$ and (c) Membership functions for output σ

3.10.4 Learning Algorithm

3.10.4.1 Consequent Parameters

In order to determine the parameters of the SO-IT3FLS, the state space representation provided below is taken into account:

$$\theta_c(t) = \theta_c(t-1) + \vartheta(t)y(t) = \xi^T\theta_c(t) + \upsilon(t) \tag{3.17}$$

where, $\vartheta(t)$ and $\upsilon(t)$ are considered to be the non-Gaussian noise. θ_c and ξ are:

$$\theta_c = \left[\underline{w}_1, \underline{w}_1, \ldots, \underline{w}_M, \underline{w}_1, \underline{w}_2, \ldots, \underline{w}_M\right]^T \tag{3.18}$$

$$\xi^T = \frac{1}{\sum_{k=1}^{K}\left(\underline{\alpha}_k + \underline{\alpha}_k\right)} \sum_{k=1}^{K}[\left(\frac{\underline{\alpha}_k \underline{z}_{\underline{\alpha}_k}^{1}}{\sum_{l=1}^{M}\left(\underline{z}_{\underline{\alpha}_k}^{l} + \underline{z}_{\underline{\alpha}_k}^{l}\right)} + \frac{\underline{\alpha}_k \underline{z}_{\underline{\alpha}_k}^{l}}{\sum_{l=1}^{M}\left(\underline{z}_{\underline{\alpha}_k}^{l} + \underline{z}_{\underline{\alpha}_k}^{l}\right)}\right), \ldots$$
$$\left(\frac{\underline{\alpha}_k \underline{z}_{\underline{\alpha}_k}^{M}}{\sum_{l=1}^{M}\left(\underline{z}_{\underline{\alpha}_k}^{l} + \underline{z}_{\underline{\alpha}_k}^{l}\right)} + \frac{\underline{\alpha}_k \underline{z}_{\underline{\alpha}_k}^{M}}{\sum_{l=1}^{M}\left(\underline{z}_{\underline{\alpha}_k}^{l} + \underline{z}_{\underline{\alpha}_k}^{l}\right)}\right),$$
$$\left(\frac{\underline{\alpha}_k \underline{z}_{\underline{\alpha}_k}^{l}}{\sum_{l=1}^{M}\left(\underline{z}_{\underline{\alpha}_k}^{l} + \underline{z}_{\underline{\alpha}_k}^{l}\right)} + \frac{\underline{\alpha}_k \underline{z}_{\underline{\alpha}_k}^{l}}{\sum_{l=1}^{M}\left(\underline{z}_{\underline{\alpha}_k}^{l} + \underline{z}_{\underline{\alpha}_k}^{l}\right)}\right), \ldots$$
$$\left(\frac{\underline{\alpha}_k \underline{z}_{\underline{\alpha}_k}^{M}}{\sum_{l=1}^{M}\left(\underline{z}_{\underline{\alpha}_k}^{l} + \underline{z}_{\underline{\alpha}_k}^{l}\right)} + \frac{\underline{\alpha}_k \underline{z}_{\underline{\alpha}_k}^{M}}{\sum_{l=1}^{M}\left(\underline{z}_{\underline{\alpha}_k}^{l} + \underline{z}_{\underline{\alpha}_k}^{l}\right)}\right)] \tag{3.19}$$

The adaptation law is calculated as:

$$\theta_c(t) = \theta_c(t-1) + \eta(t)\left(y_d - \xi^T(t-1)\theta_c(t-1)\right) \tag{3.20}$$

where,

$$\eta(t) = \left(I_M + \xi^T P(t)\xi\right)^{-1} P(t)\xi \tag{3.21}$$

$$P(t) = F_{\hat{\sigma}}\left(\left\|y_d - \xi^T(t-1)\theta_c(t-1)\right\|\right) \tag{3.22}$$

The kernel function Eq. (3.16), denoted as $F_{\hat{\sigma}}$, is determined using a basic fuzzy system, with the kernel size σ being decided by this system.

3.10.4.2 Antecedent Parameters

The number of membership functions (MFs) for inputs and the number of α-cuts α remain constant, while the centers of the MFs and the level of α-cuts are modified using the maximum correntropy unscented Kalman filter with adaptive fuzzy kernel size (MCUKF-AFKS). The details are provided below.

(1) The suggested SO-INT3FLS is stated as:

$$\theta_a(t) = \theta_a(t-1) + \vartheta(t) \;\&\; y(t) = SO - INT3FLS(x(t)|\theta_a(t)|\theta_c(t)) + \upsilon(t) \tag{3.23}$$

where,

$$\theta_a = \left[c_{\tilde{A}_1^1}, \ldots, c_{\tilde{A}_1^N}, \ldots, c_{\tilde{A}_n^1}, \ldots, c_{\tilde{A}_n^N}, \underline{\alpha}_1, \ldots, \underline{\alpha}_K, \underline{\alpha}_1, \ldots, \underline{\alpha}_K\right]^T \tag{3.24}$$

where, the center of $\tilde{A}_i^j$ is denoted by $c_{\tilde{A}_i^j}$, moreover, $\tilde{A}_i^j$ denotes the $j-th$ MF for the $i-th$ input x_i. On the other hand, $\underline{\alpha}_k$ and $\underline{\alpha}_k$ show, respectively, the lower and upper $k-th$ α-cuts α, N is the number of membership functions for inputs, and, finally, $\theta_c(t)$ shows the vector of rule parameters.

(2) Obtain the kernel size σ from the fuzzy kernel size and set the antecedent parameters vector θ_a.
(3) Create the sigma-points as below:

$$\tilde{\theta}_a^\kappa(t) = \theta_a(t) + \left(\sqrt{(d_a + \lambda)P_a(t)}\right)^\kappa, \kappa = 1, \ldots, d_a$$

$$\tilde{\theta}_a^\kappa(t) = \theta_a(t) - \left(\sqrt{(d_a + \lambda)P_a(t)}\right)^{\kappa - d_a}, \kappa = d_a + 1, \ldots, 2d_a$$

$$\tilde{\theta}_a^\kappa(t) = \theta_a(t), \kappa = 0 \tag{3.25}$$

where the dimension of the vector θ_a is denoted by d_a, P_a denotes the covariance matrix of the antecedent parameters, $\left(\sqrt{(d_a + \lambda)P_a(t)}\right)^\kappa$ shows the $\kappa - th$ column of the said parameter, and, finally, the turning parameter is denoted by λ.

(4) Determine the predicted value of $\hat{y}(t)$ using the following calculation:

$$\hat{y}(t) = \frac{1}{2(d_a + \lambda)} \sum_{\kappa=1}^{2d_a} SO - INT3FLS\left(x(t)\middle|\tilde{\theta}_a^\kappa(t)\middle|\theta_c(t)\right) + \frac{\lambda}{d_a + \lambda} SO$$
$$- INT3FLS\left(x(t)\middle|\tilde{\theta}_a^0(t)\middle|\theta_c(t)\right) \tag{3.26}$$

(5) Calculate the updated $\tilde{P}_a$ and $\tilde{R}$ as:

$$\tilde{P}_a = S_p \psi_x^{-1} S_p^T \tilde{R} = S_r \psi_y^{-1} S_r^T \tag{3.27}$$

where,

$$[P_a 00R] = \left[S_p S_p^T 00 S_r S_r^T\right] = SS^T \tag{3.28}$$

$$\psi_x = diag\left(G_\sigma(\chi_1), \ldots, G_\sigma\left(\chi_{d_a}\right)\right) \psi_y = diag\left(G_\sigma\left(\chi_{d_a+1}\right), \ldots, G_\sigma(\chi_L)\right) \tag{3.29}$$

$$\chi = S^{-1}\left[\theta_a y - \hat{y} + H\theta_a\right] - S^{-1}[IH]\theta_a \tag{3.30}$$

$$H = P_{xy}^T P^{-1} \tag{3.31}$$

$$P_{xy}^{T} = \frac{1}{2(d_a + \lambda)} \sum_{\kappa=1}^{2d_a} \tilde{\theta}_a^{\kappa} \left[SO - INT3FLS\left(x \middle| \tilde{\theta}_a^{\kappa} \middle| \theta_c\right)\right]^{T}$$
$$- \theta_a \hat{y} \frac{\lambda}{d_a + \lambda} \tilde{\theta}_a^{0} \left[SO - INT3FLS\left(x \middle| \tilde{\theta}_a^{0} \middle| \theta_c\right)\right]^{T} \quad (3.32)$$

where the dimension of χ is shown by L, and $G_\sigma(\chi)$ is defined as follows:

$$G_\sigma(\chi) = exp\left(-\frac{\chi^2}{2\sigma}\right) \quad (3.33)$$

The value of σ is decided by the suggested adaptive fuzzy kernel size.

(6) Update θ_a and P_a as:

$$\theta_a(t+1) = \theta_a(t) + \eta(t)(y(t) - \hat{y}(t)) \quad (3.34)$$

$$P_a(t+1) = (I - \eta(t)H(t))P_a(t)(I - \eta(t)H(t))^T + \eta(t)R(t)\eta^T(t) \quad (3.35)$$

where $\eta(t)$ is

$$\eta(t) = \tilde{P}_a(t)H^T(t)\left(\tilde{R}(t) + H(t)\tilde{P}_a(t)H^T(t)\right)^{-1} \quad (3.36)$$

After revising the centers of membership functions, the left and right points are modified as follows. The right point is centered on the right MF, while the left point is centered on the left MF.

3.10.4.3 Convergence Analysis

In order to do convergence analysis, the Lyapunov function is considered:

$$V(t) = \frac{1}{2}E^2 = (y_d(t) - \hat{y}(t))^2 \quad (3.37)$$

Considering a time difference of Eq. (3.37), we get:

$$\Delta V(t) = V(t+1) - V(t) = \frac{1}{2}\left((y_d(t+1) - \hat{y}(t+1))^2 - (y_d(t) - \hat{y}(t))^2\right) \quad (3.38)$$

Take into account of the equation below:

$$E(t+1) = E(t) + \Delta E(t) \cong E(t) + \left[\frac{\partial E(t)}{\partial \theta_c}\right]^T \Delta\theta_c \quad (3.39)$$

where $\theta_c(t)$ denotes the vector of rule parameters. Using Eq. (3.20), we have:

$$\theta_c(t+1) = \theta_c(t) + \Delta\theta_c(t) = \theta_c(t) + \eta(t).E(t) \tag{3.40}$$

Equation (3.40), is restated as follows:

$$\theta_c(t+1) = \theta_c(t) + \Delta\theta_c(t) = \theta_c(t) + \left(I_M + P(t)\xi^T\xi\right)^{-1}P(t)\xi.E(t) \tag{3.41}$$

By taking into account of Eqs. (3.37), (3.41) changes to:

$$\theta_c(t+1) = \theta_c(t) + \Delta\theta_c(t) = \theta_c(t) - \left(I_M + P(t)\xi^T\xi\right)^{-1}P(t).E(t)\frac{\partial E(t)}{\partial\theta_c} \tag{3.42}$$

where

$$\frac{\partial E(t)}{\partial\theta_c} = -\frac{\partial\hat{y}}{\partial\theta_c} \tag{3.43}$$

Take into account the following equation:

$$\kappa(t) = \left(I_M + \xi^T P(t)\xi\right)^{-1}P(t) \tag{3.44}$$

Next, Eq. (3.42) gets the new form of:

$$\theta_c(t+1) = \theta_c(t) + \Delta\theta_c(t) = \theta_c(t) - \kappa(t).\frac{\partial E(t)}{\partial\theta_c} \tag{3.45}$$

Then,

$$\Delta\theta_c = -\kappa(t)\frac{\partial E(t)}{\partial\theta_c}E(t) \tag{3.46}$$

Using Eq. (3.38), $\Delta V(t)$ is stated as:

$$\Delta V(t) = -\left[\frac{\partial E(t)}{\partial\theta_c}\right]^T\kappa(t)\frac{\partial E(t)}{\partial\theta_c}E(t)\left\{E(t) - \frac{1}{2}\left[\frac{\partial E(t)}{\partial\theta_c}\right]^T\kappa(t)E(t)\frac{\partial E(t)}{\partial\theta_c}\right\} \tag{3.47}$$

Using Eq. (3.39), Eq. (3.47), gets the form of:

$$\Delta V(t) = \left[\frac{\partial E(t)}{\partial\theta_c}\right]^T.\gamma.E(t).\frac{\partial\hat{y}(t)}{\partial\theta_c}.\left\{E(t) + \frac{1}{2}\left[\frac{\partial E(t)}{\partial\theta_c}\right]^T.\gamma.E(t).\frac{\partial\hat{y}(t)}{\partial\theta_c}\right\} \tag{3.48}$$

Equation (3.48), is restated as follows:

$$\Delta V(t) = -\left[\frac{\partial E(t)}{\partial \theta_c}\right]^T \kappa(t)\frac{\partial E(t)}{\partial \theta_c} E^2(t)\left\{1 - \frac{1}{2}\left[\frac{\partial E(t)}{\partial \theta_c}\right]^T \kappa(t)\frac{\partial E(t)}{\partial \theta_c}\right\} \tag{3.49}$$

Using Eq. (3.49), we have:

$$\Delta V(t) \leq -E^2(t)\lambda_{max}(\kappa(t))\left\|\frac{\partial E(t)}{\partial \theta_c}\right\|^2\left[1 - \frac{1}{2}\left[\frac{\partial \hat{y}(t)}{\partial \theta_c}\right]^T \kappa(t)\frac{\partial \hat{y}(t)}{\partial \theta_c}\right] \tag{3.50}$$

where, the maximum eigne value for matrix $\kappa(t)$ is shown by $\lambda_{max}(\kappa(t))$, and:

$$\frac{\partial E(t)}{\partial \theta_c} = -\frac{\partial \hat{y}(t)}{\partial \theta_c} \tag{3.51}$$

From the fact that:

$$\left[\frac{\partial \hat{y}(t)}{\partial \theta_c}\right]^T \kappa(t)\frac{\partial \hat{y}(t)}{\partial \theta_c} = \xi^T\left(I_M + \xi^T P(t)\xi\right)^{-1} P(t)\xi \tag{3.52}$$

The inequality (3.50), changes to:

$$\Delta V(t) \leq -E^2(t)\lambda_{max}(\kappa(t))\left\|\frac{\partial E(t)}{\partial \theta_c}\right\|^2\left[1 - \frac{1}{2}\frac{\xi^T P(t)\xi}{I_M + \xi^T P(t)\xi}\right] \tag{3.53}$$

Then, by taking the following relation into account:

$$1 - \frac{1}{2}\frac{\xi^T P(t)\xi}{I_M + \xi^T P(t)\xi} > 0 \tag{3.54}$$

The convergence is finally proved. We can apply the same approach to antecedent parameters.

3.10.5 Simulations

This part validates the effectiveness of the correntropy filter with adaptive kernel size and the intended performance of SO-IT3FLS through multiple simulations. The proposed fuzzy system and its corresponding learning algorithm are implemented under identical conditions as detailed in the referenced studies.

Example 1 *The first example considers the Automobile Miles Per Gallon data set* [55]. *The data set contains 7 inputs and 1 output. The dataset is randomly divided into training and testing sets with a percentage of 60% and 40% respectively. In order to prevent over-fitting, it is customary to designate 15% of the training set as the validation set. The SO-INT3FLS considers 5 membership functions with centres* *at -1, 0.25, 0, 0.75, and 1. The system also has a total of 5 rules. The rule takes the form of:*

$$j - th\,rule:\ if\ x_1\ is\ \tilde{A}_1^j\ and\ \cdots\ x_n\ is\ \tilde{A}_n^j\ Then\ \hat{y} \in \left[\underline{w}_j, \underline{w}_j\right] \tag{3.55}$$

The variable $\tilde{A}_i^j$ represents the $j - th$ membership function for the input x_i. There are 2 α-cuts α, and the basic parameters of SO-INT3FLS are outlined in Table 3.2. The primary covariance matrix is defined as $1e4 \cdot eye(n_p)$, where $eye(\cdot)$ denotes the identity matrix and n_p indicates the number of trainable parameters.

Table 3.3 provides a comparison between he root mean square error (RMSE) and T1FLS [56], T1FLS-SE [57], linguistic modelling [58], radial basis function (RBF) [58], RBF2 [58], and non-singleton NT2FLS [59]. The RMSE level for the recommended SO-IT3FLS scheme is observed to be superior to the other schemes. The parameter count for SO-IT3FLS is likewise lower compared to the other approaches. The SO-IT3FLS model consists of 35 membership function centres, 4 α-cuts α, and 10 rule parameters.

Figure 3.9 displays the trajectories of the root mean square error (RMSE) for both the validation and training data sets. Figure 3.10 illustrates the paths of estimation performance for the testing data and the absolute error of estimation. Figure 3.11 displays the RMSE trajectories, as well as the histogram and box plots of RMSE,

Table 3.2 Example 1: initial parameters

Parameter	Value	Equation
$c_{\tilde{A}_i^1}, c_{\tilde{A}_i^2}, c_{\tilde{A}_i^3}, c_{\tilde{A}_i^4}, c_{\tilde{A}_i^5}$	−1, 0.25, 0, 0.75, 1	(1–4)
$\underline{\alpha}_1, \underline{\alpha}_2$	0.7, 0.95	(1, 3)
$\underline{\alpha}_1, \underline{\alpha}_2$	0.1, 0.4	(2, 4)
M	5	(11–12)
$\underline{w}_l, l = 1, \ldots, M$	0	(11–12)
$\underline{w}_l, l = 1, \ldots, M$	Random number between 0 & 1	(11–12)

Table 3.3 Comparison of RMSE for various fuzzy neural networks in Ex. 1

Method	Testing RMSE	Num of paramaters
T1FLS-SE [57]	19.25	128
RBF [58]	3.26	540
RBF2 [58]	3.51	540
Linguistic [58]	3.14	540
T1FLS [56]	19.25	128
NT2FLS [59]	2.85	129
SO-IT3FLS	2.4220	49

with regard to the iterations. The anticipated output closely follows the actual output trajectory, and the root mean square error (RMSE) value is sufficiently low.

Example 2 *The second example involves the utilization of the Boston housing data collection* [60]. *The data collection has a total of 13 input values and 1 output value. The simulation condition is identical to that of Example 1. Table 3.4 displays the Root Mean Square Error (RMSE) values for the various techniques. Similar to Example 1, the suggested SO-IT3FLS demonstrates superior performance in terms of lower RMSE value and fewer trainable parameters. Figure 3.12 displays the RMSE trajectories for both the validation and training data sets. Figure 3.13 illustrates the paths of estimation accuracy for the testing data and the absolute error of the estimation. Figure 3.14 displays the RMSE trajectories, as well as the histogram and*

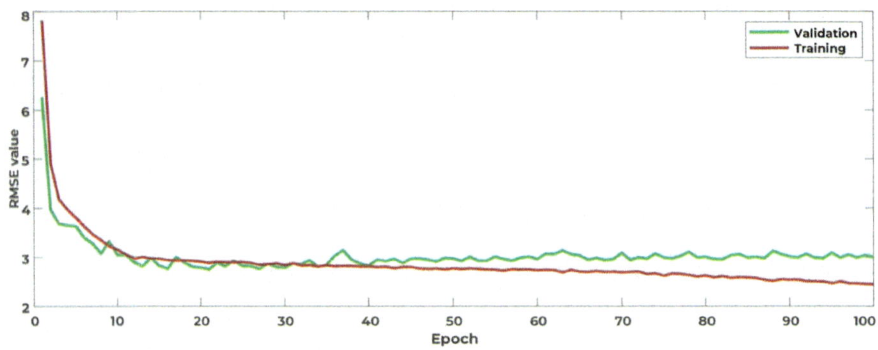

Fig. 3.9 RMSE trajectories for validation and training data sets in Ex. 1

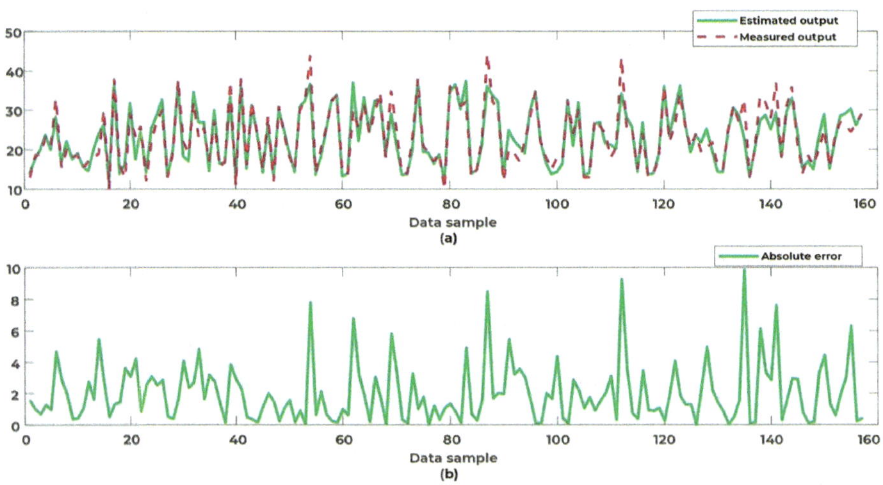

Fig. 3.10 **a** The estimation performance, for testing data; **b** Absolute error in in Ex. 1

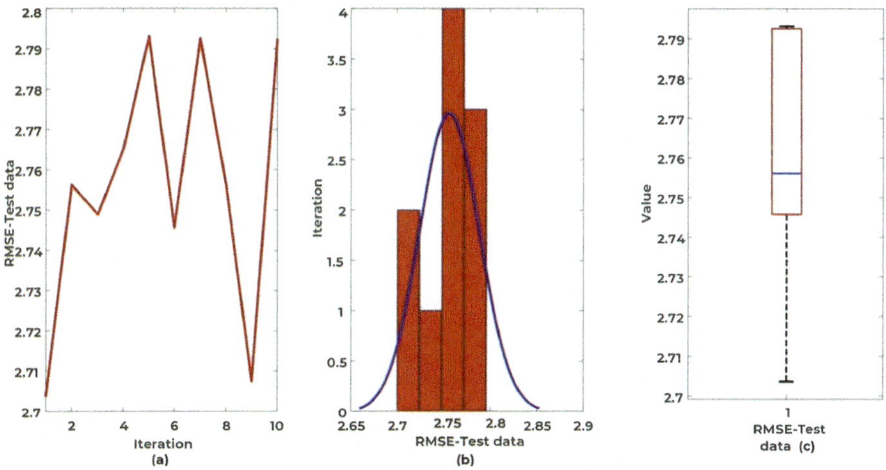

Fig. 3.11 **a** RMSE trajectory with respect to the iterations; **b** RMSE histogram plot; **b** RMSE box plot; in Ex. 1

box plots of the RMSE, with regard to the iterations. It is evident that a high level of accuracy has been attained in the estimating performance.

Example 3 *The third example focuses on the dataset of the minimum maintenance cost of electrical line* [61]. *The data set consists of 4 input values and 1 output value. The training/testing dataset consists of 400 samples for training and 200 samples for testing* [61]. *The proposed approach's capacity is being compared against IT2FLS, GT2FLS with sampling defuzzification (GT2FLS-SD), G2FLS with vertical slice defuzzification (GT2FLS-VD)* [61], *and non-singleton GT2FLS (NS-GT2FLS)* [59]. *The simulation situation and settings are identical to those of Example 1. Table 3.5 presents the RMSE comparison among the different approaches. It is observed that, as Examples 1 and 2, the suggested SO-IT3FLS yields superior results in terms of a lower RMSE magnitude. Figure 3.15 shows the RMSE trajectories for both the validation and training data sets. Figure 3.16 illustrates the paths of estimate*

Table 3.4 RMSE for various fuzzy neural networks in Ex. 2

Method	Testing RMSE	Number of paramaters
T1FLS-SE [57]	24.38	8192
RBF [58]	6.94	1332
RBF2 [58]	6.91	1332
Linguistic [58]	5.32	1332
T1FLS [56]	24.38	8192
NT2FLS [59]	5.17	28
SO-IT3FLS	3.1051	51

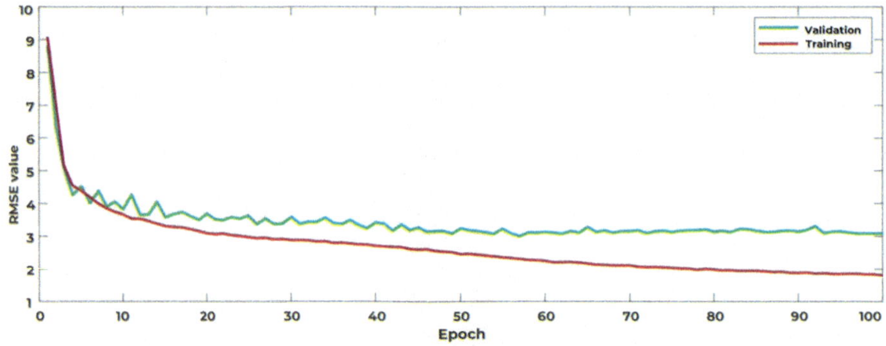

Fig. 3.12 RMSE trajectories for validation and training data sets

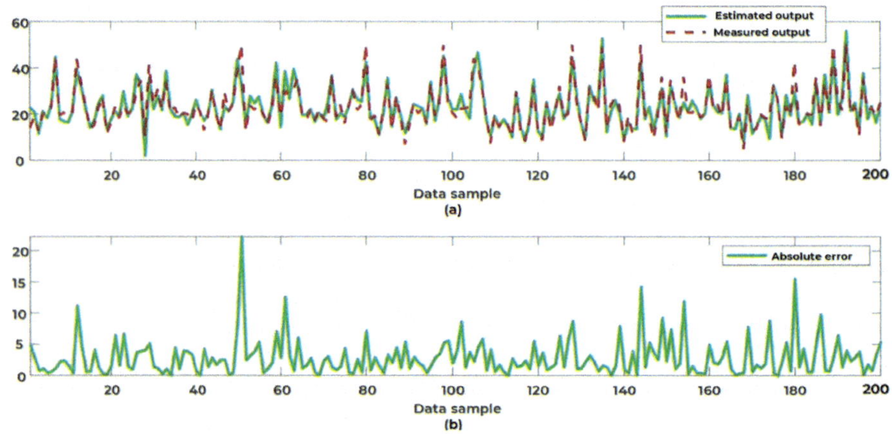

Fig. 3.13 **a** The estimation performance, for testing data; **b** Absolute error; in Ex. 2

accuracy for the testing data and the absolute error of estimation. Figure 3.7 displays the RMSE trajectories, as well as the histogram and box plots of RMSE, with regard to the iterations. It is evident that the estimated output trajectory has closely aligned with the actual output trajectory.

Example 4 *This section evaluates the performance of the proposed SO-IT3FLS and learning approach in identifying a nonlinear dynamic system* [62]*:*

$$y(t) = y(t-1)y(t-2)(y(t-1)-0.5)/\left[1+y^2(t-1)+y^2(t-2)\right]+u(t-1) \tag{3.56}$$

The input signal, denoted as $u(t)$, is defined as $sin(2\pi n/25)$. The input variables utilised in the SO-IT3FLS are considered to be $y(t-1)$, $y(t-2)$ and $u(t-1)$.

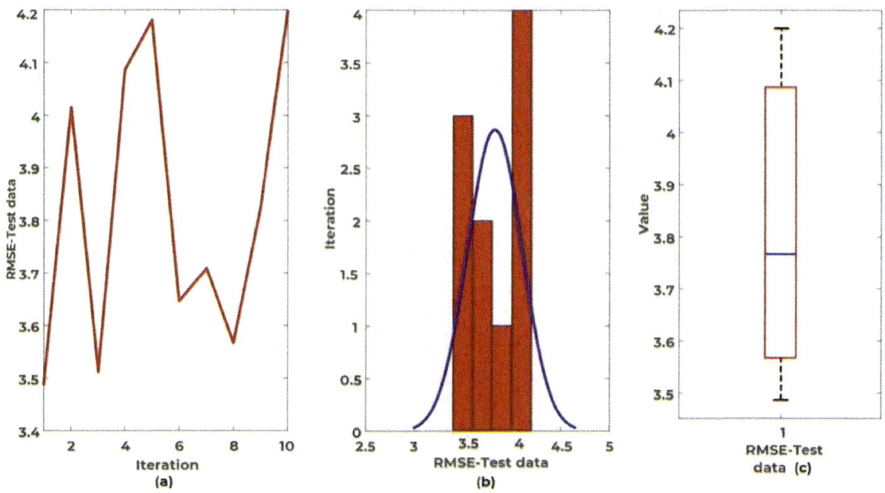

Fig. 3.14 **a** RMSE trajectory with respect to the iterations; **b** RMSE histogram plot; (b): Box plot of RMSE; in Ex. 2

Table 3.5 RMSE for various fuzzy neural networks in Ex. 3

Method	Testing RMSE
NS-GT2FLS	198.2255
GT2FLS-SD	424.139295
GT2FLS-VD	317.43325
IT2FLS	353.99755
SO-IT3FLS	182.7507

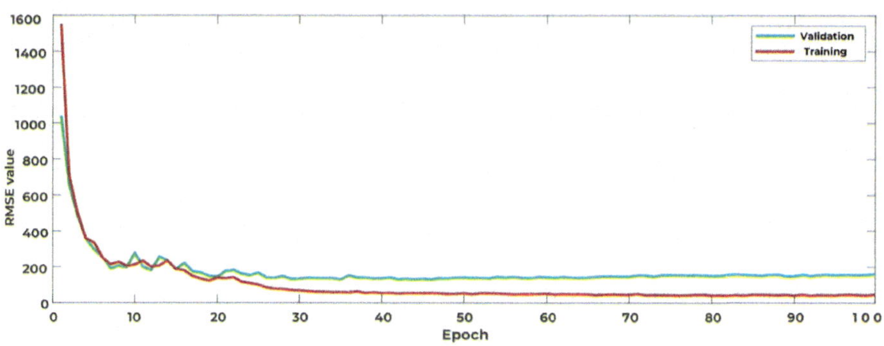

Fig. 3.15 RMSE trajectories for validation and training data sets in Ex. 3

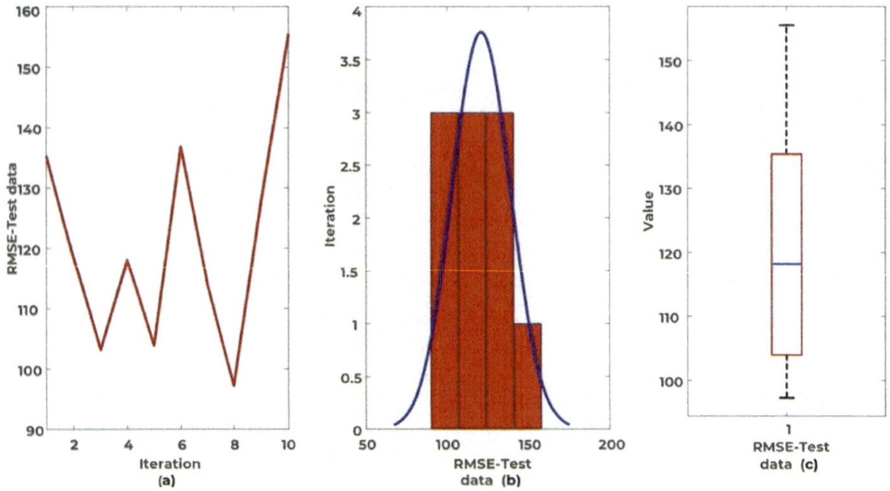

Fig. 3.16 **a** The estimation performance, for testing data; **b** Absolute error; in Ex. 3

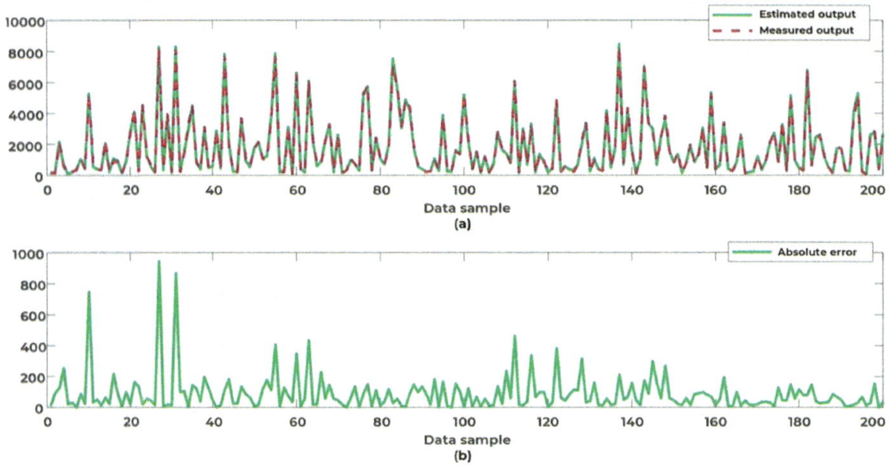

Fig. 3.17 **a** RMSE trajectory with respect to the iterations; **b** RMSE histogram plot; **b** RMSE box plot; in Ex. 3

The training dataset consists of 5000 data points, generated by $t \in [0, 5000]$. The testing dataset consists of 300 data points, generated by $t \in [5001, 5300]$ [63]. Figure 3.18 displays the RMSE trajectories for both the validation and training data sets. Figure 3.19 illustrates the paths of estimation performance for the testing data and the absolute error of estimation. Figure 3.20 displays the RMSE trajectories, as well as the histogram and box plots of the RMSE, with regard to the iterations.

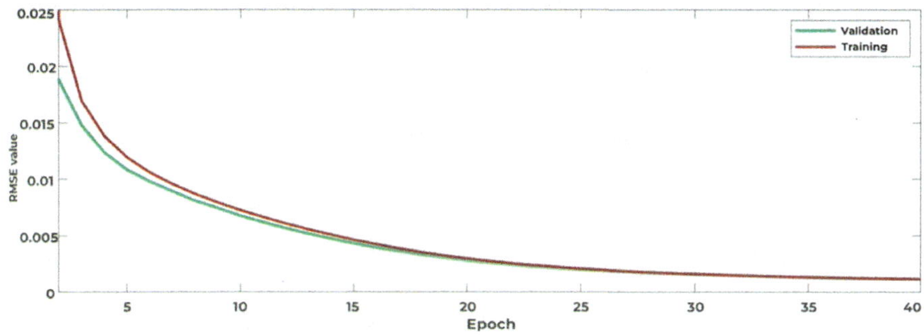

Fig. 3.18 RMSE trajectories for validation and training data sets in Ex. 5

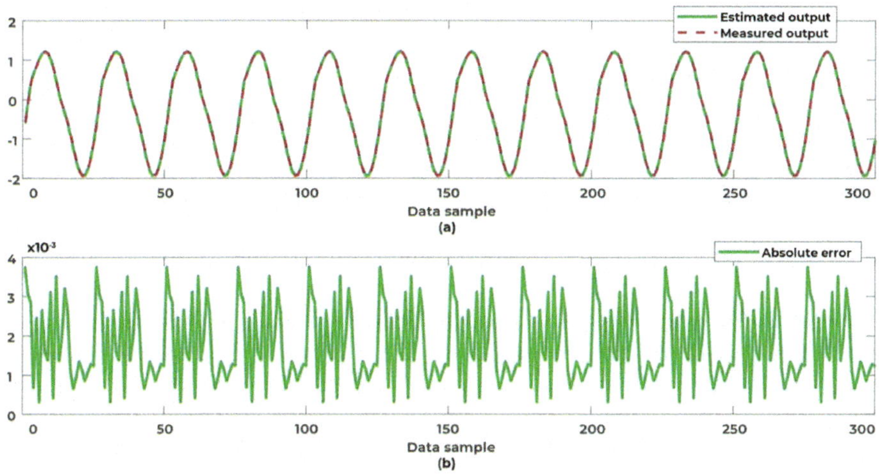

Fig. 3.19 **a** The estimation performance, for testing data; **b** Absolute error; in Ex. 5

A comparison of the performance of SO-IT3FLS is made with several other fuzzy logic systems, namely sequential evolving Takagi–Sugeno FLS (ET-FLS) [64], FLS (S-FLS) [62], resource allocating network (RAN) [65], radial basis function (RBF) [66], type-1 FLS (T1-FLS) [67], type-1 FLS with Takagi–Sugeno structure (T1-FLS-TS) [68], FLS with evolving-construction (FLS-EC) [63], and self-structuring FLS (S-FLS) [69]. Additionally, the proposed learning algorithm is compared to the recursive least square (LS) [70], extended Kalman filter (KF) [71], hybrid learning algorithms [62, 63], and genetic algorithm [72].

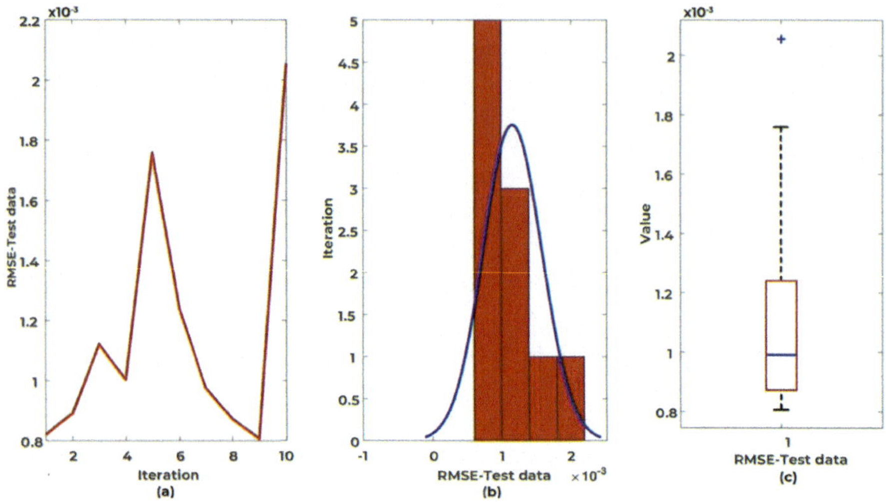

Fig. 3.20 **a** RMSE trajectory with respect to the iterations; **b** RMSE histogram plot; **b** RMSE box plot; in Ex. 5

The comparison results are presented in Table 3.6, which clearly illustrates the superiority of the suggested SO-IT3FLS and learning algorithm.

Example 5 *Learning algorithm performance is assessed in the last case, and its ability to withstand non-Gaussian noise is examined. A comparison is made between the performance of the suggested correntropy filter, which has an adaptable kernel size, and three other filters: the traditional Kalman filter (KF), the maximum correntropy criterion Kalman filter (MCC-KF), and the unscented Kalman filter (UKF).*

The subsequent navigation issue is regarded as [73]:

Table 3.6 RMSE comparison for various learning methods and neural structure in Ex. 4

Structures	Testing RMSE	Learning method	Rule numbers
SO-IT3FLS	0.0046	MCKF + MCUKF	9
T1-FLS-TS [68]	0.0225	Recursive LS	22
S-FLS [62]	0.0221	EKF + self-structuring technique	17
S-FLS [69]	0.0201	modified LS	5
RBF [66]	0.0271	EKF + self-structuring algorithm	22
T1-FLS [67]	0.0244	Genetic algorithm + KF	28
RAN [65]	0.0297	EKF	35
FLS-EC [63]	0.0056	LS + evolving-construction method	20
ET-FLS [64]	0.0212	Clustering + LS	4

$$\begin{aligned}&\left[y_1(t+1)y_2(t+1)y_3(t+1)y_4(t+1)\right] =\\ &[1\,0\,\Delta\,k\,0\,1\,0\,0\,0\,0\,0\,0\,1\,0\,0\,\Delta\,k\,0\,1]\left[y_1(t)y_2(t)y_3(t)y_4(t)\right]\\ &+[00\Delta ksin(\theta)\Delta ksin(\theta)]u(t)+\vartheta(t)\end{aligned}$$

$$z(t+1) = [1\,0\,0\,1\,0\,0\,0\,0]\left[y_1(t)\,y_2(t)y_3(t)y_4(t)\right]^T + \upsilon(t) \tag{3.57}$$

In this context, $\Delta k = 3s$ denotes the step size, $u(t)$ represents the input, θ indicates the direction of the vehicle, and $\vartheta(t)$ and $\upsilon(t)$ are Gaussian noise variables with a mean of zero and a covariance of 0.1. A sudden and unplanned sound, as depicted in Fig. 3.21, is combined with $\vartheta(t)$ and $\upsilon(t)$. Figure 3.22 illustrates the variations in correntropy kernel size when impulsive noise is present. It is evident that the proposed fuzzy system effectively adjusts the size of the correntropy kernel in response to the presence of impulsive noise in the system. The graph in Fig. 3.23 displays the RMSE trajectory for the third state. The values of RMSE for the Kalman Filter (KF), Unscented Kalman Filter (UKF), Modified Covariance Correction-Kalman Filter (MCC-KF), and the suggested approach can be found in Table 3.7. As observed, the suggested technique demonstrates a significant improvement in performance. By referring to Fig. 3.23, it is evident that the RMSE trajectory of the proposed technique rapidly converges following the occurrence of impulsive noise. When a sudden loud noise is applied to the system, the size of the Kernel is adjusted accordingly. It is important to mention that a big kernel size reduces the accuracy of the estimation, whereas a small kernel size reduces the resilience against non-Gaussian disturbances. In traditional correntropy-based Kalman filters, the kernel size is treated as a constant value. The kernel size is initialised to a value of $\sigma = 6$. We select an initial kernel size that is relatively large since our assumption is that the system is not affected by sudden and brief bursts of noise at the start. When impulsive noise is absent, the MCKF exhibits behaviour that closely resembles that of a regular Kalman filter. It is important to mention that our proposed method for adaptive kernel size does not incur any computational cost in terms of calculating adaptive sigma. Additionally, the value of the kernel size is modified with a single iteration when the impulsive noise is applied [27].

Note 1 *The proposed type-3 FLS exhibits enhanced uncertainty modelling capabilities since:*

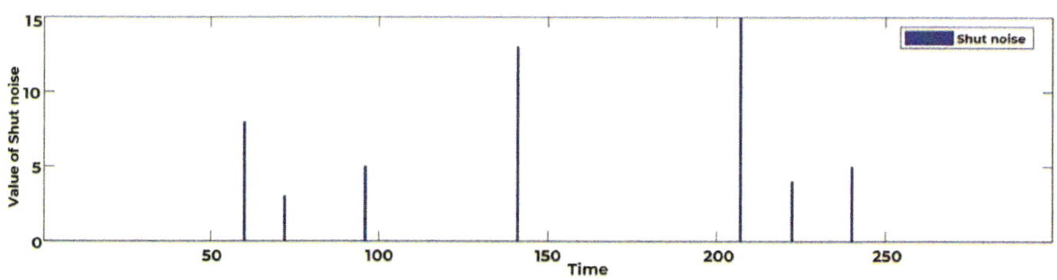

Fig. 3.21 The shot noise in outputs and states

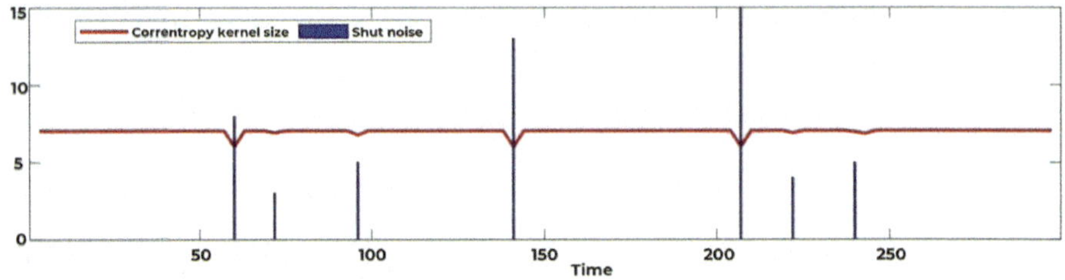

Fig. 3.22 The alterations of the correntropy kernel size in presence of shot noise

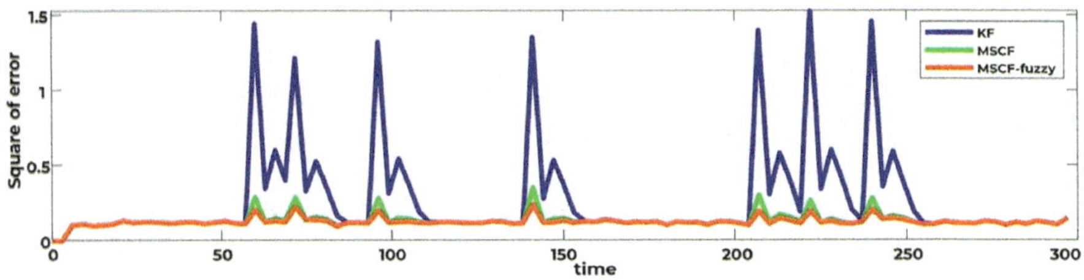

Fig. 3.23 RMSE trajectory for third state

Table 3.7 RMSE comparison for different Kalman filter in Ex. 5

	States			
Methods	y_1	y_2	y_3	y_4
UKF	0.83183	0.83472	0.26235	0.26159
KF	0.83183	0.83472	0.26235	0.26159
MCC-KF	0.2749	0.26881	0.13093	0.13273
Proposed method	0.22131	0.219	0.12229	0.12433

1. *the utilisation of fuzzy sets for secondary memberships rather than crisp values.*
2. *The upper and bottom bounds of FOU are fuzzy sets, rather than precise values.*
3. *The optimisation process focuses on the value of horizontal slices, MF parameters, and rules.*
4. *The recommended learning strategy possesses a favourable characteristic that enhances its ability to withstand non-Gaussian disturbances.*
5. *It is not necessary to compute the output derivative of FLSs with regard to the MF parameters. Furthermore, the leaning strategy ensures the preservation of the nonlinear structure.*

The suggested FLS and its related learning approach demonstrate superiority and required accuracy in three modelling tasks using real-world data sets. Furthermore, it has been shown that the proposed learning strategy has superior resilience to non-Gaussian noise. These results ensure the successful implementation of the proposed approach in both practical and educational systems [26].

3.11 Adaptive General Type-2 Fuzzy Systems for Online Frequency Regulation

3.11.1 Problem Formulation

The MG system comprises wind-turbine/diesel-engine generators (WTG/DEG), battery energy storage system (BESS), photovoltaic (PV) panel, flywheel energy storage system (FESS), and fuel cells (FCs). The recommended control configuration is presented in Fig. 3.1. The controller is designed to ensure that the frequency fluctuation ($\Delta\omega$) is brought to a negligible level under the demanding conditions. The recommended model represents the dynamics of the system using a fractional-order fuzzy dynamic system.

$$D_t^q \hat{y} = \hat{f}(\theta|x) + \varepsilon + u \tag{3.1}$$

where the derivative of the Caputo fractional-order is shown by $D_t^q \hat{y}$ and it is considered to be:

$$D_t^q \hat{y}(t) = \frac{1}{\Gamma(1-q)} \int_0^t (t-\upsilon)^{-q} \frac{d}{dt} \hat{y}(\upsilon) d\upsilon \tag{3.2}$$

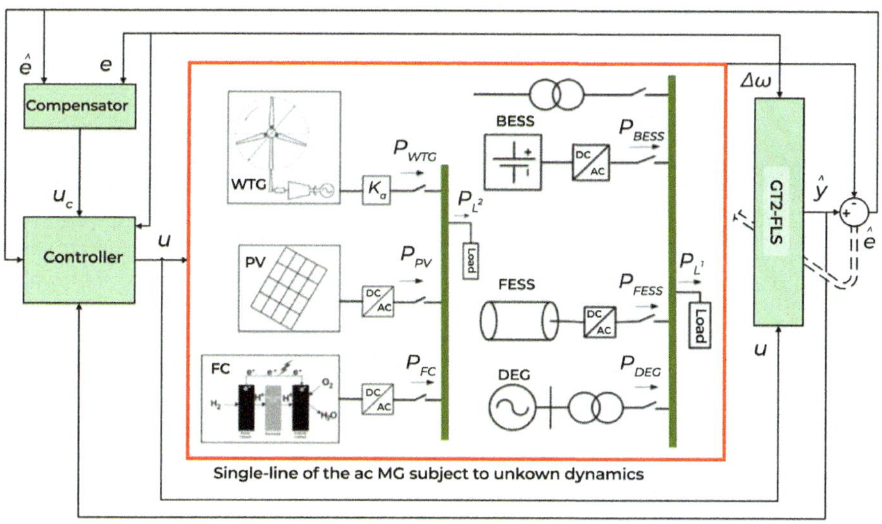

Fig. 3.1 GT2-FLS based control strucure

where $\hat{y}$ denotes the estimated value of $\Delta\omega$, the approximation error is shown by ε, the inputs of $\hat{f}$ is denoted by x so that it is determined as $x = \left[\Delta\omega, \frac{d}{dt}\Delta\omega, \int_0^t \Delta\omega(\tau)d\tau\right]^T$, the control signal is shown by u, and, finally, θ denotes the adjustable parameters of $\hat{f}$. The tunable parameters consist of rule parameters which shows the secondary membership value.

Assumption 1 *There is an assumed optimal fuzzy system, denoted as $\hat{f}^*(\theta^*|x)$, with an optimal vector θ^*. The approximation error ε is considered to be sufficiently minimal, and the following condition holds:*

$$D_t^q y = \hat{f}^*\left(\theta^*|x\right) + u \tag{3.3}$$

where, y is the output $(\Delta\omega)$.

3.11.2 General Type-2 FLS

This section provides a comprehensive overview of the proposed GT2-FLS, including in-depth information and specifications (Fig. 3.2).

(1) Obtain the inputs.
(2) Calculate the memberships, taking into account the secondary membership at level α. The membership function (MF) is illustrated in Fig. 3.3. Next, there is:

$$\underline{\mu}_{\tilde{A}^h_{i,\alpha}}(x_i) = \frac{1}{2}\left(\underline{\mu}_{\tilde{A}^h_{i,0}}(x_i) + \underline{\mu}_{\tilde{A}^h_{i,0}}(x_i)\right) + \frac{1}{2ln(1/\delta)}\left(\underline{\mu}_{\tilde{A}^h_{i,0}}(x_i) - \underline{\mu}_{\tilde{A}^h_{i,0}}(x_i)\right)\sqrt{ln(1/\alpha)} \tag{3.4}$$

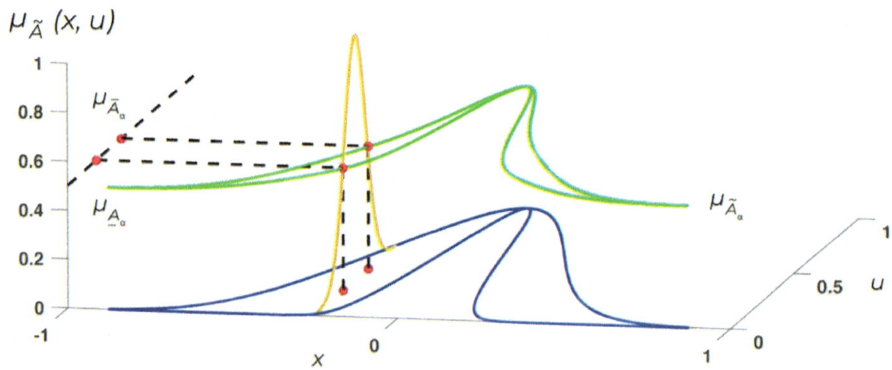

Fig. 3.2 Suggested MF

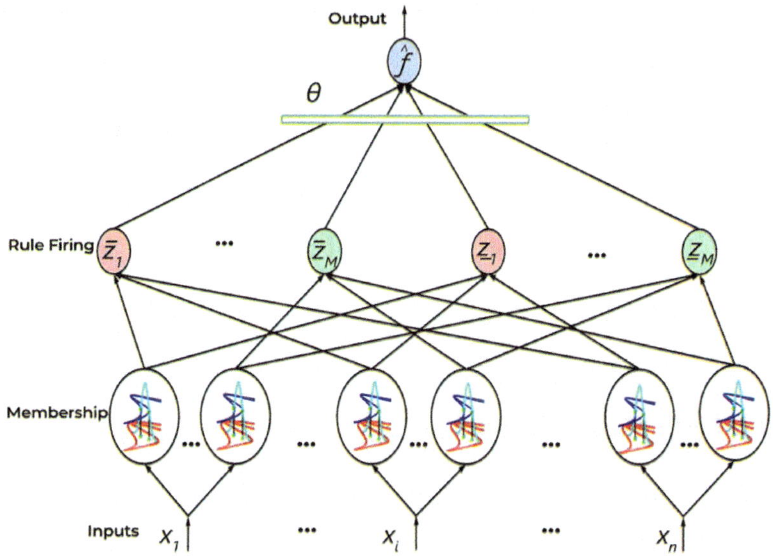

Fig. 3.3 The suggested GT2-FLS structure

$$\underline{\mu}_{\tilde{A}^h_{i,\alpha}}(x_i) = \frac{1}{2}\left(\underline{\mu}_{\tilde{A}^h_{i,0}}(x_i) + \underline{\mu}_{\tilde{A}^h_{i,0}}(x_i)\right) - \frac{1}{2ln(1/\delta)}\left(\underline{\mu}_{\tilde{A}^h_{i,0}}(x_i) - \underline{\mu}_{\tilde{A}^h_{i,0}}(x_i)\right)\sqrt{ln(1/\alpha)} \tag{3.5}$$

where,

$$\underline{\mu}_{\tilde{A}^h_{i,0}}(x_i) = exp\left(-\frac{\left(x_i - c_{\tilde{A}^h_{i,0}}\right)^2}{\underline{\sigma}^2_{\tilde{A}^h_{i,0}}}\right) \tag{3.6}$$

$$\underline{\mu}_{\tilde{A}^h_{i,0}}(x_i) = exp\left(-\frac{\left(x_i - c_{\tilde{A}^h_{i,0}}\right)^2}{\underline{\sigma}^2_{\tilde{A}^h_{i,0}}}\right) \tag{3.7}$$

where $\tilde{A}^h_{i,0}$ represents the $h-th$ membership function (MF) for input x_i at secondary membership $\alpha = 0$, $c_{\tilde{A}^h_{i,0}}$ denotes the centre of $\tilde{A}^h_{i,0}$, δ is a tiny positive constant, and $\underline{\sigma}_{\tilde{A}^h_{i,0}}$ and $\underline{\sigma}_{\tilde{A}^h_{i,0}}$ represent the lower and upper bounds of the standard deviation at secondary membership $\alpha = 0$.

(3) The rule firings are given as below:

$$\underline{z}_{h,\alpha} = \underline{\mu}_{\tilde{A}^h_{1,\alpha}}(x_i) \times \cdots \underline{\mu}_{\tilde{A}^h_{n,\alpha}}(x_i) \tag{3.8}$$

$$\underline{z}_{h,\alpha} = \underline{\mu}_{\tilde{A}^h_{1,\alpha}}(x_i) \times \cdots \underline{\mu}_{\tilde{A}^h_{n,\alpha}}(x_i) \tag{3.9}$$

where, $h = 1, \ldots, \varphi$ in which φ denotes the rules and MFs number.

(4) Based on the simple type-reduction, the output is inferred as:

$$\hat{f} = \frac{\sum_{h=1}^{\varphi}\left(\underline{z}_{h,\alpha}\underline{w}_h + \underline{z}_{h,\alpha}\underline{w}_h\right)}{\sum_{h=1}^{\varphi}\underline{z}_{h,\alpha} + \underline{z}_{h,\alpha}} \tag{3.10}$$

where the rule parameters is denoted by $\underline{w}_h$.

3.11.3 Main Results

This section presents the formulation of the learning algorithm for the consequent parameters and the value of secondary membership.

Theorem 1 *Asymptotic stability is guaranteed when the control signal (u), the compensator (u_c), and the GT2-FLS parameters tuning rules (θ) are taken into account in the following manner:*

$$u = D_t^q r - \hat{f}(\theta|x) - \lambda e + u_c \tag{3.11}$$

$$u_c = -\underline{\varepsilon}\left(tanh(e) + \frac{tanh(e)|\hat{e}|}{|e| + \delta}\right) \tag{3.12}$$

$$D_t^q = -\eta\hat{e}\xi \tag{3.13}$$

In this context, η represents the tuning rate, whereas λ and δ are positive constants. The variables e and $\hat{e}$ correspond to the tracking and estimation errors, respectively, and are defined as the differences between $\hat{y}$ and y, and between $\hat{y}$ and r. The reference signal is represented by r. The approximation error is denoted by ε, and $\underline{\varepsilon}$ is the upper bound of ε. The suggested GT2-FLS is represented by $\hat{f}(\theta|x)$, where θ is the tuneable parameters of $\hat{f}$. The derivative of $\hat{f}$ with respect to θ is represented by ξ. Both the rule parameters and the value of the secondary membership are included in the vector θ.

Proof Using Eqs. (3.1) and (3.3) the dynamics of $\hat{e}$ (estimation error) are given below:

$$D_t^q\hat{e} = -\hat{f}^*(\theta^*|x) + \hat{f}(\theta|x) + \varepsilon \tag{3.14}$$

Applying the Taylor expansion, Eq. (3.14) can be stated as:

$$D_t^q \hat{e} = \tilde{\theta}^T \xi + \varepsilon \tag{3.15}$$

where, $\tilde{\theta} = \theta - \theta^*$, $\theta^T = \left[\underline{w}_1, \ldots, \underline{w}_\varphi, w_1, \ldots, \underline{w}_\varphi, \alpha\right]$, $\underline{w}_j, \underline{w}_j, j = 1, \ldots, \varphi$ denote the successive parameters, the magnitude of the secondary membership is denoted by α and ξ is a vector that contains derivative of $\hat{f}$ with respect to $\underline{w}_j, \underline{w}_j, j = 1, \ldots, \varphi$ and α. Then we can write:

$$\xi = \left[\frac{\partial \hat{f}}{\partial \underline{w}_1}, \ldots, \frac{\partial \hat{f}}{\partial \underline{w}_\varphi}, \frac{\partial \hat{f}}{\partial \underline{w}_1}, \ldots, \frac{\partial \hat{f}}{\partial \underline{w}_\varphi}, \frac{\partial \hat{f}}{\partial \alpha}\right]^T \tag{3.16}$$

where,

$$\frac{\partial \hat{f}}{\partial \underline{w}_h} = \frac{\sum_{h=1}^{\varphi} \underline{z}_{h,\alpha}}{\sum_{h=1}^{\varphi} \underline{z}_{h,\alpha} + \underline{z}_{h,\alpha}}, h = 1, \ldots, \varphi \tag{3.17}$$

$$\frac{\partial \hat{f}}{\partial \underline{w}_h} = \frac{\sum_{h=1}^{\varphi} \underline{z}_{h,\alpha}}{\sum_{h=1}^{\varphi} \underline{z}_{h,\alpha} + \underline{z}_{h,\alpha}}, h = 1, \ldots, \varphi \tag{3.18}$$

In order to calculate $\partial \hat{f}/\partial \alpha$, both the upper and lower memberships $\underline{\mu}_{\tilde{A}_{i,\alpha}^h}(x_i)$ and $\underline{\mu}_{\tilde{A}_{i,\alpha}^h}(x_i)$ in Eqs. (3.4, 3.5) can be reformulated as:

$$\underline{\mu}_{\tilde{A}_{i,\alpha}^h}(x_i) = exp\left(-\frac{\left(x_i - c_{\tilde{A}_i^h}\right)^2}{\underline{\sigma}_{\tilde{A}_{i,\alpha}^h}^2}\right) \tag{3.19}$$

$$\underline{\mu}_{\tilde{A}_{i,\alpha}^h}(x_i) = exp\left(-\frac{\left(x_i - c_{\tilde{A}_i^h}\right)^2}{\underline{\sigma}_{\tilde{A}_{i,\alpha}^h}^2}\right) \tag{3.20}$$

where, $\underline{\sigma}_{\tilde{A}_{i,\alpha}^h}/\underline{\sigma}_{\tilde{A}_{i,\alpha}^h}$ show the upper/lower standard division for $\tilde{A}_{i,\alpha}^h$ ($h - th$ MF for input x_i at secondary level α). Using Eqs. (3.4, 3.5) and Eqs. (3.19, 3.20), $\underline{\sigma}_{\tilde{A}_{i,\alpha}^h}$ and $\underline{\sigma}_{\tilde{A}_{i,\alpha}^h}$ are given to be:

$$\underline{\sigma}_{\tilde{A}_{i,\alpha}^h}^2 = \left(x_i - c_{\tilde{A}_i^h}\right)^2 / ln \left(\frac{1}{\left[\frac{1}{2}\left(\underline{\mu}_{\tilde{A}_{i,0}^h}(x_i) + \underline{\mu}_{\tilde{A}_{i,0}^h}(x_i)\right) + \frac{1}{2ln(1/\delta)}\left(\underline{\mu}_{\tilde{A}_{i,0}^h}(x_i) - \underline{\mu}_{\tilde{A}_{i,0}^h}(x_i)\right)\sqrt{ln(1/\alpha)}\right]}\right) \tag{3.21}$$

$$\underline{\sigma}_{\tilde{A}_{i,\alpha}^h}^2 = \left(x_i - c_{\tilde{A}_i^h}\right)^2 / ln$$

$$\left(\frac{1}{\left[\frac{1}{2}\left(\underline{\mu}_{\tilde{A}^h_{i,0}}(x_i)+\underline{\mu}_{\tilde{A}^h_{i,0}}(x_i)\right)-\frac{1}{2ln(1/\delta)}\left(\underline{\mu}_{\tilde{A}^h_{i,0}}(x_i)-\underline{\mu}_{\tilde{A}^h_{i,0}}(x_i)\right)\sqrt{ln(1/\alpha)}\right]}\right) \tag{3.22}$$

Then using Eqs. (3.10, 3.19–3.22), $\partial\hat{f}/\partial\alpha$ is calculated to be:

$$\frac{\partial\hat{f}}{\partial\alpha}=\sum_{h=1}^{\varphi}\left(\sum_{i=1}^{n}\frac{\partial\underline{z}_{h,\alpha}}{\partial\underline{\sigma}_{\tilde{A}^h_{i,\alpha}}}\frac{\partial\underline{\sigma}_{\tilde{A}^h_{i,\alpha}}}{\partial\alpha}\right)(\partial\hat{f}/\partial\underline{z}_{h,\alpha})+\sum_{h=1}^{\varphi}\left(\sum_{i=1}^{n}\frac{\partial\underline{z}_{h,\alpha}}{\partial\underline{\sigma}_{\tilde{A}^h_{i,\alpha}}}\frac{\partial\underline{\sigma}_{\tilde{A}^h_{i,\alpha}}}{\partial\alpha}\right)(\partial\hat{f}/\partial\underline{z}_{h,\alpha}) \tag{3.23}$$

where, $\frac{\partial\hat{f}}{\partial\underline{z}_{h,\alpha}}$ and $\frac{\partial\hat{f}}{\partial\underline{z}_{h,\alpha}}$ are considered as:

$$\frac{\partial\hat{f}}{\partial\underline{z}_{h,\alpha}}=\frac{\underline{w}_h}{\sum_{h=1}^{\varphi}\underline{z}_{h,\alpha}+\underline{z}_{h,\alpha}}-\hat{f},\, h=1,\ldots,\varphi \tag{3.24}$$

$$\frac{\partial\hat{f}}{\partial\underline{z}_{h,\alpha}}=\frac{\underline{w}_h}{\sum_{h=1}^{\varphi}\underline{z}_{h,\alpha}+\underline{z}_{h,\alpha}}-\hat{f} \tag{3.25}$$

$\frac{\partial\underline{z}_{h,\alpha}}{\partial\underline{\sigma}_{\tilde{A}^h_{i,\alpha}}}$ and $\frac{\partial\underline{z}_{h,\alpha}}{\partial\underline{\sigma}_{\tilde{A}^h_{i,\alpha}}}$ have the following computation:

$$\frac{\partial\underline{z}_{h,\alpha}}{\partial\underline{\sigma}_{\tilde{A}^h_{i,\alpha}}}=\frac{2\left(x_i-c_{\tilde{A}^h_i}\right)^2}{\underline{\sigma}^3_{\tilde{A}^h_{i,\alpha}}}\underline{z}_{h,\alpha} \tag{3.26}$$

$$\frac{\partial\underline{z}_{h,\alpha}}{\partial\underline{\sigma}_{\tilde{A}^h_{i,\alpha}}}=\frac{2\left(x_i-c_{\tilde{A}^h_i}\right)^2}{\underline{\sigma}^3_{\tilde{A}^h_{i,\alpha}}}\underline{z}_{h,\alpha} \tag{3.27}$$

$\frac{\partial\underline{\sigma}_{\tilde{A}^h_{i,\alpha}}}{\partial\alpha}$ and $\frac{\partial\underline{\sigma}_{\tilde{A}^h_{i,\alpha}}}{\partial\alpha}$ are calculated to be:

$$\frac{\partial\underline{\sigma}_{\tilde{A}^h_{i,\alpha}}}{\partial\alpha}=\left(x_i-c_{\tilde{A}^h_i}\right)^2\Big/\left[ln\left(\frac{1}{\underline{\mu}_{\tilde{A}^h_{i,\alpha}}(x_i)}\right)\right]^3\cdot\frac{-1/\left(\alpha\sqrt{ln(1/\alpha)}\right)}{\underline{\mu}_{\tilde{A}^h_{i,\alpha}}(x_i)^2} \tag{3.28}$$

$$\frac{\partial\underline{\sigma}_{\tilde{A}^h_{i,\alpha}}}{\partial\alpha}=\left(x_i-c_{\tilde{A}^h_i}\right)^2\Big/\left[ln\left(\frac{1}{\underline{\mu}_{\tilde{A}^h_{i,\alpha}}(x_i)}\right)\right]^3\cdot\frac{-1/\left(\alpha\sqrt{ln(1/\alpha)}\right)}{\underline{\mu}_{\tilde{A}^h_{i,\alpha}}(x_i)^2} \tag{3.29}$$

Using Eqs. (3.11, 3.1), the dynamics e are given as:

$$D_t^q e=-\lambda e+\varepsilon+u_c \tag{3.30}$$

Now, we consider the following Lyapunov function for the purpose of studying stability:

$$V = \frac{1}{2}e^2 + \frac{1}{2}\hat{e}^2 + \frac{1}{2\eta}\tilde{\theta}^T\tilde{\theta} \tag{3.31}$$

The fractional time derivative of Eq. (3.31) gets the following form:

$$D_t^q V \leq eD_t^q e + \hat{e}D_t^q\hat{e} + \frac{1}{\eta}\tilde{\theta}^T D_t^q\theta \tag{3.32}$$

Inserting $D_t^q e$ and $D_t^q\hat{e}$ from Eqs. (3.30, 3.15) into Eq. (3.32), the recent inequality changes to:

$$D_t^q V \leq e(-\lambda e + \varepsilon + u_c) + \hat{e}\left(\tilde{\theta}^T\xi + \varepsilon\right) + \frac{1}{\eta}\tilde{\theta}^T D_t^q\theta \tag{3.33}$$

Simplifying Eq. (3.33), gives:

$$D_t^q V \leq -\lambda e^2 + e\varepsilon + \hat{e}\varepsilon + eu_c + \tilde{\theta}^T\left(\frac{1}{\eta}\tilde{\theta}^T D_t^q\theta + \hat{e}\xi\right) \tag{3.34}$$

Taking the tuning law Eq. (3.13) into account, the inequality (3.34) obtains the following form:

$$D_t^q V \leq -\lambda e^2 + e\varepsilon + \hat{e}\varepsilon + eu_c \tag{3.35}$$

The inequality (3.35) is restated as:

$$D_t^q V \leq -\lambda e^2 + |e||\varepsilon| + |\hat{e}||\varepsilon| + eu_c \tag{3.36}$$

Inserting u_c from Eq. (3.12) into Eq. (3.36), gives:

$$D_t^q V \leq -\lambda e^2 + |e||\varepsilon| + |\hat{e}||\varepsilon| - e\underline{\varepsilon}\left(tanh(e) + \frac{tanh(e)|\hat{e}|}{|e| + \delta}\right) \tag{3.37}$$

Considering the fact that $etanh(e) \leq |e|$, the inequality (3.37) is reformulated as follows:

$$D_t^q V \leq -|\hat{e}|\left(-|\varepsilon| + \underline{\varepsilon}|e|/(|e| + \delta)\right) - \lambda e^2 - |e|\left(-|\varepsilon| + \underline{\varepsilon}\right) \tag{3.38}$$

By taking into account the facts that $\underline{\varepsilon} \geq |\varepsilon|$ and $|e|/(|e| + \delta) \approx 1$, Eq. (3.38) implies that $D_t^q V \leq 0$. According to the theory of the invariant sets, it is evident that $D_t^q V = 0$ when $e = 0$ and the proof is completed.

Note 1 *The suggested GT2-FLS has a lower computational cost compared to the standard version. In the proposed method, there is just one α-cut. However, the α-cut is not fixed but rather determined in an adaptive manner.*

3.11.4 Simulations

The efficacy of the suggested controller is evaluated in several scenarios. The diagram illustrating the structure of the managed MG is presented in Fig. 3.4, while the specific parameters used for the simulation are provided in Table 3.1. When considering inputs of GT2-FLS, two membership functions are used. These functions have centres at –1 and 1, and are divided into upper and lower standard divisions with a ratio of 2:1. The initial α-cut is defined as 0.5.

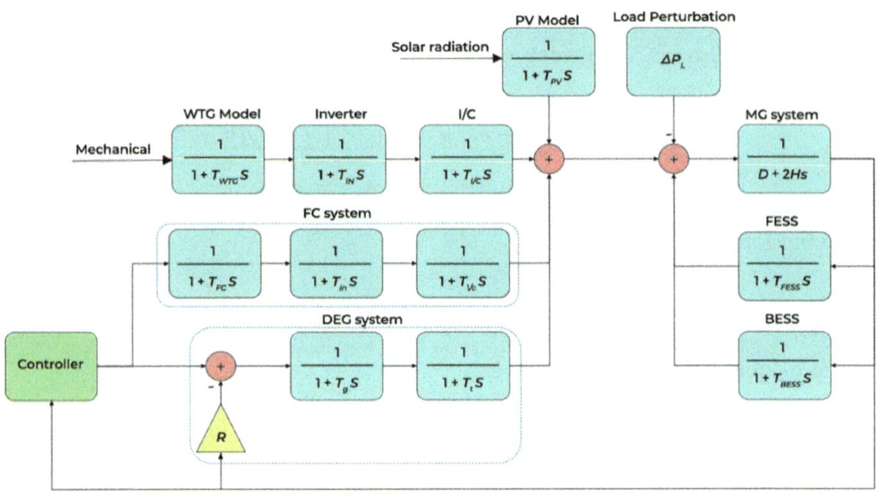

Fig. 3.4 The dynamics of the case study

Table 3.1 MG system simulation parameters

Parameter description				MG description		
T_{PV}	1.80 (s)	T_{WTG}	1.50 (s)	FC	70 (KW)	$P_{L_1} =$ 210 (KW)
T_{IN}	0.040 (s)	D	0.0150 (pu/Hz)	WTG	100 (KW)	
H	0.170 (pu s)	$T_{I/c}$	0.004 (s)	PV	30 (KW)	
T_{BESS}	0.100 (s)	T_g	0.080 (s)	DEG	160 (KW)	$P_{L_2} =$ 200 (KW)
T_t	0.40 (s)	T_{FESS}	0.10 (s)	FESS	45 (KW)	
T_{FC}	0.260 (s)	R	0.33 (pu/Hz)	BESS	45 (KW)	

Scenario 1:

In the first scenario, a modest change in load disturbance (Δp_L) is considered, while assuming constant values for solar radiation, mechanical power, and parameters. The path of Δp_L and the effectiveness of the frequency change under this Δp_L are illustrated in Fig. 3.5.

Figure 3.6 displays the results of the online dynamic approximation approach, while Fig. 3.7 illustrates controller u trajectory. It is observed that $\Delta\omega$ is converging towards zero and the control signal has a uniform form. Furthermore, the identification performance is commendable.

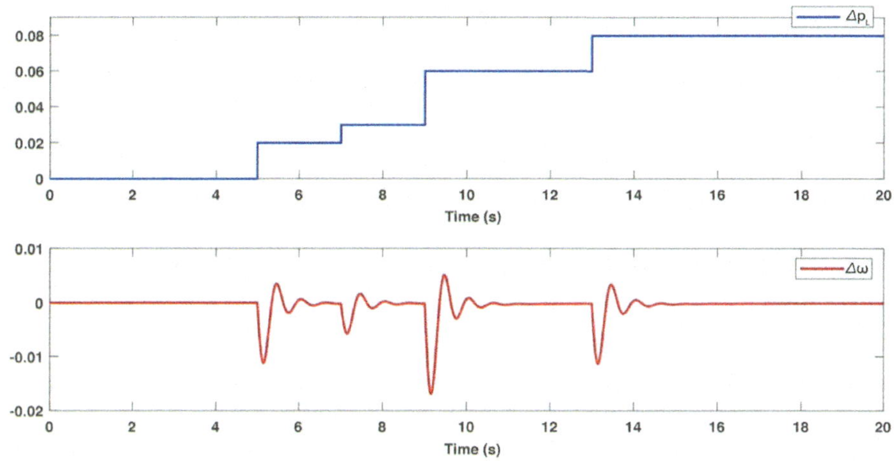

Fig. 3.5 Δp_L and $\Delta\omega$ trajectories in Scenario 1

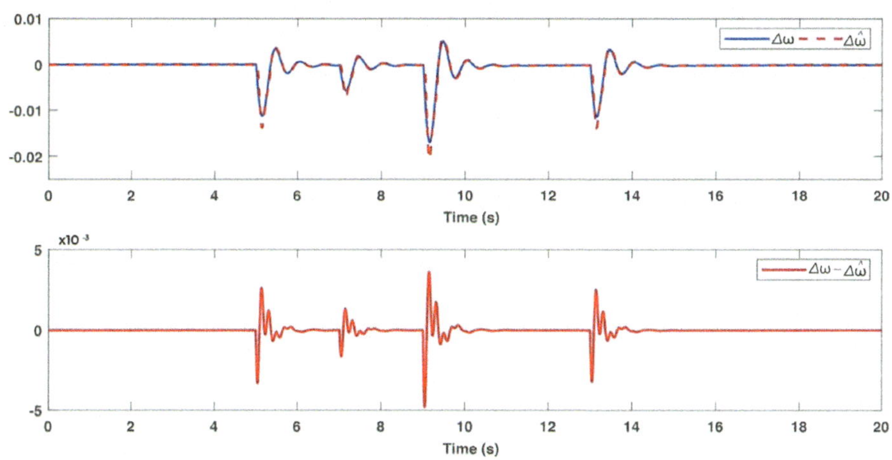

Fig. 3.6 Online identification performance in Scenario 1

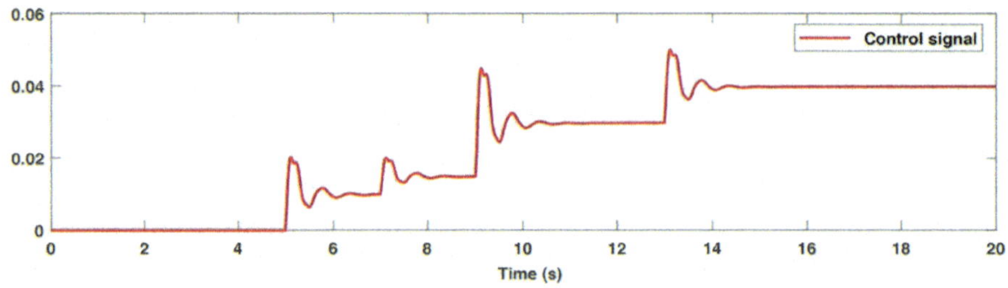

Fig. 3.7 The control signal in Scenario 1

Scenario 2:

The second scenario examines the robustness in relation to a significant change in Δp_L. Similarly to the first example, the value of Δp_L is altered at time $t = 1s$. Figure 3.8 illustrates the path of Δp_L and the change in frequency caused by this disturbance. Figure 3.9 illustrates the online identification proficiency. Figure 3.10 illustrates the controller u. We see that a high level of robustness proficiency is achieved when dealing with a significant load disruption.

Scenario 3:

In this case, the parameters' values are variable, but their changes are described by the equation $p = (sin(t) + 1)p$. The value of Δp_L is assumed to be 0.1 per unit ($p.u.$). Figure 3.11 illustrates the path of Δp_L and the fluctuations in frequency. Figure 3.12 displays the performance of online identification, whereas Fig. 3.13 illustrates controller u trajectory. It can be observed that the identification and control technique operate well, and the trajectory of Δω reaches zero within a limited time duration.

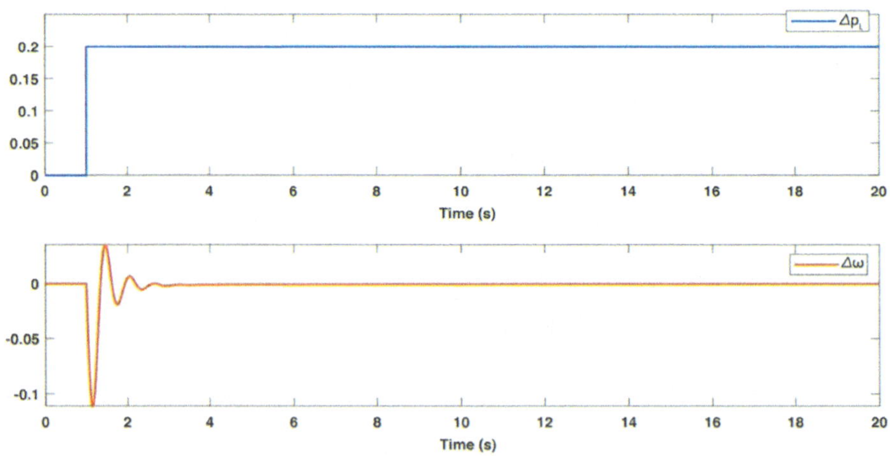

Fig. 3.8 Δp_L and $\Delta\omega$ trajectories in Scenario 2

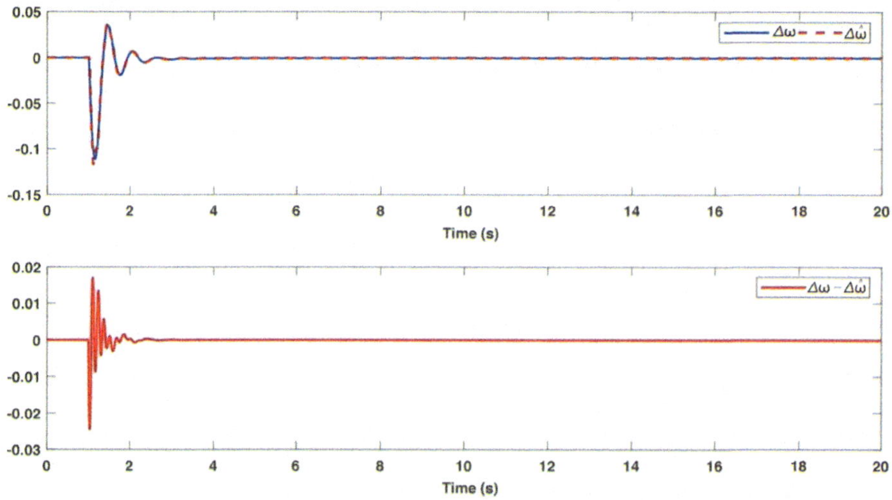

Fig. 3.9 Online identification performance in Scenario 2

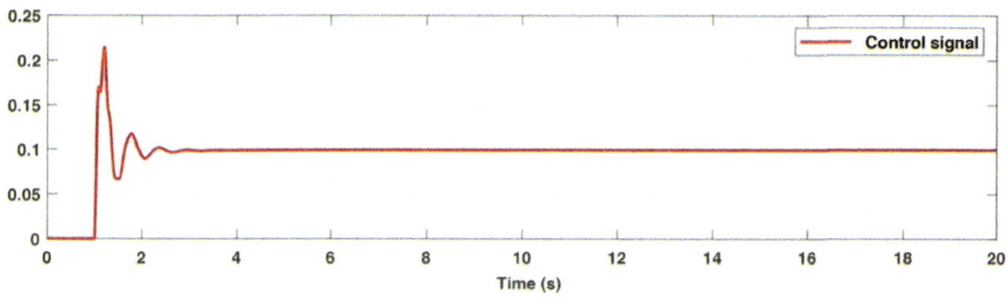

Fig. 3.10 The control signal in Scenario 2

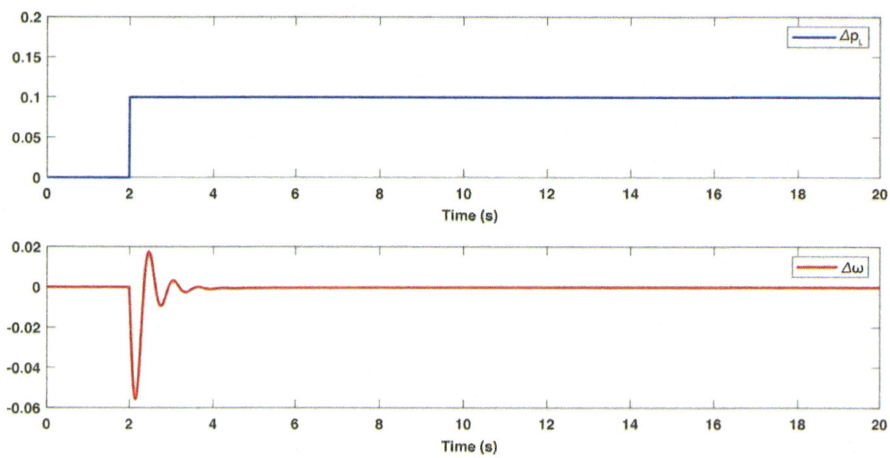

Fig. 3.11 Δp_L and $\Delta\omega$ trajectories in Scenario 3

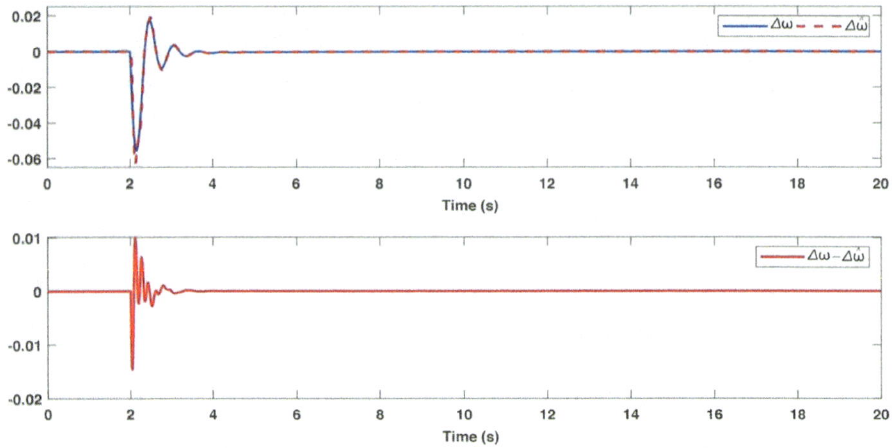

Fig. 3.12 Online identification performance in Scenario 3

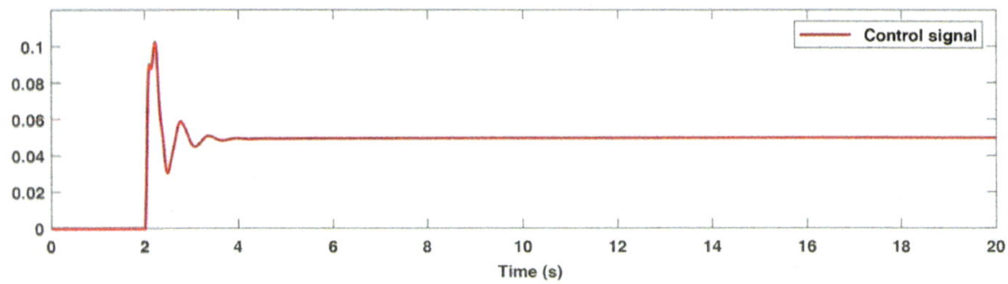

Fig. 3.13 The control signal in Scenario 3

Scenario 4:

Here, the solar radiation and mechanical power are considered to change over time, as shown in Fig. 3.14. All other conditions remain the same as for the third scenario. Figure 3.15 illustrates the frequency variation path. Figure 3.16 displays the online identification proficiency, whereas Fig. 3.17 illustrates the path of controller u. Based on the analysis of Figs. 3.15 and 3.16, it can be concluded that the suggested controller and identification approach exhibit robustness against fluctuations in parameters and mechanical power, various load disturbances, and changes in solar radiation.

Scenario 5:

To enhance the assessment of the identification system, a white noise is introduced into the control signal. The noise is considered to have a mean of zero and a variance of 0.1. The outcome of the identification is presented in Fig. 3.17. It is shown that the dynamic GT2-FLS, when combined with the optimised secondary membership, results in a strong capability of estimation. The estimated signal $\Delta\hat{\omega}$ closely follows the path of the actual signal $\Delta\omega$.

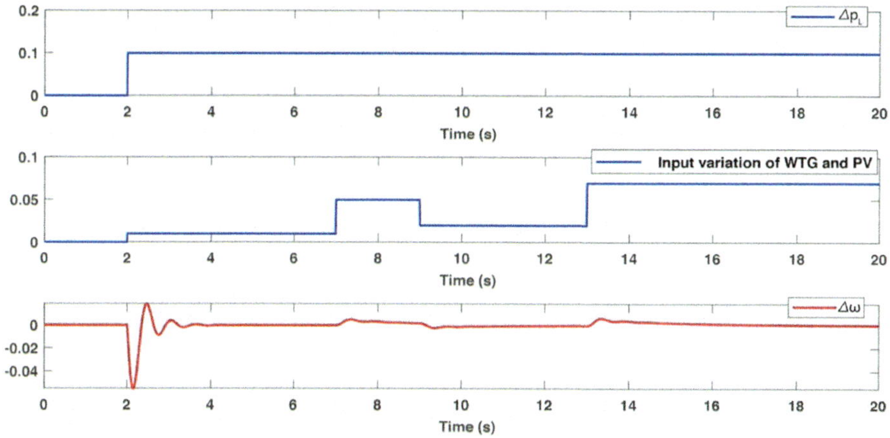

Fig. 3.14 Δp_L and $\Delta\omega$ trajectories in Scenario 4

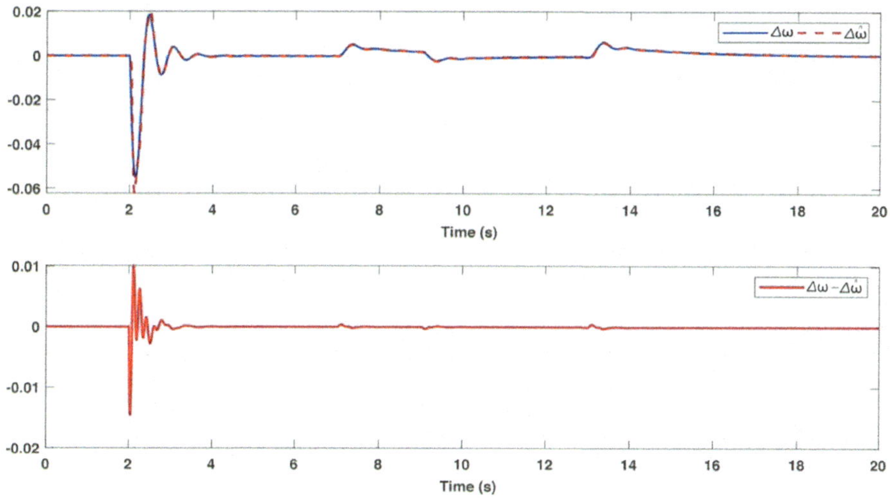

Fig. 3.15 Online identification performance in Scenario 4

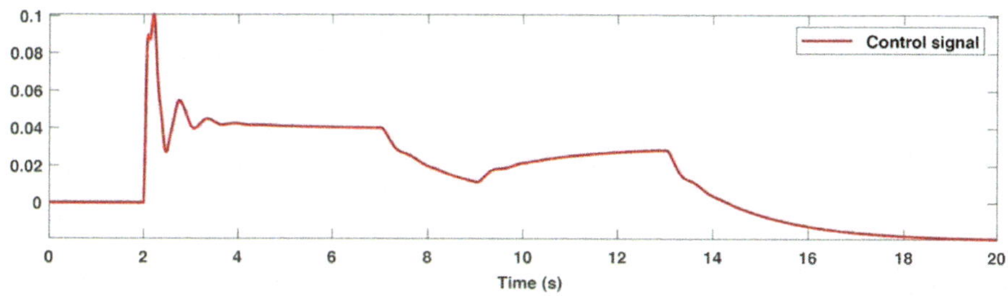

Fig. 3.16 The control signal in Scenario 4

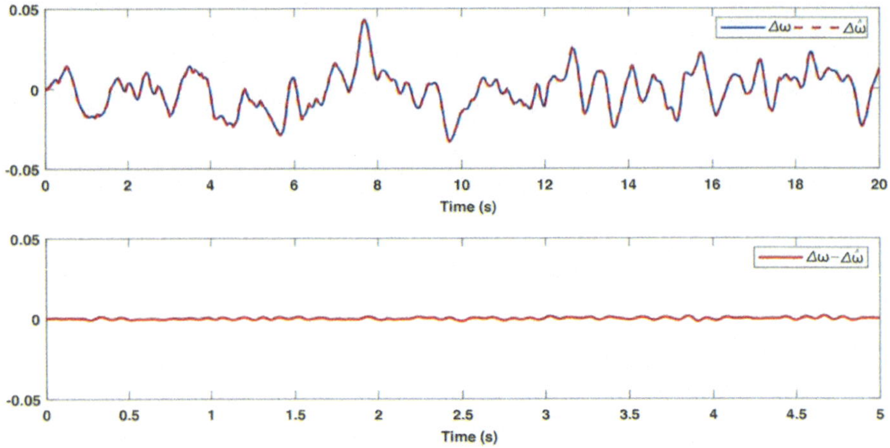

Fig. 3.17 Online identification performance in Scenario 5

Comparison 1:

This section demonstrates the effectiveness and superiority of the recommended scheme compared to other strategies such as PI controller [74], Fuzzy-PI control [75], Fuzzy-PI-PSO controller [75], fractional-order sliding mode control system (FSMC) [76, 77], and the fractional PID controller (FPID), which utilises fractional calculus principles [78]. The simulation circumstances are identical to those of scenario 1. The table provides the values of the integral $\int_0^T \Delta\omega^2 dt$, where T represents the duration of the simulation, for various control methods (as indicated in Table 3.2). It is evident that the proposed approach results in a more effective control, despite the fact that the dynamics are completely unknown.

Comparison 2:

In order to demonstrate the effectiveness of the recommended GT2-FLS, a comparison is provided between T1-FLS and interval T2-FLS (IT2-FLS). To assess the identification performance, a control signal is subjected to white noise with different variances and a mean of zero. The table presents the root-mean-square of error, denoted as $\Delta\omega - \Delta\hat{\omega}$ (RMSE), in Table 3.3. It is evident that the suggested GT2-FLS

Table 3.2 Performance comparison for various control techniques

	Methods					
	PI	Fuzzy-PI	Fuzzy-PI-PSO	FSMC	FPID	Proposed method
$\int_0^T \Delta\omega^2 dt$	24e-3	20e-3	15e-3	12e-3	15e-3	10e-3

Table 3.3 Comparison of the identification performance for various FLSs

	RMSE for $\Delta\omega - \Delta\hat{\omega}$		
Noise Variance	T1-FLS	IT2-FLS	GT2-FLS
0.0141	0.0723	0.0517	0.0501
0.05	0.1241	0.0931	0.0627
0.1	0.2180	0.1341	0.0731

performs better than IT2-FLS and T1-FLS in terms of identification performance, particularly in situations with high levels of noise [78].

References

1. A. Mohammadzadeh, M.H. Sabzalian, W. Zhang, An interval type-3 fuzzy system and a new online fractional-order learning algorithm: theory and practice. IEEE Trans. Fuzzy Syst. **28**, 1940–1950 (2019). https://doi.org/10.1109/TFUZZ.2019.2928509
2. Renewable Energy and Energy Efficiency Organization. http://www.satba.gov.ir/en/home.
3. Y. Cao, A. Raise, A. Mohammadzadeh, S. Rathinasamy, S.S. Band, A. Mosavi, Deep learned recurrent type-3 fuzzy system: application for renewable energy modeling/prediction. Energy Rep. **7**, 8115–8127 (2021)
4. B. Lin, Conceptual design and modeling of a fuel cell scooter for urban asia. J. Power. Sources **86**(1–2), 202–213 (2000)
5. S.S. Band, A. Mohammadzadeh, P. Csiba, A. Mosavi, A.R. Varkonyi-Koczy, Voltage regulation for photovoltaics-battery-fuel systems using adaptive group method of data handling neural networks (gmdh-nn). IEEE Access **8**, 213 748–213 757 (2020)
6. Mohammadzadeh, M.H. Sabzalian, W. Zhang, An interval type-3 fuzzy system and a new online fractional-order learning algorithm: theory and practice. IEEE Trans. Fuzzy Syst (2019)
7. Ma, A. Mohammadzadeh, H. Turabieh, M. Mafarja, S.S. Band, A. Mosavi, Optimal type-3 fuzzy system for solving singular multipantograph equations. IEEE Access **8**, 225 692–225 702 (2020)
8. S.W. Sung, J. Lee, I.-B. Lee, Process identification and PID control. Wiley Online Library, vol. 6 (2009)
9. D.E. Kirk, Optimal control theory: an introduction. Courier Corporation (2012)
10. A. Tofighi, M. Kalantar, Power management of pv/battery hybrid power source via passivity-based control. Renew. Energy **36**(9), 2440–2450 (2011)
11. M.R. Mojallizadeh, M. Badamchizadeh, S. Khanmohammadi, M. Sabahi, Designing a new robust sliding mode controller for maximum power point tracking of photovoltaic cells. Sol. Energy **132**, 538–546 (2016)
12. Z. Liu, A. Mohammadzadeh, H. Turabieh, M. Mafarja, S.S. Band, A. Mosavi, A new online learned interval type-3 fuzzy control system for solar energy management systems. IEEE Access **9**, 10498–10508 (2021)
13. P.T. Krein, J. Bentsman, R.M. Bass, B.L. Lesieutre, On the use of averaging for the analysis of power electronic systems. IEEE Trans. Power Electron. **5**, 182–190 (1990)
14. B. Lin, Conceptual design and modeling of a fuel cell scooter for urban Asia. J. Power. Sources **86**, 202–213 (2000)
15. A. Mohammadzadeh, M.H. Sabzalian, W. Zhang, An interval type-3 fuzzy system and a new online fractional-order learning algorithm: theory and practice. IEEE Trans. Fuzzy Syst (2019). https://doi.org/10.1109/TFUZZ.2019.2928509

16. A. Mosavi, S.N. Qasem, M. Shokri, S.S. Band, A. Mohammadzadeh, Fractional-order fuzzy control approach for photovoltaic/battery systems under unknown dynamics, variable irradiation and temperature. Electronics **9**(9), 1455 (2020)
17. X. Xu, D. Niu, L. Peng, S. Zheng, J. Qiu, Hierarchical multi-objective optimal planning model of active distribution network considering distributed generation and demand-side response. Sust. Energy Technol. Assess. **53**, 102438 (2022)
18. M. Davari, M.P. Aghababa, F. Blaabjerg, M. Saif, An innovative, adaptive faulty signal rectifier along with a switching controller for reliable primary control of GC-VSIs in CPS-based modernized microgrids. IEEE Trans. Power Electron. **36**(7), 8370–8387 (2021)
19. C. Zhong, Y. Zhou, J. Chen, Z. Liu, Dc-side synchronous active power control of two-stage photovoltaic generation for frequency support in islanded microgrids. Energy Rep. **8**, 8361–8371 (2022)
20. E. Gershon, U. Shaked, I. Yaesh, H-infinity control and estimation of state-multiplicative linear systems, Vol. 318, Springer Science & Business Media (2005)
21. H. Liu, F. Zhao, X. Chen, H. Hou, J. Qiu, Observer-based finite-time H_{∞} sliding mode control of stochastic nonlinear singular systems and its applications. Nonl. Dynam. 1–10 (2022)
22. K. Zhou, S.-K. Oh, J. Qiu, Design of ensemble fuzzy-rbf neural networks based on feature extraction and multi-feature fusion for gis partial discharge recognition and classification. J. Electr. Eng. Technol. **17**(1), 513–532 (2022)
23. S. Gholami, S. Saha, M. Aldeen, Fault tolerant control of electronically coupled distributed energy resources in microgrid systems. Int. J. Electr. Power Energy Syst. **95**, 327–340 (2018)
24. G. Cao, G. Lou, W. Gu, L. Sheng, $H\infty$ robustness for distributed control in autonomous microgrids considering cyber disturbances. CSEE J. Power Energy Syst. (2021)
25. S.R. Yan, W. Guo, A. Mohammadzadeh, S. Rathinasamy, Optimal deep learning control for modernized microgrids. Appl. Intell. **53**(12), 15638–15655 (2023)
26. Sultan Noman Qasem, Ali Ahmadian, Ardashir Mohammadzadeh, Sakthivel Rathinasamy, Bahareh Pahlevanzadeh, A type-3 logic fuzzy system: Optimized by a correntropy based kalman filter with adaptive fuzzy kernel size. Inform. Sci. **572**, 424–443 (2021)
27. G. Hua, F. Wang, J. Zhang, K.A. Alattas, A. Mohammadzadeh, M. The Vu, A new type-3 fuzzy predictive approach for mobile robots. Mathematics **10**(17), 3186 (2022)
28. Y. Wu, H. Sheng, Y. Zhang, S. Wang, Z. Xiong, W. Ke, Hybrid motion model for multiple object tracking in mobile devices. IEEE Intern. Things J. **10**(6), 4735–4748 (2022)
29. M. Liu, Q. Gu, B. Yang, Z. Yin, S. Liu, L. Yin, W. Zheng, Kinematics model optimization algorithm for six degrees of freedom parallel platform. Appl. Sci. **13**(5), 3082 (2023)
30. X. Hou, L. Zhang, Y. Su, G. Gao, Y. Liu, Z. Na, Q. Xu, T. Ding, L. Xiao, L. Li et al., A space crawling robotic bio-paw (scrbp) enabled by triboelectric sensors for surface identification. Nano Energy **105**, 108013 (2023)
31. A. Mohammadzadeh, S. Ghaemi, O. Kaynak et al., Robust predictive synchronization of uncertain fractional-order time-delayed chaotic systems. Soft. Comput. **23**(16), 6883–6898 (2019)
32. B. Li, Y. Tan, A.-G. Wu, G.-R. Duan, A distributionally robust optimization based method for stochastic model predictive control. IEEE Trans. Autom. Control **67**(11), 5762–5776 (2021)
33. A. Mohammadzadeh, R.H. Vafaie, A deep learned fuzzy control for inertial sensing: Micro electro mechanical systems. Appl. Soft Comput. **109**, 107597 (2021)
34. R.H. Vafaie, A. Mohammadzadeh, M. Piran et al., A new type-3 fuzzy predictive controller for mems gyroscopes. Nonlinear Dyn. **106**(1), 381–403 (2021)
35. A.S. Alkabaa, O. Taylan, M. Balubaid, C. Zhang, A. Mohammadzadeh, A practical type-3 Fuzzy control for mobile robots: predictive and Boltzmann-based learning. Compl. Intell. Syst. **9**(6), 6509–6522 (2023)
36. R. Calegari, G. Ciatto, V. Mascardi, A. Omicini, Logic-based technologies for multi-agent systems: a systematic literature review. Auton. Agent. Multi-Agent Syst. **35**, 1–67 (2021)
37. M. Nie, W.W. Tan, Towards an efficient type-reduction method for interval type-2 fuzzy logic systems. 2008 IEEE International Conference on Fuzzy Systems (IEEE World Congress on Computational Intelligence). IEEE, pp. 1425–1432 (2008)

38. S. Yan, A.A. Aly, B.F. Felemban, M. Gheisarnejad, M. Tian, M.H. Khooban, A. Mohammadzadeh, S. Mobayen, A new event-triggered type-3 fuzzy control system for multi-agent systems: optimal economic efficient approach for actuator activating. Electronics **10**(24), 3122 (2021)
39. H. Tirandaz, S.S. Aminabadi, H. Tavakoli, Chaos synchronization and parameter identification of a finance chaotic system with unknown parameters, a linear feedback controller. Alex. Eng. J. **57**(3), 1519–1524 (2018)
40. M. Jun-hai, C. Yu-Shu, Study for the bifurcation topological structure and the global complicated character of a kind of nonlinear finance system (i). Appl. Math. Mech. **22**(11), 1240–1251 (2001)
41. X. Zhao, Z. Li, S. Li, Synchronization of a chaotic finance system. Appl. Math. Comput. **217**(13), 6031–6039 (2011)
42. A. Sambas, S. He, H. Liu, S. Vaidyanathan, Y. Hidayat, J. Saputra et al., Dynamical analysis and adaptive fuzzy control for the fractional-order financial risk chaotic system. Adv. Difference Equ. **2020**(1), 1–12 (2020)
43. S. Wang, S. Bekiros, A. Yousefpour, S. He, O. Castillo, H. Jahanshahi, Synchronization of fractional time-delayed financial system using a novel type-2 fuzzy active control method. Chaos Solitons Fractals **136**, 109768 (2020)
44. M.H. Sabzalian, A. Mohammadzadeh, W. Zhang, K. Jermsittiparsert, General type-2 fuzzy multi-switching synchronization of fractional-order chaotic systems. Eng. Appl. Artif. Intell. **100**, 104163 (2021)
45. M.W. Tian, Y. Bouteraa, K.A. Alattas, S.R. Yan, A.K. Alanazi, A. Mohammadzadeh, S. Mobayen, A type-3 fuzzy approach for stabilization and synchronization of chaotic systems: applicable for financial and physical chaotic systems. Complexity **2022**(1), 8437910 (2022)
46. Norelys Aguila-Camacho, Manuel A. Duarte-Mermoud, Javier A. Gallegos, Lyapunov functions for fractional order systems. Commun. Nonlin. Sci. Numer. Simul. **19**(9), 2951–2957 (2014)
47. Stephen P. Boyd, Laurent El Ghaoui, Eric Feron, Venkataramanan Balakrishnan, Linear matrix inequalities in system and control theory, vol. 15. SIAM (1994)
48. L. Xie, Output feedback H-infinity control of systems with parameter uncertainty. Int. J. Control. **63**(4), 741–750 (1996)
49. Jun-Guo Lu, Yang-Quan Chen, Robust stability and stabilization of fractional-order interval systems with the fractional order α: The $0 \leq \alpha \leq 1$ case. Autom. Contr. IEEE Trans. **55**(1), 152–158 (2010)
50. Ardashir Mohammadzadeh, Mohammad Hosein Sabzalian, Weidong Zhang, An interval type-3 fuzzy system and a new online fractional-order learning algorithm: theory and practice. IEEE Trans. Fuzzy Syst. (2019)
51. Mohammad Hosein Sabzalian, Ardashir Mohammadzadeh, Shuyi Lin, Weidong Zhang, Robust fuzzy control for fractional-order systems with estimated fraction-order. Nonlin. Dyn. **98**(3), 2375–2385 (2019)
52. Saeed Mirzajani, Mohammad Pourmahmood Aghababa, Aghileh Heydari, Adaptive t–s fuzzy control design for fractional-order systems with parametric uncertainty and input constraint. Fuzzy Sets Syst. **365**, 22–39 (2019)
53. A. Mohammadzadeh, O. Castillo, S.S. Band, A. Mosavi, A novel fractional-order multiple-model type-3 fuzzy control for nonlinear systems with unmodeled dynamics. Int. J. Fuzzy Syst. **23**(6), 1633–1651 (2021)
54. Auto-mpg data, Machine learning databases (1996). https://archive.ics.uci.edu/ml/machine-learning-databases/auto-mpg/
55. Di Wang, Xiao-Jun Zeng, John A. Keane, A simplified structure evolving method for Mamdani fuzzy system identification and its application to high-dimensional problems. Inform. Sci. **220**, 110–123 (2013)
56. Di Wang, Xiao-Jun Zeng, John A. Keane, A structure evolving learning method for fuzzy systems. Evolv. Syst. **1**(2), 83–95 (2010)

57. W. Pedrycz, K.-C. Kwak, Linguistic models as a framework of user-centric system modeling. Inf. Sci. **220**, 110–123 (2013)
58. A. Mohammadzadeh, E. Kayacan, A non-singleton type-2 fuzzy neural network with adaptive secondary membership for high dimensional applications. Syst. Man Cybern. Part A: Syst. Humans, IEEE Trans. **36**(4), 727–745 (2006)
59. Housing data, Machine learning databases (1996). https://archive.ics.uci.edu/ml/machine-learning-databases/housing/
60. M. Almaraashi, R. John, A. Hopgood, S. Ahmadi, Learning of interval and general type-2 fuzzy logic systems using simulated annealing: theory and practice. Inf. Sci. **360**, 21–42 (2016)
61. Hai-Jun Rong, N. Sundararajan, Guang-Bin Huang, P. Saratchandran, Sequential adaptive fuzzy inference system (SAFIS) for nonlinear system identification and prediction. Fuzzy Sets Syst. **157**(9), 1260–1275 (2006)
62. Di Wang, Xiao-Jun Zeng, John Keane, et al., An evolving-construction scheme for fuzzy systems. Fuzzy Syst. IEEE Trans. **18**(4), 755–770 (2010)
63. Plamen P. Angelov, Dimitar P. Filev, An approach to online identification of Takagi-Sugeno fuzzy models. Syst. Man Cybern. Part B: Cybern. IEEE Trans. **34**(1), 484–498 (2004)
64. V. Kadirkamanathan, M. Niranjan, A function estimation approach to sequential learning with neural networks. Neural Comput. **5**(6), 954–975 (1993)
65. Lu Yingwei, Narasimhan Sundararajan, Paramasivan Saratchandran, A sequential learning scheme for function approximation using minimal radial basis function neural networks. Neural comput. **9**(2), 461–478 (1997)
66. L. Wang, J. Yen, Extracting fuzzy rules for system modeling using a hybrid of genetic algorithms and Kalman filter. Fuzzy Sets Syst. **101**(3), 353–362 (1999)
67. Plamen Angelov, Dimitar Filev. Simpl_eTS: a simplified method for learning evolving Takagi-Sugeno fuzzy models. In *Fuzzy Systems, 2005. FUZZ'05. The 14th IEEE International Conference on*, pages 1068–1073. IEEE (2005)
68. José de Jesús Rubio, SOFMLS: online self-organizing fuzzy modified least-squares network. Fuzzy Syst. IEEE Trans. **17**(6), 1296–1309 (2009)
69. Mar 13053'*f* a de los Angeles Hernandez, Patricia Melin, Gerardo M Méndez, Oscar Castillo, Ismael López-Juarez, A hybrid learning method composed by the orthogonal least-squares and the back-propagation learning algorithms for interval A2-C1 type-1 non-singleton type-2 TSK fuzzy logic systems. Soft Comput **19**(3), 661–678 (2015)
70. Frederick E. Daum, Extended Kalman Filters. Encyclop. Syst. Contr., 411–413 (2015)
71. Ricardo Mart 13053' *f* nez-Soto, Oscar Castillo, and Juan R Castro. Genetic algorithm optimization for type-2 non-singleton fuzzy logic controllers. In *Recent Advances on Hybrid Approaches for Designing Intelligent Systems*, pp. 3–18. Springer (2014)
72. Reza Izanloo, Seyed Abolfazl Fakoorian, Hadi Sadoghi Yazdi, Dan Simon, Kalman filtering based on the maximum correntropy criterion in the presence of non-gaussian noise. In *2016 Annual Conference on Information Science and Systems (CISS)*, pp. 500–505. IEEE (2016)
73. H. Bevrani, A. Ghosh, G. Ledwich, Renewable energy sources and frequency regulation: survey and new perspectives. IET Renew. Power Gener. **4**(5), 438–457 (2010)
74. Hassan Bevrani, F. Habibi, P. Babahajyani, Masayuki Watanabe, Yasunori Mitani, Intelligent frequency control in an ac microgrid: Online pso-based fuzzy tuning approach. IEEE Trans. Smart Grid **3**(4), 1935–1944 (2012)
75. Minghao Zhou, Yong Feng, Chen Xue, Fengling Han, Deep convolutional neural network based fractional-order terminal sliding-mode control for robotic manipulators. Neurocomputing (2019)
76. J. Chen, C. Li, X. Yang, Global mittag–leffler projective synchronization of nonidentical fractional-order neural networks with delay via sliding mode control. Neurocomputing **313**, 324–332 (2018)
77. Lin Xu, Baoye Song, Maoyong Cao, Yihui Xiao, A new approach to optimal design of digital fractional-order piλdμ controller. Neurocomputing **363**, 66–77 (2019)
78. A. Mohammadzadeh, M.H. Sabzalian, A. Ahmadian, N. Nabipour, A dynamic general type-2 fuzzy system with optimized secondary membership for online frequency regulation. ISA Trans. **112**, 150–160 (2021)

Index

R. Sabetahd et al., *Applied Type-3 Fuzzy Logic Systems and Controllers*, Studies in Fuzziness and Soft Computing 417,
https://doi.org/10.1007/978-3-031-75988-8

The manufacturer's authorised representative in the EU is Springer
Nature Customer Service Centre GmbH, Europaplatz 3, 69115 Heidelberg,
Germany. If you have any concerns regarding our products, please
contact ProductSafety@springernature.com

Printed and bound by CPI Group (UK) Ltd, Croydon, CR0 4YY
30/09/2026
02225563-0005